B. Graham
28 - 10 - 93

Computation and Neural Systems

Computation and Neural Systems

Edited by:

Frank H. Eeckman
Lawrence Livermore National Laboratory

James M. Bower
California Institute of Technology

KLUWER ACADEMIC PUBLISHERS
BOSTON/DORDRECHT/LONDON

Distributors for North America:
Kluwer Academic Publishers
101 Philip Drive
Assinippi Park
Norwell, Massachusetts 02061 USA

Distributors for all other countries:
Kluwer Academic Publishers Group
Distribution Centre
Post Office Box 322
3300 AH Dordrecht, THE NETHERLANDS

Library of Congress Cataloging-in-Publication Data

Computation and neural systems / edited by Frank H. Eeckman, James Bower.
     p.   cm.
    Includes bibliographical references and index.
    ISBN 0-7923-9349-X
    1. Neural networks (Neurobiology)--Congresses. 2. Neural networks
(Computer science)--Congresses. I. Eeckman, Frank H. II. Bower,
James (James M.)
QP363.3.C635   1993
591.1'88--dc20   93-4227
                                 CIP

Printed on acid-free paper.

Printed in the United States of America

TABLE OF CONTENTS

Introduction xi

Part I - Analysis and Modeling Tools and Techniques

Section 1: Analysis 1

1 Self-Organising Ion Channel Densities: The Rationale for 'anti-Hebb', *Anthony J. Bell* 3

2 A Modified Hodgkin-Huxley Spiking Model With Continuous Spiking Output, *Frédéric E. Theunissen, Frank H. Eeckman and John P. Miller* 9

3 The Neurone as a Nonlinear System: A Single Compartment Study, *P.G. Hearne, J. Wray, D.J. Sanders, E. Agar and G.G.R. Green* 19

4 Unsupervised Hebbian Learning and the Shape of the Neuron Activation Function, *Jonathan L. Shapiro and Adam Prügel-Bennett* 25

5 A Method for Estimating the Neural Input to a Neuron Using the Ionic Current Model, *Yoshimi Kamiyama, Hiroyuki Ishii and Shiro Usui* 31

6 A Model of a Thalamic Relay Cell Incorporating Voltage-Clamp Data on Ionic Conductances, *Diana K. Smetters* 37

7 An Entropy Measure for Revealing Deterministic Structure in Spike Train Data, *Garrett T. Kenyon and David C. Tam* 43

8 A Multi-Neuronal Vectorial Phase-Space Analysis for Detecting Dynamical Interactions in Firing Patterns of Biological Neural Networks, *David C. Tam* . 49

9 The 'Ideal Homunculus': Statistical Inference from Neural Population Responses, *Peter Földiák* 55

10 The Effect of Slow Synaptic Coupling on Populations of Spiking Neurons, *C.A. van Vreeswijk, L.F. Abbott and A. Treves* 61

11 Neural Geometry Revealed by Neurocomputer Analysis of Multi-Unit Recordings, *Andras J. Pellionisz, David L. Tomko and James R. Bloedel* 67

12 Signal Delay in Passive Dendritic Trees, *Hagai Agmon-Snir and Idan Segev* 73

Section 2: Modeling **79**

13 Matching Neural Models to Experiment, *William R. Foster, Julian F.R. Paton, John J. Hopfield, Lyle H. Ungar and James S. Schwaber* 81

14 Neuronal Model Predicts Responses of the Rat Baroreflex, *James S. Schwaber, Julian F.R. Paton, Robert F. Rogers and Eliza B. Graves* 89

15 Simulations of Synaptic Integration in Neocortical Pyramidal Cells, *Paul C. Bush and Terrence J. Sejnowski* 97

16 Modeling Stimulus Specific Habituation: The Role of the Primordial Hippocampus, *DeLiang Wang* 103

17 Effects of Single Channel Kinetics upon Transmembrane Voltage Dynamics, *Adam F. Strassberg and Louis J. DeFelice* 109

18 An Object-Oriented Paradigm for the Design of Realistic Neural Simulators, *David C. Tam and R. Kent Hutson* 115

19 Neural Network Simulation in a CSP Language for Multicomputers, *F. Bini, M. Mastroianni, S. Russo, and G. Ventre* 121

20 Designing Networks of Spiking Silicon Neurons and Synapses, *Lloyd Watts* 127

21 RallPacks: A Set of Benchmarks for Neuronal Simulators, *Upinder S. Bhalla, David H. Bilitch and James M. Bower* 133

Part II - Sensory Systems

Section 3: Visual Systems **141**

22 Invariant Contrast Adaptation in the Primate Outer Plexiform Layer, *Josef Skrzypek and George Wu* 143

23 A Computational Model of the *Limulus* Eye on the Connection Machine, *Daryl R. Kipke, Erik D. Herzog and Robert B. Barlow Jr* 151

24 Does Synaptic Facilitation Mediate Motion Facilitation in the Retina ? *Norberto M. Grzywacz, Franklin R. Amthor, and Lyle J. Borg-Graham* 159

25 Fast-Most Information Extraction by Retinal Y-X Cells, *Zhaoping Li* 165

26 An Adaptive Neural Model of Early Visual Processing, *Harry G. Barrow and Alistair J. Bray* 171

27 A Local Model for Transparent Motions Based on Spatio-Temporal Filters, *James A. Smith and Norberto M. Grzywacz* 177

28 A Neural Model of Illusory Contour Perception, *Brian Ringer and Josef Skrzypek* 183

29 Adaptive Receptive Fields for Textural Segmentation, *Edmond Mesrobian and Josef Skrzypek* 189

30 Cortical Mechanisms for Surface Segmentation, *Paul Sajda and Leif H. Finkel* 195

31 Direction Selectivity Using Massive Intracortical Connections, *Humbert Suarez, Christof Koch and Rodney Douglas* 201

32 Correlational-Based Development of Disparity Sensitivity, *Gregory S. Berns, Peter Dayan and Terrence J. Sejnowski* 207

33 Dynamical Control of Visual Attention Through Feedback Pathways: A Network Model, *Janani Janakiraman and K. P. Unnikrishnan* 215

34 The Gating Lattice: A Neural Substrate for Dynamic Gating, *Eric O. Postma, H. Jaap van den Herik and Patrick T. W. Hudson* 221

Section 4: Auditory Systems **227**

35 Hair Cell Modelling to Explore The Physiological Basis of Tuning in the Lower Vertebrate Ear, *David Egert* 229

36 Mechanisms of Horizontal Sound Localization at Low and High Frequency in Avian Nuc. Laminaris, *W. Edward Sullivan* 235

37 A Neuronal Modeling System for Generating Brainstem Maps of Auditory Space, *Bruce R. Parnas and Edwin R. Lewis* 241

Section 5: Other Sensory Systems **247**

38 Modeling Chemotaxis in the Nematode *C. elegans, Shawn R. Lockery, Steven J. Nowlan and Terrence J. Sejnowski* 249

39 Formal Model of the Insect Olfactory Macroglomerulus, *C. Linster, C. Masson, M. Kerszberg, L. Personnaz, and G. Dreyfus* 255

40 Dynamic Activity, Learning, and Memory in a Model of the Olfactory Bulb, *Tamás Gröbler and Péter Érdi* 261

41 Differential Effects of Norepinephrine on Synaptic Transmission in Layers 1A and 1B of Rat Olfactory Cortex, *Michael C. Vanier and James M. Bower* .. 267

42 Cholinergic Modulation of Associative Memory Function in a Realistic Computational Model of Piriform Cortex, *Ross E. Bergman, Michael Vanier, Gregory Horwitz, James M. Bower and Michael E. Hasselmo* 273

43 Numerical Simulations of the Electric Organ Discharge of Weakly Electric Fish, *Christopher Assad, Brian Rasnow and James M. Bower* 281

Part III - Motor Systems

Section 6: Pattern Generators **287**

44 Identification of Leech Swim Neurons Using a Resonance Technique, *Richard A. Gray and W. Otto Friesen* 289

45 The Use of Genetic Algorithms to Explore Neural Mechanisms that Optimize Rhythmic Behaviors: Quasi-Realistic Models of Feeding Behavior in *Aplysia* , *I. Kupfermann, D. Deodhar, S.R. Rosen, and K.R. Weiss* 295

46 Undulatory Locomotion - Simulations with Realistic Segmental Oscillator, *T. Wadden, S. Grillner, T. Matsushima, and Anders Lansner* 301

47 Network Model of the Respiratory Rhythm, *Eliza B. Graves, William C. Rose, Diethelm W. Richter, and James S. Schwaber* 307

48 Model of the Peristaltic Reflex, *A.D. Coop and S.J. Redman* 313

Section 7: Cortex, Cerebellum, and Spinal Cord **321**

49 Simulation of the Muscle Stretch Reflex by a Neuronal Network, *Bruce P. Graham and Stephen J. Redman* 323

50 Modeling and Simulation of Compartmental Cerebellar Networks for Conditioning of Rabbit Eyeblink Response, *P.M. Khademi, E.K. Blum, P.K. Leung, D.G. Lavond, R.F. Thompson, D.J. Krupa and J. Tracy* 331

51 Recurrent Backpropagation Models of the Vestibulo-Ocular Reflex Provide Experimentally Testable Predictions, *Thomas J. Anastasio* 337

52 Prolonged Activation with Brief Synaptic Inputs in the Purkinje Cell: Intracellular Recording and Compartmental Modeling, *Dieter Jaeger, Erik De Schutter, and James M. Bower* 343

53 Electrophysiological Dissection of the Excitatory Inputs to Purkinje Cells, *John H. Thompson and James M. Bower* 349

54 Integration of Synaptic Inputs in a Model of the Purkinje Cell, *Erik De Schutter and James M. Bower* 355

55 Unsupervised Learning of Simple Speech Production Based on Soft Competitive Learning, *Georg Dorffner and Thomas Schönauer* 363

56 Dopaminergic Modulation and Neural Fatigue in Discrete Time Sigmoidal Networks, *Nur Arad, Eytan Ruppin and Yehezkel Yeshurun* 369

Part IV - Cerebral Cortex

Section 8: Development and Map Formation 375

57 Volume Learning: Signaling Covariance Through Neural Tissue, *P. R. Montague, P. Dayan, and T.J. Sejnowski* 377

58 Hebbian Learning in Feedback Networks: Development Within Visual Cortex, *Dawei W. Dong* 383

59 The Role of Subplate Feedback in the Development of Ocular Dominance Columns, *Harmon S. Nine and K.P. Unnikrishnan* 389

60 A Comparison of Models of Visual Cortical Map Formation, *Edgar Erwin, Klaus Obermayer, and Klaus Schulten* 395

61 Field Discontinuities and Islands in a Model of Cortical Map Formation, *F. Wolf, H.-U. Bauer, and T. Geisel* 403

62 Orientation Columns from First Principles, *Ernst Niebur and Florentin Wörgötter* 409

63 A Model of the Combined Effects of Chemical and Activity-Dependent Mechanisms in Topographic Map Formation, *Martha J. Hiller* 415

Section 9: Associative Memory and Learning 423

64 NMDA-Activated Conductances Provide Short-Term Memory for Dendritic Spine Logic Computations, *R.V. Jensen and Gordon M. Shepherd* 425

65 A Model of Cortical Associative Memory Based on Hebbian Cell Assemblies, *Erik Fransén, Anders Lansner and Hans Liljenström* 431

66 Self-Teaching Through Correlated Input, *Virginia R. de Sa and Dana H. Ballard* 437

67 Multi-Layer Bidirectional Auto-Associators, *Dimitrios Bairaktaris* 443

68 A Neural Model for a Randomized Frequency-Spatial Transformation, *Yossi Matias and Eytan Ruppin* 449

69 Outline of a Theory of Isocortex, *Mark James and Doan Hoang* 455

70 A Network for Semantic and Episodic Associations Showing Disturbances Due to Neural Loss, *Michael Herrmann, Eytan Ruppin and Marius Usher* 461

71 Synaptic Deletion and Compensation in Alzheimer's Disease: A Neural Model, *David Horn, Eytan Ruppin, Marius Usher and Michael Herrmann* 467

Section 10: Cortical Dynamics **473**

72 Effects of Input Synchrony on the Response of a Model Neuron, *Venkatesh N. Murthy and Eberhard E. Fetz* 475

73 Analysis and Simulation of Synchronization in Oscillatory Neural Networks, *Xin Wang, E.K. Blum and P.K. Leung* 481

74 Alternating Predictable and Unpredictable States in Data from Cat Visual Cortex, *K. Pawelzik, H.-U. Bauer, and T. Geisel* 487

75 Temporal Structure of Spike Trains from MT Neurons in the Awake Monkey, *Wyeth Bair, Christof Koch, William Newsome, and Kenneth Britten* 495

76 Temporal Structure Can Solve the Binding Problem for Multiple Feature Domains, *Thomas B. Schillen and Peter König* 503

77 Assembly Formation and Segregation by a Self-Organizing Neuronal Oscillator Model, *Peter König, Bernd Janosch, and Thomas B. Schillen* 509

78 Inter-area Synchronization in Macaque Neocortex During a Visual Pattern Discrimination Task, *Steven L. Bressler and Richard Nakamura* 515

79 Acetylcholine and Cortical Oscillatory Dynamics, *Hans Liljenström and Michael E. Hasselmo* 523

80 Dissipative Structures and Self-Organizing Criticality in Neural Networks with Spatially Localized Connectivity, *Robert W. Kentridge* 531

Index 537

INTRODUCTION

This volume includes papers presented at the First Annual Computation and Neural Systems meeting held in San Francisco, California July 26 - July 29 , 1992. This meeting was the natural outgrowth of two previous workshops on Analysis and Modeling of Neural Systems also held in San Francisco in the previous two summers (see Analysis and Modeling of Neural Systems, 1992 and Neural Systems: Analysis and Modeling, 1993 by Kluwer Academic Publishers). The expansion of these earlier meetings into CNS*92 reflects the continuing dramatic growth of interest in computational neuroscience. The papers in this volume were originally submitted in January of 1992. Each paper was then peer reviewed by at least two referees before it was accepted for the meeting. The papers appearing here were prepared in final form after the meeting and resubmitted in October 1992. This collection represents 80 of the 107 papers actually presented at the meeting. The present volume is organized in the same fashion as the Analysis and Modeling volumes.

As can be seen from the Table of Contents, CNS*92 was intended to showcase current work in computational neuroscience, broadly defined. While often most closely associated with computer modeling, in fact, computational neuroscience is best defined by its focus on understanding the nervous system as a computational device rather than by a particular experimental technique. Accordingly, while the majority of these papers describe analysis and modeling efforts, other papers describe the results of new biological experiments explicitly placed in the context of computational issues.

In addition to the scientific results presented here, numerous papers also describe the ongoing technical developments that are critical for the continued growth of computational neuroscience. To a large extent, this involves the use of ever more powerful computers and computer algorithms for modeling. However, equally important, and also represented here, are papers describing improvements in experimental techniques and neuronal data analysis.

Finally, we believe that the distribution of subjects in these papers reflects the current state of the field well. Clearly, the visual system continues to predominate as the most studied sensory system. Similarly, the mammalian cerebral cortex is the focus for many current computational efforts. If one works on the mammalian visual cortex one has a lot of company in the neurosciences. At the same time, however, we are pleased by the diversity of systems and brain regions represented in these papers. This diversity enhanced CNS*92 immeasurably and bodes well for the vitality of computational neuroscience in the future.

xii

We acknowledge the many professional scientists that helped put this meeting together. Not listed are the numerous students at the University of California at Berkeley who performed many important tasks during the meeting itself. However, no meeting of this size and complexity could possibly be successful without substantial support from professional staff. For this reason we are extremely grateful for the very hard work performed by Chris Ploegaert and Patti Bateman at Caltech and Chris Ghinazzi at Lawrence Livermore Laboratory. We are especially thankful for their competent tolerance of our frequent miscalculations. We also want to acknowledge John Uhley and Maneesh Sahani from Caltech for setting up and maintaining the meeting's computer facilities. With respect to this volume in particular, we acknowledge Katherine Butterfield at UC Berkeley for expert help in typesetting, and Alex Greene and Stephanie Faulkner at Kluwer for their editorial assistance. Many thanks also to Barbara Liepe for overall help and support.

CNS*92 was regarded by the organizers as an experiment to determine if current biological research could support an annual meeting focussed on computational neuroscience. We hope that the quality and diversity of papers in this volume will lead the reader, as it has the organizers, to the conclusion that this is in fact the case. If so, we hope to see you at the Second Annual Computation and Neural Systems Meeting (CNS*93) to be held in Washington DC from July 31 to August 7, 1993.

Jim Bower
Caltech

Frank Eeckman
LLNL

CNS*92 Organizing and Program Committee:

James M. Bower, Caltech;
Frank H. Eeckman, Lawrence Livermore Laboratory.;
Edwin R. Lewis, University of California at Berkeley;
John P. Miller, University of California at Berkeley;
Muriel D. Ross, NASA Ames Research Center;
Nora Smiriga, Lawrence Livermore Laboratory.

CNS*92 Reviewers:

Larry Abbott , Brandeis University; Charles Anderson, Washington University; Pierre Baldi, Caltech; William Bialek , NEC; James Bower, Caltech; Ronald Calabrese, Emory University; Erik De Schutter, Caltech; Rodney Douglas, Oxford University; Frank Eeckman, LLNL; Bard Ermentrout, University of Pittsburgh; Michael Hasselmo, Harvard University; William Holmes , Ohio University; Christof Koch, Caltech; Nancy Kopell, Boston University; Giles Laurent, Caltech; Edwin Lewis, University of California at Berkeley; Gerald Loeb, Queen's University, Kingston; Eve Marder, Brandeis University; Barlett Mel, Caltech; John Miller, University of California at Berkeley; Kenneth Miller, Caltech; Mark Nelson, University of Illinois at Champaign Urbana; Ernst Niebur,Caltech; Ed Posner, JPL; Idan Segev, Hebrew University, Jerusalem; Roger Traub, IBM Watson Research Labs; Marius Usher, Caltech; Jack Gallant, Washington University; Taichi Wang, Rockwell International Science Center; Charles Wilson, University Tennesee; Alan Yuille, Harvard University; Anthony Zador, Yale University

SECTION 1

ANALYSIS

The topics in this section range from self-organization of ion channels at the subcellular level, to methods for analysis of multi-unit recordings at the systems level. Mathematical techniques focus on information theory, dynamical systems theory, and bayesian statistics. Readers may find related work in chapters 40 (section 5), 56 (section 7) and 80 (section 10).

1

SELF-ORGANISING ION CHANNEL DENSITIES: THE RATIONALE FOR 'ANTI-HEBB'

Anthony J. Bell

AI-lab, Vrije Universiteit Brussel,
Pleinlaan 2, B-1050 Brussels, BELGIUM

1.1 LOCAL DENDRITIC PROCESSING

The neuron is a finely calibrated electrical system, and not only at well-studied structures, like synapses and nodes of Ranvier. Such calibration must involve activity-dependent adjustment of internal parameters. Yet current 'learning algorithms' remain pre-occupied with alterations of synaptic weights alone.

Recently, Mel 1992 and Brown et al 1991 [2] have argued that when voltage-dependent factors are at play, the locations of correlated synapses on the dendritic membrane are critical. These studies have suggested that complex processing occurs locally in dendrites. Even in electrotonically compact cells, correlated synaptic inputs in one dendritic compartment may cooperate in reaching thresholds for local voltage-dependent conductances. The crucial question is how might one organise the distribution of such active conductances in an activity-dependent fashion. If we can answer this, we will also understand the integrative role of these channels in the dendrites.

'Colocality' arguments may be applied to channels as well as to synapses. Channels also carry currents which may be correlated with those of neighbouring channels, and we know they cooperate in reaching thresholds—the obvious example being at the spike initiation zone. Intriguingly, voltage-dependent channels have been found to have a diverse range of thresholds, and time scales ranging from 1 to 10^{-4} seconds [1]. Evidence is also accumulating that their distributions in the dendrites are inhomogeneous.

The clear implication is that if the distribution of channels is important for single neuron 'computation', we ought to be able to define learning rules for the formation of these distributions.

1.2   THE RATIONALE FOR 'ANTI-HEBB'

In search of such a framework, we start with a basic problem in neural signal processing: how to faithfully transmit a signal along a cable. In any passive cable, leakage currents will reduce the amplitude of a signal and capacitance will cause it to lose its spatio-temporal structure. Consider Figure 1.1(a), depicting a chain of six 'compartments'. Compartment 0 is transmitting a 'signal'—a simple non-linear spiking behaviour generated by the Morris-Lecar equations [7]. The other compartments are passive (RC) cable. The result (Figure 1.1b) is that the signal dies away in the cable the way we might expect in an axon with its active channels blocked.

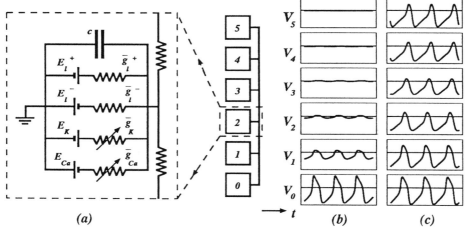

(a)   (b)   (c)

Figure 1.1   A cable which learns to conduct a signal. (a) Circuit diagram of a compartment. (b) Voltage trajectories in the 6 compartments when only 0 carries an active signal, and 1-5 are passive. (c) After learning, 1-5 have become active (through changing their $\bar{g}$'s), and so faithfully conduct the signal.

Associated with the degradation of the signal in (b) is a *voltage curvature* (or current density) at each point on the cable. This reflects the distorting capacitive and leakage cross-membrane currents at that point. If we could somehow cancel out these currents, the voltage-curvature (and thus the distortion) would be zero, and the signal would be faithfully transmitted. So we have a measure of the distortion occurring at a point x along the cable, and it is: $\partial^2 V(x)/\partial x^2$.

In order to cancel the distortion at this point, we can allow currents to be steadily introduced there which are anti-correlated with the distorting currents. The full equation for a point along the cable is then the following (with V and

i_j functions of x and t, and c and $\overline{g}_j$ at this stage functions of x alone):

$$\frac{\partial^2 V}{\partial x^2} = c\frac{\partial V}{\partial t} - \sum_j \overline{g}_j i_j \tag{1.1}$$

Here, c is the membrane capacitance and $\overline{g}_j$ is the maximum conductance of a channel species j. Both these variables are normalised by the axial conductance g which thus doesn't appear: Also, i_j is the specific *inward* current (per unit $\overline{g}_j$) through channel species j. This variable covers all non-capacitive currents, be they synaptic, intrinsic or leakage.

It is by altering the values of the $\overline{g}_j$'s that we can cancel distortion. This is done by raising the value of $\overline{g}_j$ according to minus the time-integrated correlation of the local distortion with the channel species' current. This gives us the following rule (now $\overline{g}_j$, V and i_j are all functions of x and t):

$$\frac{d\overline{g}_j}{dt} = -\epsilon\frac{\partial^2 V}{\partial x^2}i_j \tag{1.2}$$

If this rule is run during numerical integration of the system dynamics, and if the 'learning rate' ϵ is small enough that the $\overline{g}_j$'s change much more slowly than the V's, then the spatial voltage curvatures occurring in the course of spike generation will be steadily reduced by the changes in $\overline{g}_j$'s. The rule is not strictly 'anti-Hebbian' in the sense that it involves an anti-correlation of two *currents* rather than the *'outputs'* of two 'neurons', as in more abstract models. However the name is used in order to stress the anti-correlation aspect.

The simplest test of this rule is to make the same kinetic systems available in the passive cable as in the signal-generating compartment. As shown in the electrical circuit in Figure 1.1(a), these consist of 4 conductances: the Morris-Lecar conductances $\overline{g}_{Ca}$ and $\overline{g}_K$ and inward and outward ohmic leakage conductances, $\overline{g}_l^+$ and $\overline{g}_l^-$. The latter two are designed to remove any DC components of the voltage curvature, rather in the way that 'bias' weights in error correction algorithms are designed to make make the errors zero-mean.

The system was then integrated numerically, with the $\overline{g}$'s in compartments 1-5 starting on zero and following equation 1.2. When the final state was reached, the time-integrated voltage curvature was close to zero in each compartment, the signals were well-propagated (Figure 1.1c) and the conductances had converged to values such that each compartment's $\overline{g}$-vector, given by $[c, \overline{g}_{Ca}, \overline{g}_K, \overline{g}_l^+, \overline{g}_l^-]$, was pointing in roughly the same direction. The length of each $\overline{g}$-vector was then determined by the fixed parameter c, the capacitance of each compartment.

1.3   IMPLICATIONS

What has been shown is that we can use the learning rule in equation 1.2 to adjust the dynamics of compartments so that they are *tuned* with their neighbours. The term 'tuned' is used loosely here to denote the fact that the internal dynamics of the compartment matches the dynamics it is accustomed to experiencing from its surroundings. In [1], however, the case of *tuning* a linear system to be resonant at a stimulating frequency is also addressed—it is another application of the learning rule of equation 1.2.

When a system is 'tuned', voltage curvatures, representing transmission distortions, are minimised; and correspondingly, the transmitted *power* is maximised. One measure which decreases as the system learns, is Φ, the rate of *dissipation* of free energy in the cable:

$$\Phi(t) = \int g(x) \left(\frac{\partial V}{\partial x}\right)^2 dx \qquad (1.3)$$

where $g(x)$ is the spatially-varying axial conductance. Intriguingly, Landauer [5] has linked dissipation with loss of information. The erasure of a 'bit' in a computer memory, for example, is the only fundamental operation which requires heat generation. This connection between information, thermodynamics and a learning rule suggests that dendritic information processing may have arisen out of the natural biological prerogative not to waste energy.

Whether or not this turns out to be useful way of looking at things, certain problems still have to be addressed. One problem is the 'shunting' of signals. Compartments carrying a signal (whether synaptically or intrinsically generated), must be stopped from learning, because the simplest way to remove a voltage curvature is to remove the current causing it, or to shunt the signal away using high leakage. Another problem is that 'hotspots'—clusters of channels—cannot develop, since they would quickly disintegrate under the force of their own voltage perturbations. These effects of the learning rule run counter to empirically observed channel clustering [4]. They also run counter to our intuition, for they suggest that the best way to avoid distortion of a signal is to have no signal to begin with.

These problems turn out to be artifacts of compartmental modelling, in which a dendrite, for example, is treated as a set of discrete-space points, while the channel densities are treated as continuous variables (like a 'fluid'). In reality, the converse is true: cell membrane is a continuous 2D fluid while channels themselves are discrete particles. A leakage channel can never really be 'at the same place' as a synaptic one. With this shift in perspective, we can see that the 'shunting scenario' never really occurs. Furthermore, calculations show that a *particle* version of the learning rule enables channels with correlated currents to cluster, while a *fluid* rule would predict their dispersion.

Finally, implicit in this discussion, though ignored thus far, is the issue of membrane-level biological mechanisms for such self-organisation. The rule in equation 1.2 suggests experiments which could confirm a self-organisational model. There are two membrane-level mechanisms which might implement such a rule. The first, *activity-dependent motion* of the channels in the 2D plane of the membrane, would require that an individual channel move according to the time-integrated product of its current with the local dendritic electrical field. This is possible. Experiments [6] and models [3] have documented the electrokinetic redistribution of channels, though no-one has shown this to be dependent on channel activity. The second mechanism, *differential degradation* of the channel proteins, is based on the fact that membrane proteins have limited life-times. A channel could be chemically 'labelled' for recycling with a probability proportional to the correlation in equation 1.2. Such 'sorting' of proteins using 'molecular filters' is known to occur. Again, the crucial experimental point is to demonstrate the dependence of such degradation on channel activity.

These and other issues, including interactions with network dynamics, are addressed in greater detail in a forthcoming PhD thesis [1].

Acknowledgements Supported by a ESPRIT Basic Research Action 3234 on subsymbolic computing. Thanks to Luc Steels, Terry Sejnowski and Tim Smithers.

References.

[1] Bell A.J. 1992. in Moody J. et al (eds), *Adv. Neur. Inf. Proc. Sys 4*, Morgan-Kaufmann 1992.   Bell A.J. 1993. *PhD thesis*, in preparation.

[2] Brown T. et al 1991. in Lippmann R. et al. (eds), *Adv. neur. inf. proc. sys. 3*, Morgan Kaufmann.   Mel B. 1991. *Neural Computation*, 4, 502-517.

[3] Fromherz P. 1988. *Proc. Natl. Acad. Sci. USA* 85, 6353-6357.
Poo M-M. & Young S. 1990. *J. Neurobiol.* 21, 1, 157-168.

[4] Jones O.T. 1989. *Science*, 244, 1189-1193.

[5] Landauer R. 1961. *IBM J. Res. Dev. 5, 183-91.*

[6] Lo Y-j. & Poo M-m. 1991. *Science*, 254, 1019-1022.
Stollberg J. & Fraser S.E. 1990. *J. Cell Biol.*, 111, 2029-2039.

[7] Rinzel J. & Ermentrout G. 1989. in Koch C. & Segev I. (eds) 1989, *Methods in Neuronal Modeling*, MIT Press.

2

A MODIFIED HODGKIN-HUXLEY SPIKING MODEL WITH CONTINUOUS SPIKING OUTPUT

Frédéric E. Theunissen[1]
Frank H. Eeckman[2]
John P. Miller[1]

[1] *Dept. of Mol. and Cell Biology, University of California, Berkeley CA 94720 and* [2]*ISCR, Lawrence Livermore Lab, Livermore CA 94551*

ABSTRACT

We have used a modified Hodgkin and Huxley (HH) description of voltage dependent channels to model the currents involved in spike generation in the cricket cercal sensory system. We included this spiking model in a compartmental neuron model in order to obtain the correct scaling factor between the current actually injected in an experiment and the current arriving at the spiking initiating zone. After fixing the parameters for our spiking model so that we were able to match the spike shape of the real neuron, we found ourselves with a model which showed an almost continuous intensity versus rate characteristic. The rate of firing would increase smoothly from 0 as the intensity of the stimulus was increased. This is result is radically different from the typical bifurcation one finds in a traditional HH model. Our results were reminiscent of the characteristics of neurons with additional A-currents. We found that neither an additional A-current nor a consideration of sources of current noise is needed to obtain the continuous firing rate. Noise is, however, necessary to model the variability in the firing rate and therefore to investigate neural coding.

2.1 INTRODUCTION

Realistic modeling of spiking neurons requires not only an accurate description of the passive electroanatomy of the cell (as it relates to the effective integration of presynaptic inputs) but also an accurate description of the spiking mechanism, as the spike train is the ultimate output of the unit. An accurate spiking model is of particular importance when the modeled cell is capable of repetitive firing and possibly "codes" by modulating its spike rate or statistics. Many such neurons (in particular sensory and motor neurons) are capable of firing rates which vary continuously from 0 to their saturating value. For example, we can record from interneurons in the cricket cercal sensory system which respond to increasing steps of current by firing repetitively with steady rates which vary continuously from 0 to 200 spikes per second. Similar behaviors have been observed in crustacean motor neurons (Connor et al, 1977), cat motor neurons (Calvin & Stevens, 1968) and in auditory sensory neurons (Yu & Lewis, 1989) to mention a few.

However, the original (and universal) spiking model based on the kinetics of each of the different classes of channels, the Hodgkin-Huxley (HH) model, shows a strict non-linearity as the neuron is driven into repetitive firing, with the firing rate jumping from zero to a value very close to the saturating level. This phenomenon is well understood mathematically and is called a Hopf bifurcation (Rinzel & Ermentrout, 1989).

Connor et al. (Connor, 1977) modified the HH model to match the kinetics of Na and K channels observed in the crustacean motor neuron and added a description for an additional channel which carried a transient potassium current (A current). With their model, they were able to not only match the currents of their specific channels, but also to match the low firing rates that were observed experimentally. In another line of studies, Yu and Lewis (1989) showed that the Hopf bifurcation in the HH model dynamics could be linearized simply by adding noise to the spiking model. Using the full HH description with an additional gaussian current noise, they were able to match the large dynamic firing range of a frog saccular neuron as well as its statistics as described by the cycle histograms. Confronted with these two alternatives, we attempted, in previous work, to match both the frequency versus current injection (FI) plot and its variance in interneurons of the cricket cercal sensory system, starting with the Fitz-Hugh Nagumo reduction of the HH model (FitzHugh, 1961). We showed that we could match the FI plot by either adding a transient shunting A- like current or by adding noise. However to match both the FI plot and the variance a unique combination of shunting current and noise would work (Theunissen et al, 1991).

In this work, we attempted to repeat these results by restricting ourselves to a modified HH model. We were inspired by theoretical work from two teams. On one hand, Rinzel & Ermentrout (1989) showed that a continuous FI plot can be obtained with two dimensional dynamical systems (HH is a 4 dimensional dynamical system). In fact, both the Hopf bifurcation and an arbitrarily slow firing rate (Such as Connor et al.) can be obtained from two-dimensional models, and the difference between the two behaviors becomes clear in a phase plane analysis which displays the nullclines and singular points. On the other hand, Kepler et al (1991) derived a rigorous mathematical methodology to reduce the number of variables of HH like equations by grouping variables which have similar time constants. They showed that a reduced two-dimensional model displayed most of the complex observed phenomena and demonstrated that a two dimensional version of the HH model plus A current could give the correct spike shape as well as the correct continuous FI plot. Since two dynamical variables are sufficient to obtain the continuous FI plot, we attempted to determine parameters for the Na and K current in the HH model so that the shape of the fast spikes and the continuous FI plot of giant interneurons in the cricket cercal sensory system were accurately simulated. We first modified the parameters to match the spike shape and, to our surprise, these initial modifications yielded a continuous firing model. We then reduced our model using the Kepler and Abbot methodology. The phase plane analysis of the reduced model showed the characteristic signature of a continuous FI plot spiking model.

2.2 METHODS

2.2.1 Experimental Data.

FI plots were obtained experirmentally by intracellular injections of current steps into interneurons in the cercal sensory system. Currents up to 5 nA were injected in 0.5 nA increments and voltage traces were recorded simultaneously. From the size of the action potential we could deduce that the electrode was not in the axon of the cell but still relatively close to the SIZ. We also minimized the synaptic input to the interneuron by isolating the sensory input to the cercal system . The cercal system is an air current detection system and the recordings were performed in a sound proof chamber. In addition, we did a series of FI recordings in which we covered the cerci with Vaseline and in this manner eliminated any remaining noise from brownian motion of the air particles. Finally, to obtain the exact shape (i.e. not attenuated) of the action potential, additional recordings were obtained from the axon of the cell.

2.2.2 Model Simulations.

We simulated our electrophysiological experiments with two different models, each allowing for a different level of detail. In the first set of calculations the multi-compartmental simulator NeMoSys (Eckman 1993) which allows for a detailed representation of both the anatomy and the electrophysiological parameters of the cell. We used a 1000 compartment model neuron obtained from reconstructions of dyed filled interneurons. In this model, the presumed location of the electrode site at the SIZ was verified by matching the observed attenuated amplitude of the spikes to the model values. In this manner, the multi-compartment model allowed us to compare directly the FI plots of model vs. data, since the model correctly predicted the relationship between the current injected at the electrode and the depolarization at the SIZ.

The second model was a patch of active membrane which had the same electrophysiological characteristics as the active compartments in the full 1000 comparment model, the only difference being that the equations for the voltage dependent channels were reduced with the Kepler et al methodology to a two dimensional set. With the reduced model, the correct spiking dynamics were preserved both in terms of spike shape and continuous firing rates, were preserved. In addition, the reduced model allowed us to analyze the dynamical behavior of our active membrane using two dimensional phase plane analysis. The phase plane calculations were done with the UNIX version of PHSPLAN (Developed by Bard Ermentrout).

a. Passive compartments.

The electrophysiological parameters for the passive properties of our neuron were set from previous complex input impedance experiments. The values are: $R_m = 415 \ \Omega \cdot cm^2$; $R_i = 100 \ \Omega \cdot cm$; $Cm = 1\mu F/cm^2$.

b. Spike-generating currents.

The active compartments had additional excitatory and inhibitory currents which were modified versions of HH sodium and potassium currents. Starting with the classic HH equations we modified the slope and the location on the voltage axis of the activation (m and n) and inactivation (h) variables of the model.

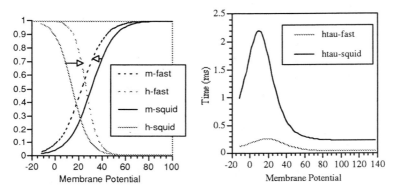

Figure 1.

Na channel m(inf) and h(inf) parameters and time constant t(h). Comparison for the squid and our fast model. Note that the curves for m and h are shifted in opposite directions : the inactivation is shifted upwards with respect to the activation. m was set to be instantaneous in our model.

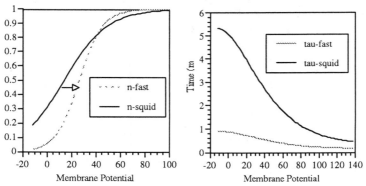

Figure 2.

K channel n(inf) parameter and time constant $\tau(n)$.

The modifications were done recursively, by using NeMoSys interactively, until the spike shape in the model matched the spike shape recorded in the axon. The parameters in the model corresponding to the fast spikes of these interneurons are referred as m-fast, h-fast, etc.

As seen in figs 1 and 2, the major differences between our model and the classic HH kinetics are 1) the shift in the inactivation, both in terms of h and n, in the positive direction on the voltage axis with respect to the activation and 2) much shorter time constants. The exact parameter values and the form of the equations for n, m, and h are shown in the appendix. These simple modifications yielded a very good fit for the spike shape and also changed the behavior of our model spiking neuron from one displaying a Hopf bifurcation to a neuron with continuous firing rate.

2.3 RESULTS

2.3.1. Matching the spike shape at the axon.

Both the multi-compartmental full HH model and the patch of membrane with the reduced two dimensional model gave very good fits for the shape of an action potential as shown on figure 3.

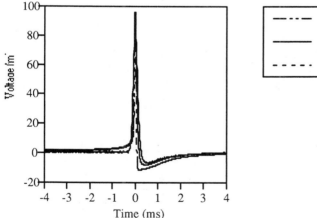
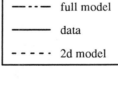

Figure 3.
Shape of an action potential as recorded in the axon (data) and as predicted by the models: the NeMoSys simulation (full model) and the single compartment calculation (2d model).

2.3.2. Matching the FI plot.

The model FI plot obtained with NeMoSys matched the experimental curves surprisingly well without any further modifications in the active compartments. The slopes are

practically identical as shown in Figure 4. To align the curves on the voltage axis the
shunting at the model injection site was varied. Figure 4 shows a characteristic FI plot for a
single cell in the two experimental protocols: with and without the cerci covered. By
shunting off all receptor noise, we effectively increase the threshold of the cell. The effect
of noise on threshold and encoding of information will be addressed in future work.

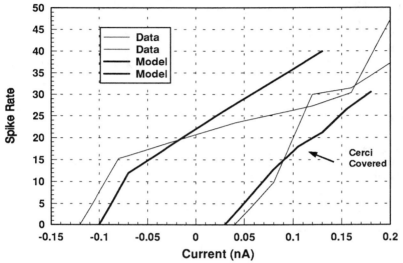

Figure 4.
Frequency of firing versus current injected for the giant interneuron (data) and for the model. The curves are obtained
for two levels of background noise : a. for the animal exposed to the background noise inside a sound proof box and
b. for the animal with its sensory receptors shut off. The only noise present in the second case is the intrinsic noise of
the receptors.

c. Phase plane analysis.

To achieve a better understanding of the effects of our modifications on the dynamics of the
spiking membrane and to see how we obtained a continuous firing rate, we applied phase
plane analysis to our model. We reduced the model as mentioned above to a two
dimensional model where V and W are the independent variables. V is the voltage of the
membrane and the voltage which sets m through minf(V). W is a "retarded' potential which
sets the time course of the "slower" variables n and h. The differential equation for W
depends on ninf, hinf, ntau and htau (Kepler et al). The time course of V and W were
calculated for both the reduced classical HH and for the reduced model of the modified
(fast) HH we use to model our spikes. We have plotted W versus V during a spike for both
cases in the phase planes of figure 5. Also show in the phase planes are the nullclines, that
is the points where the derivative dV/dt and dW/dt are zero. Below the W nullcline dW/dt is
positive meaning that the trajectories are pushed up in the phase plane. Below the V
nullcline dV/dt is positive meaning that the trajectories are pushed to the right. Where the

two nullclines intersect, we have a singular point where both derivatives are zero. These singular points can be stable or unstable (saddle points or spirals). The shape of the derivative surface (suggested here from our nullclines) will set the behavior of the dynamical system. A clear difference arises between the two phase planes. In the fast HH model (Figure 5A), the V nullcline has an area where its positive slope is steeper than the slope of the W nullcline. This allows for multiple singular points. As the V nullcline is lower and raised by decreasing or increasing the input current, one goes from 3 singular points to 2 to 1.

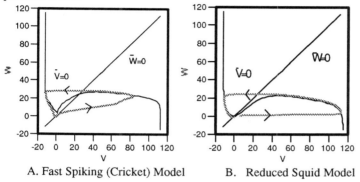

A. Fast Spiking (Cricket) Model   B. Reduced Squid Model

Figure 5.
Phase plane trajectory of a spike using a 2 dimensional reduction of the full HH description. Graph A shows the trajectory for a spike with the fast HH kinetics and Graph B shows the trajectory for a spike with classical HH kinetics. The phase planes also show the nullclines for V and W.

The 3 singular points are stable, unstable and unstable. As the V nullcline is raised, the stable and the first unstable singular point (a saddle point) coalesce, yielding zero frequency firing (Rinzel and Ermentrout, 1988). This particular situation is the one we chose to display in figure 5A. In the traditional HH model, there is only a single singular point which jumps from stable to unstable as the V nullcline is raised. As for the small width of our spike (about 0.15 ms at half width), we can observe that the spike raises and falls faster in the fast HH model. As seen from the phase plane, the trajectory is further away from the V nullcline yielding larger dV/dt values and also the dW/dt values are larger as suggested by the greater displacement in the W direction. The phase plane analysis of the reduced model confirms and explains the results obtained in the NeMoSys simulations.

2.4 CONCLUSION

The Hopf bifurcation in a FI plot obtained from a HH model can be eliminated by complementing the model with an additional A-current (Connor et al, 1977) or additional current noise (Yu & Lewis, 1989). Here, we have shown that the Hopf bifurcation can also be eliminated by changing the HH parameters for the Na and K channel without any further modifications. It would be of great interest to investigate if the characteristics of the channels in our model have a physiological correlate. A-channels are not needed to obtain continuous firing rates, but the authors don't know of any case where repetitive firing has been observed with only HH like Na and K channels.

In this study, we have also ignored the effect of noise in the generation of spikes. Our model as described here could not take into account the variability of the firing rate. This variability is present even in the case where the sensory receptors are shut off and will have an effect on the spike generation strategy to maximize the encoding of neural information.

APPENDIX.

The equations in the fast HH model are :
1. The differential equations

$$Cm\frac{dV}{dt} = I - g_{Na}m_\infty^3 h(V - E_{Na}) - g_k n^4(V - E_k) - gl(V - E_l)$$

$$\frac{dh}{dt} = \frac{h_\infty(V) - h}{\tau_h(V)}, \text{ and similarly for } n.$$

The conductance values are g_{Na} = 117.4 mS/cm^2, g_K = 622 mS/cm^2, g_l = 2.33 mS/cm^2. The reversal potentials are E_{Na} = 115 mV, E_K = -12 mV, E_l = -3.47 mV,

2. The values of m, h and n at infinity.

$$m_\infty(V) = \frac{1}{1 + e^{-\frac{V - 22.72}{9.55}}}; \quad h_\infty(V) = \frac{1}{1 + e^{\frac{V - 26.84}{5.07}}}; \quad n_\infty(V) = \frac{1}{1 + e^{-\frac{V - 26.32}{9.45}}}$$

where the voltage is expressed in mV with 0 mV being rest.

3. The time constants

$$\tau_h(V) = \frac{0.4}{e^{\frac{V - 24.06}{14.04}} + e^{-\frac{V - 19.54}{20.0}}} + 0.03; \quad \tau_n(V) = \frac{1.76}{e^{\frac{V + 7.49}{34.83}} + e^{-\frac{V - 8.46}{52.56}}} + 0.16$$

where the time is expressed in ms and voltage in mV with 0 mV being rest.

REFERENCES.

1. L.F. Abbott and T.B. Kepler. (1990) Model Neurons: From Hodgkin to Hopfield. In: *Statistical Mechanics of Neural Networks*. Garrido L. (ed). Springer Verlag, Berlin.

2. J.A. Connor, D. Walter and R. McKown. (1977) Neural Repetitive Firing. *Biophysical Journal* **18**:81-102.

3. F.H. Eeckman, F.E. Theunissen and J.P. Miller (1993) NeMoSys: a system for realistic neural modeling. In Neural Systems: Analysis and Modeling. Eeckman (ed). Kluwer Academic Publishers, Norvell MA.

4. R. FitzHugh. (1961) Impulses and physiological states in theoretical studies of nerve membrane. *Biophysical Journal* **1**: 445-466.

5. T.B. Kepler, L. F. Abbott. and E. Marder (1992). Reduction of Conduction-Based Neuron Models. Biol. Cybern. 66:381-387.

6. J. Rinzel and G.B. Ermentrout. (1988) Analysis of neural excitability and oscillations in *Methods in Neural Modeling*. Koch and Segev (eds). MIT Press Cambridge MA.

7. F.E. Theunissen, F.H. Eeckman and J.P. Miller.(1993) Linearization by noise and/or additional shunting current of a modified FitzHugh Nagumo spiking model. In *Neural Systems: Analysis and Modeling*. Eeckman (ed). Kluwer Academic Publishers, Norvell MA.

8. X. Yu and E.R. Lewis. (1989) Studies with spike initiators: Linearization by noise allows continous signal modulation in neural networks. *IEEE Trans. on Biomed. Eng* **36**:36-43.

THE NEURONE AS A NONLINEAR SYSTEM : A SINGLE COMPARTMENT STUDY

P.G. Hearne, J. Wray, D.J. Sanders
E. Agar and G.G.R. Green

Neural Network and Sensory Systems Group, Department of Physiological Sciences, The Medical School, Newcastle upon Tyne, NE2 4HH UK

ABSTRACT

A neurone is described as a single-input single-output nonlinear system. The Volterra kernels of a single compartmental model are extracted using a feedforward network. We then demonstrate that the derived kernels can represent features of the compartmental model. This technique can also be applied to intracellular recordings to allow the construction of improved models.

1 INTRODUCTION

A neurone can be regarded as a nonlinear system because the membrane resistance is a nonlinear function of the membrane potential. The characterisation of this system is one of finding a mathematical description which both captures its main features and is meaningful. Two alternative approaches have been used in the study of neurones. First, the specification of a mechanistic model followed by parameter estimation and, second, the use of *black box* identification techniques.

Using the former approach, the work of Hodgkin and Huxley [1] has developed into the compartmental modelling method. This involves the application of steady-state signals to construct sets of coupled ordinary differential equations. The model is said to describe the system modelled because each variable is thought to represent a feature of the mechanism underlying the reponse of the cell. This class of model has been very successful in both clarifying theoretical issues and in reproducing experimental data. There are, however, problems with their application, particularly concerning the estimation of parameters. With stimuli that are typically used in experimental neuroscience a correct fit of data is insufficient to uniquely justify a particular, complex model.

Black box techniques [2] are used to explicitly represent the nonlinear nature of the system. The output of a time-invariant, analytic system with finite memory can be represented as a Volterra series whose terms characterise the dynamic response of its steady-state, linear and nonlinear components. The use of stochastic stimuli to probe the system means that its functional characteristics can be fully described. The Volterra approach to modelling characterises a system as a mapping between its input and output spaces. For a causal system the discrete-time Volterra series is given as,

$$
v(t) = h_0 + \sum_{\tau_1=0}^{T} h_1(\tau_1)i(t-\tau_1) + \sum_{\tau_1=0}^{T}\sum_{\tau_2=0}^{T} h_2(\tau_1,\tau_2)i(t-\tau_1)i(t-\tau_2) + \cdots
$$
$$
+ \sum_{\tau_1=0}^{T} \cdots \sum_{\tau_n=0}^{T} h_n(\tau_n,\cdots,\tau_n)i(t-\tau_1)\cdots i(t-\tau_n) + \cdots
$$

where $i(t)$ and $v(t)$ are the input and output at time t respectively, T is the total memory of the system and $h_n(\tau_1 \cdots \tau_n)$ is the nth order Volterra kernel.

In this paper we use an artificial neural network to extract the Volterra kernels of a lumped soma model and show how these kernels can be interpreted in terms of the variables of the compartmental model.

2   COMPUTATION OF VOLTERRA KERNELS

A major difficulty in the application of Volterra series to system modelling concerns the estimation of system kernels. The measurement of the Volterra kernels has only been possible where the contributions of each of the terms in the series can be separated from the total systems response, for example for a finite order system [2]. Neural networks can be used for time-series prediction when the inputs are regarded as a window in time. Previously we have shown how a network has a Taylor series expansion which can be expanded and rearranged to give the discrete Volterra kernels of the system [5, 4], and the general $n'th$ order kernel is given by,

$$
h_n(j_1, j_2, \cdots, j_n) = \sum_{k=1}^{K} c_k a_{nk} w_{j_1 k} w_{j_2 k} \cdots w_{j_n k}
$$

where c_k is the weight of the connection from the kth hidden unit to the output unit, a_{nk} is the coefficient of the nth term of the Taylor series expansion of the

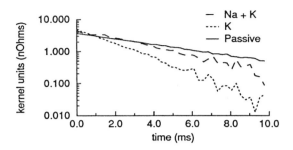

Figure 1   First order kernels

output function for the kth hidden unit, w_{jk} is the weight from the jth input unit to the kth hidden unit and K is the number of hidden units.

3   METHOD

The output, $v(t)$, represents the membrane potential of a lumped soma model with the following characteristics: radius, $r = 20\mu m$; specific capacitance, $C_m = 1\mu F.cm^{-2}$; leak resistance, $R_m = 5K\Omega.cm^2$; resting potential, $E_{rest} = 0mV$; leak reversal potential, $E_{leak} = 0mV$; potassium reversal potential, $E_K = -5mV$; sodium reversal potential, $E_{Na} = 115mV$. Equations were solved by the Euler method using software written in our laboratory. A feed-forward network, 127 input units; 8 hidden units; 1 output unit, was trained using 8192 input-output mappings from the $2sec$ simulation. The input vector represented a time window of 127 values of the past input. The input signal was Gaussian white-noise with a variance of $1nA$ sampled every $0.244ms$. The network was trained using a backpropagation algorithm until a fit was obtained. Three input-output mappings were learnt; a passive compartment, a compartment with a potassium current(I_K) and a compartment with an additional sodium current (I_{Na}). Kernels of the system were then extracted using the method outlined above.

4   RESULTS

Figure 1 shows the first order kernel obtained from the three simulations. For the passive compartment, when the system is linear, it shows the impulse response of the system. The time constant, τ, for this system can be estimated from the value on the abscissa which corresponds to a drop in the impulse response to $1/e^{th}$ of its value; this is $5ms$ which matches the value of R_mC_m given above. The slope of the first order kernel was increased by the addition of I_K

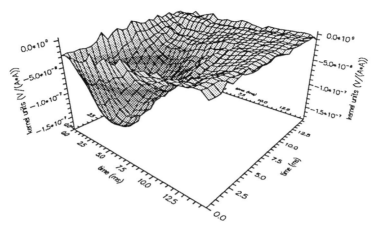

Figure 2   Second order kernel for I_K model

[3], $G_{K(max)} = 400mS.cm^{-2}$. This would be expected solely from the addition of another conductance. Nonlinear effects are also present, as indicated by the second order kernel in figure 2; the negative values of the kernel indicate that the conductance is outward rectifying. The lag at which the minimum occurs indicates the time delay associated with I_K. The effects of I_K can be shown to be restricted to low order because the offset, first and second order kernels are sufficient to predict the membrane potential.

When I_{Na} [3], $G_{Na(max)} = 200mS.cm^{-2}$, and I_K, $G_{K(max)} = 40mS.cm^{-2}$, are added the first order kernel is affected because the sodium current has a significant linear component at a faster time scale. Examination of the second order, and slices through the third order kernel reveal that there is an almost instantaneous nonlinearity which is an order of magnitude greater than that in the potassium-only case. Figure 3 shows the predicted (a) and actual (b) time-series for all orders of nonlinearity. In fact, because the higher order powers of the weights are small, good fits are obtained using the kernels upto fourth order.

5   CONCLUSION

We have shown it is possible to obtain the kernels of a discrete-time Volterra system using a neural network method. We have used it to identify features in the kernel representations of a model neurone. Because the kernel extraction is a general method, employing few assumptions, it can be used to give the kernels of real cells which can subsequently lead to improved models by locating

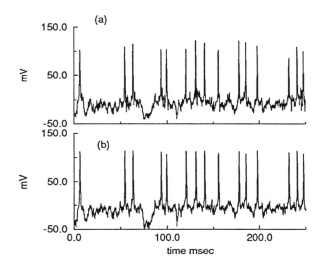

Figure 3   Predicted (a) and actual (b) time series

and labelling features in the Volterra kernels by manipulation of the model parameters.

REFERENCES

[1] A. L. Hodgkin and A. F. Huxley. A quantitative description of membrane current and its application to conduction and excitation in nerve. *Journal of Physiology*, 117:500–544, 1952.

[2] M. Schetzen. *The Volterra and Weiner theories of nonlinear systems*. John Wiley and Sons, New York, NY, 1980.

[3] R. D. Traub. Motorneurons of different geometry and the size principle. *Biological Cybernetics*, 25:163–176, 1977.

[4] J. Wray. *Theory and application of artificial neural networks*. PhD thesis, University of Newcastle-upon-Tyne, 1992.

[5] J. Wray and G. G. R. Green. Analysis of networks that have learnt control problems. In *Proceedings of the International Conference Control '91, Herriot-Watt, UK*, pages 261–265. Institution of Electrical Engineers, 1991.

UNSUPERVISED HEBBIAN LEARNING AND THE SHAPE OF THE NEURON ACTIVATION FUNCTION

Jonathan L. Shapiro and Adam Prügel-Bennett

Department of Computer Science, The University, Manchester, Oxford Road, Manchester, M13 9PL U.K.

ABSTRACT

A model describing unsupervised Hebbian learning in a neuron whose output is a continuous, nonlinear function of its input is presented. The model can be solved exactly for simple pattern sets. The processing function which the neuron learns is strongly dependent on the shape of the activation function. When the activation function is linear, the neuron learns to compute a statistical property of the collection of input patterns. When it grows more slowly than linear, the neuron learns to the average of the pattern set. When the activation function grows sufficiently faster than linear, the neuron forms a grandmother cell representation of the patterns.

4.1   INTRODUCTION

This paper reports results of mathematical analysis on unsupervised Hebbian learning in an extremely simplified model of a single neuron. We show that the shape of the neuron activation function can control what the neuron learns. In particular, the processing function which the neuron performs after learning can be affected in a crucial way by the degree of nonlinearity of this function.

The ingredients of the model are standard idealizations. The input signal at the ith synapse is represented as a continuous number x_i. The post-synaptic potential (PSP) of the cell, V, is modelled as the weighted sum of inputs $V = \sum_i x_i w_i$, where the weight w_i represents the synaptic strength of the ith synapse. The neuron fires in response to the post-synaptic potential with a frequency which

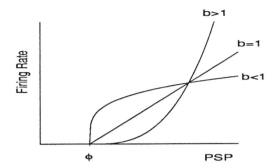

Figure 4.1   The neuron activation function for different values of b.
Below the threshold ϕ, the neuron does not fire.

is determined by a function of the PSP; this function is called the activation
function, and is denoted $A(V)$.

The activation function is assumed to take a very simple form – it is zero below
a threshold and increases as a power above the threshold,

$$A(V) = \left\{ \begin{array}{ll} (V - \phi)^b & V > \phi \\ 0 & V \leq \phi \end{array} \right. . \tag{4.1}$$

This is characterized by two parameters: b measures the non-linearity of the
response, and ϕ is the threshold. This is shown in figure 1.

The synaptic weights are modified according to a Hebbian rule with an additional term which prevents the weights from growing too large. The change in
the weights for each pattern and learning step is,

$$\delta w_i \propto A(V)(x_i - Vw_i). \tag{4.2}$$

The first term is a form of Hebbian learning — the change in weight is proportional to the product of the input and output activations. The second term
prevents unbounded growth of the weights. Learning is unsupervised: the neuron starts in an arbitrary state, it then repeatedly responds to patterns, and
the weights are changed via equation 2.

4.2   LINEAR VERSUS NONLINEAR NEURONS

Hebbian learning with linear activation functions has been widely studied [1, 2].

This single neuron model (equation 4.2) with $A(V) = V$ was proposed by Oja [3]. He showed that under these dynamics the weight vector converges to the eigenvector with the largest eigenvalue (i.e. the principal component) of the correlation matrix $M_{ij} = < x_i x_j >$, where $< \ldots >$ denotes average over the patterns. This has an important statistical interpretation – it is the first stage of principal component analysis (PCA), which is a well known method of data compression. Using these neurons as building blocks, Oja and other researchers have developed neural network architectures which perform principal component analysis [1, 4]. Because the linear neuron learns the principal component of the correlation matrix which is a statistical property of the ensemble of patterns, it learns about the pattern *set*, not about individual patterns.

We have solved the nonlinear generalization of Oja's model for a range of different pattern sets and find that these neurons learn very differently from linear neurons. The neurons have two modes of operation. The weights after learning are either close to one of the patterns or to some mixture of patterns. When the neuron learns close to a single pattern, it functions as a discriminator or matched filter; it has learned to recognize one pattern in the presence of the other patterns. When it learns to the mixture, it is more like the principal component (in some cases it *is* the principal component); it cannot recognize any individual pattern but has learned a property of the pattern set.

4.3   RESULTS

The effect of the nonlinearity is well illustrated by the case of a neuron learning two normalized input patterns, $\vec{x}^1$ and $\vec{x}^2$ with correlation $c = \vec{x}^1 \cdot \vec{x}^2$. When the correlation is small, there are two stable fixed points, one near each pattern. As the correlation increases, these move away from the patterns. There is a third fixed point midway between the two patterns but this point is unstable. For highly correlated patterns, the only stable fixed point is halfway between the patterns. There is a critical correlation, c^*, which depends upon the nonlinearity through $c^* = (b - 1)/(b + 1)$, at which the two stable fixed points coalesce with the unstable fixed point. The fixed points for $b = 2$ are shown in figure 2.

Similar behaviour occurs with many patterns; the neuron either learns to discriminate one pattern from all the others, or learns a mixture of all patterns. To illustrate this, we have simulated a neuron learning six characters. This is shown in figure 3. Figure 3(a) shows the patterns. Figure 3(b) and 3(c) show the weight vector at various learning times for $b = 2$ and $b = 1.3$ respectively. In the first case, the neuron learns to recognize the 'U' which is the pattern

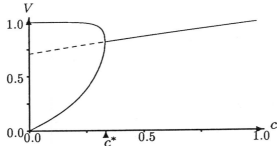

Figure 4.2 The PSP of the neuron after learning when a pattern in input versus the correlation between patterns. Here $b = 2$, there are two patterns, and $c^* = 1/3$. The dashed curve shows the unstable fixed point.

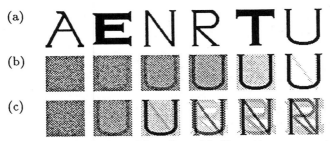

Figure 4.3 Simulation of learning. (a) Six patterns presented to the neuron, consisting of 4096 pixels. (b) The synaptic weight vector after 0, 80, 84, 86, 90 and 100 learning steps for $b = 2$. The neuron has learned to recognize U in the presence of other patterns. (c) The synaptic weight vector after 0, 15, 20, 100, 150, and 200 learning steps for $b = 1.3$. The neuron has learned a mixture of the letters.

it was closest to initially. In the second, although the synapses start to learn the 'U', ultimately they converge to a mixture of some of the patterns. This is because b was not sufficiently large to distinguish patterns with this correlation.

We have also solved the model for a large number of random patterns when the number of synapses N diverges. This was done by applying methods similar to those used by Amit *et al.*[5] and Gardner [6] to study the Hopfield model and perceptrons. Details are presented elsewhere [7].

The neuron learns to distinguish one random pattern from the others if the activation function grows faster than linear ($b > 1$) and the total number of patterns P is not too large. The maximum number of patterns which can be input to the neuron without destroying the discrimination ability is a complicated function of N and b, but it can be large compared with N. For example, if $b = 2$ a neuron with 10 000 synapses can learn to distinguish a single pattern in from 124 000 patterns.

We have modeled the neuron using standard idealizations, only the form of the activation function is novel, and the results are surprisingly simple. The processing function which the neuron learns is determined by the nonlinearity parameter b and the correlation between the patterns. The neuron can either learn to compute a statistical mixture of the patterns, like the principal component of the correlation matrix in the case of $b = 1$ or the average, or can form a grandmother cell representation of one of the patterns. In the latter case, the pattern which is represented depends on the initial weights. Of course a realistic set of patterns may contain a range of two-pattern correlations. In this case the neuron will perform a kind of cluster analysis, forming an average representation of closely cluster data but able to distinguish different clusters. It is also possible to extend these results to cover more complex activation functions, such as sigmoid functions, by approximating the function the by appropriate power law in the region where the inter-pattern correlations occurs.

REFERENCES

[1] Oja, E., "Neural Networks, Principal Components, and Subspaces," Int. J. of Neural Systems, 1(1):61–68, 1989.

[2] Linsker, R., "Self-Organization in a Perceptual Network," Computer, March 1988:105–117, 1988.

[3] Oja, E., "A Simplified Neuron Model As a Principal Component Analyzer," J. Math. Bio., 15:267-273, 1982.

[4] Sanger, T. D., "Optimal Unsupervised Learning in a Single-Layer Linear Feedforward Neural Network," Neural Networks, 2:459-473, 1989.

[5] Amit, D. J., Gutfreund, H., and Sompolinsky H, "Statistical Mechanics of Neural Networks near Saturation," Ann. Phys, 173:30–67, 1987.

[6] Gardner, E., "The space of interactions in neural network models," J. Pys. A, 21:257–270, 1988.

[7] Prugel-Bennett A. and Shapiro, J. L., "Statistical mechanics of unsupervised hebbian learning," Submitted to J. Phys. A., 1992.

A METHOD FOR ESTIMATING THE NEURAL INPUT TO A NEURON USING THE IONIC CURRENT MODEL

Yoshimi Kamiyama[†], Hiroyuki Ishii[‡] and Shiro Usui[‡]

[†] *Department of Knowledge-based Information Engineering*
[‡] *Department of Information and Computer Sciences*
Toyohashi University of Technology
Toyohashi 441, Japan

1 INTRODUCTION

The ionic current properties of a neuron play a fundamental role for generating the voltage responses such as post synaptic potential and action potential. The voltage- and time-dependent characteristics of each ionic current can be measured by applying voltage clamp techniques which lead to the precise measurement of the membrane ionic currents underlying the voltage response. From these measurements it is possible to obtain the exact mathematical descriptions of the ionic current in response to electrical and chemical stimuli similar to Hodgkin-Huxley equations[1]. The ionic current model provides a basis for understanding how the individual ionic current flows during the response and/or predicts the blocking effect of a particular ionic channel.

In this study, we have developed a method for estimating input signals to a neuron using the ionic current model. Although voltage response of a neuron is generated by chemical neurotransmitter and/or direct current input through gap junction in the real neural system, these signals can not be measured directly by the conventional electrophysiological experimental method. The proposed method enables us to estimate the total input current generating the experimentally recorded voltage responses of a neuron.

2 COMPUTATIONAL METHODS

The component of each ionic current and membrane capacitance of a neuron give a following differential equation:

$$C\frac{dV}{dt} = I_{inp} - \sum_{ion} I_{ion}(V, t) \qquad (1)$$

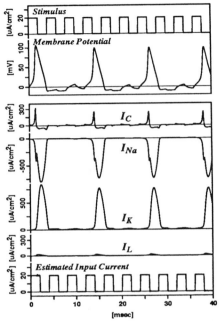

Figure 1   Estimation of Input
Current using the Hodgkin-
Huxley Model.   Voltage responses were generated by applying pulse current stimuli
(upper panel). Capacitive current (I_C), each ionic current
(I_{Na}, I_K, I_L) and "Estimated
Input Current" were numeri-
cally estimated from the membrane potentials by the proposed method.

where, I_{inp} is a total input current to the cell. The conventional voltage clamp
method is achieved to control membrane potential constant ($dV/dt = 0$) and
measures the resulted I_{inp}. However, from mathematical point of view with
the model, if the membrane is clamped by the experimentally recorded voltage
response, then the estimated I_{inp} is the current generating the response. That
is, the proposed method is based on an idea of applying "response clamp" to
the model. Since the first derivative and each ionic current are function of
the potential change, the total sum of these currents represents the signals
transmitted from the external neurons.

In the case of Hodgkin-Huxley model, the input current is described as

$$I_{inp} = C\frac{dV}{dt} + I_{Na}(V,t) + I_K(V,t) + I_L(V,t). \qquad (2)$$

In order to evaluate the method, we first calculated the action potential data
by computer and used them as a test data. Fig. 1 shows the calculated action
potential to the pulse stimuli and each current component was estimated by
the proposed method. Although it is difficult to find the input current from
the spike trains, the estimated I_{inp} agreed well with the input current which
produced the test data.

3   RESPONSE DYNAMICS OF RETINAL HORIZONTAL CELL

In order to show the validity of the method to a real neural system, we analyzed the light response properties of the retinal horizontal cell with the ionic current model. Horizontal cells are second-order neuron in the retina, and they make synaptic connections with photoreceptors and gap junctional connections with neighboring horizontal cells. The ionic currents of the horizontal cell are identified by the voltage clamp study[2, 3] and we reconstructed the currents by Hodgkin-Huxley type equations[4].

3.1   Ionic Current Model of a Solitary Horizontal Cell

In the solitary horizontal cell from the goldfish retina, there are at least four major voltage-dependent currents[2]: calcium current (I_{Ca}), anomalous rectifying potassium current (I_{Ka}), delayed rectifying potassium current (I_{Kv}), transient outward potassium current (I_A), and one pharmacologically activated current[3]: glutamate-induced current (I_{glu}). Glutamate is a neurotransmitter from cone to horizontal cell. The component of each ionic current, membrane capacitance and gap junctional current give the following differential equation:

$$C\frac{dV}{dt} = I_{gap} - (I_{Ca} + I_A + I_{Ka} + I_{Kv} + I_{glu} + I_l) \qquad (3)$$

Since I_{gap} and I_{glu} are activated by the external input, the input current is defined as follows:

$$I_{inp} \equiv I_{glu} - I_{gap} = C\frac{dV}{dt} + I_{Ca} + I_A + I_{Ka} + I_{Kv} + I_l \qquad (4)$$

Here, we cannot separate I_{inp} into each component, but if we stimulate the retina by a full field stimulus of light, then I_{gap} can be neglected due to the uniform polarization of the horizontal cells.

3.2   Ionic Current Characteristics in Horizontal Cell during a Light Response

Intracellular recordings were made on the L-type horizontal cell in the carp retina[4]. The L-type horizontal cell is believed to receive the main input from red-sensitive cones via glutamate and send a GABAergic negative feedback signal to the cone[5]. The L-type horizontal cell responds with hyperpolarization to any stimulus condition. However, the response waveform of the cell strongly

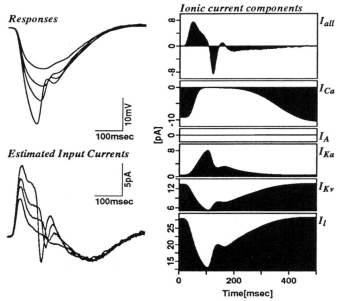

Figure 2   Estimated Ionic Current Dynamics in the L-type Horizontal Cell during the Light Response. Each ionic current component is calculated using the model from the voltage response (upper left). I_{Ca} slowly changes with hyperpolarization. I_A is almost constant. I_{Ka}, I_{Kv} and I_l flow passively to the response. Note that the estimated input current, I_{inp} shows faster dynamics and the result indicates that the horizontal cell membrane acts like a low-pass filter (modified from [4]).

depends on stimulus wavelength, size and intensity, and shows nonlinear dynamical behaviors[6].

Fig. 2 (right half) shows the ionic current components for an L-type horizontal cell light response (top left) calculated by the ionic current model. I_{Ca} gradually changes by the membrane hyperpolarization. I_A is negligible. I_{Ka}, I_{Kv} and I_l show the similar waveform to the response and thus these currents are passive components. However, the estimated input current (left bottom), I_{inp}, shows faster dynamics than the voltage response, and note that the inflection like a notch after the peak is more pronounced. This result clearly demonstrate that the L-type horizontal cell membrane acts like a low-pass filter and indicates the advantage for analyzing the input current to the cell response.

4 DISCUSSION

In this study, we proposed a method for estimating the input current using the ionic current model and applied to analyze the retinal horizontal cell responses. Horizontal cell behaves like a low-pass filter because of their large capacitance and the electrical coupling among neighboring cells, i.e., the voltage response of the cell is smoothed both in time and space. These properties have made the study of synaptic mechanism difficult. However, the present study provides us an useful method and clues for understanding the synaptic mechanism. The estimated input current with a faster dynamics and a nonlinear behaviors were more pronounced in the input current level which were never seen in the voltage recordings. Thus, the analysis of the input current provides a view of how visual information is coded in cone-horizontal cell network. In conclusion, merging the proposed method and the experimental techniques such as a selective blocking of the neural pathway with drugs will make further detailed study of the real neural systems.

REFERENCES

[1] Hodgkin, A.L. and Huxley, A.F., "A quantitative description of membrane current and its application to conduction and excitation in nerve", J. Physiol., 117, 500–544 (1952)

[2] Tachibana, M., "Ionic currents of solitary horizontal cells isolated from goldfish retina", J. Physiol., 345, 329–351 (1983)

[3] Tachibana, M., "Permeability changes induced by L-glutamate in solitary retinal horizontal cells isolated from *Carassius auratus*", J. Physiol., 358, 153–167 (1985)

[4] Usui, S., Ishii, H. and Kamiyama, Y., "An analysis of retinal L-type horizontal cell responses by the ionic current model", Neuroscience Research, Suppl. 15, S91–S105 (1991)

[5] Kaneko, A., "The functional role of retinal horizontal cells", Jpn. J. Physiol., 37, 341–358 (1987)

[6] Usui, S., Mitarai, G. and Sakakibara, M., "Discrete nonlinear reduction model for horizontal cell response in the carp retina", Vision Res., 23, 4, 413–420 (1983)

(Correspondence should be addressed to Shiro USUI)

A MODEL OF A THALAMIC RELAY CELL INCORPORATING VOLTAGE-CLAMP DATA ON IONIC CONDUCTANCES

Diana K. Smetters

Department of Brain and Cognitive Science
Massachusetts Institute of Technology, Cambridge, Massachusetts 02139

ABSTRACT

We have constructed a model of a thalamic relay cell based on voltage clamp data obtained in thalamic slices and isolated relay cells. Analysis of the model provides insight into the role of the various voltage-dependent conductances in the response of these cells to current injection. The model also provides useful information to guide future experiments.

1 INTRODUCTION

Thalamic relay cells have been extensively studied anatomically and physiologically, yet their functional role is still not completely understood. These cells show complex behavior, responding with tonic firing to a stimulus when depolarized, but when the same stimulus is applied from a hyperpolarized potential, they respond with a "low-threshold spike" (LTS) – a large Ca^{++}-mediated action potential. This state-dependency has fundamental implications for the response of these cells to synaptic input.

Switching between these modes is controlled by a large number of voltage- and ion-dependent conductances [9]. As a result, understanding of the functional behavior of thalamic cells would be greatly facilitated by a detailed understanding of the role of these currents in cell behavior. Additionally, such an understanding may provide new insight into the role of membrane nonlinearities in neuronal function in general. To facilitate such an understanding, I have created a model of a thalamic neuron based as closely as possible on voltage-clamp data recently available in these cells, supplemented where necessary with data from other cell types.

2   METHODS

To simplify interpretation, the model utilizes a simple 14-compartment equivalent cylinder model of an average LGN X-cell.[1] The somatic diameter was fixed at the average for cat LGN cells [2]. The dendritic membrane resistivity $(40\ k\Omega - cm^2)$ was chosen based on similarity to the value measured with perforated patch electrodes in hippocampal pyramidal cells [18]. The length and diameter of the dendritic cable and the somatic membrane resistivity were chosen to match the average input resistance $(60\ M\Omega)$ and primary time constant (16.5 msec) of cat LGN cells as recorded with KAc electrodes [4], and to be within the normal range of dendritic length and surface area for LGN cells. The resulting somatic shunt (approximately 10nS) was considered to represent both electrode-induced leak and a potassium-selective voltage-independent current known to be present in these cells. Calcium accumulation in the submembrane space is controlled by a simple first order decay with a time constant of 1 msec. Most simulations were performed using NEURON, developed by Michael Hines. A previous, simplified version of the model was constructed in Genesis.

Ionic conductance models

Three potassium currents have recently been described in thalamic neurons [6, 7, 13], and models of these currents were derived from published data for use in this study. Two of these currents can be seen in Figure 1. Ik2 is a slowly-inactivating K^+ current with fairly slow activation [7]. Iaf is a rapidly inactivating current very similar to Ia as seen in many cell types, and showing two time constants of inactivation in isolated thalamic cells [6]. Ias [13] is a rapidly activating, slowly inactivating current very similar to Id in the hippocampus [19]. The steady-state voltage-dependence and two time constants of inactivation used in the model are based on data from thalamic neurons [13], but activation kinetics were not available and were adapted from recordings in sensory ganglion cells [14].

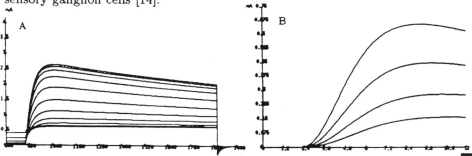

Figure 1   A) The modeled Ik2 at 23degC as current evoked by steps to 0mV from various holding potentials (holding period not shown). B) The modeled Iaf, also at 23degC in response to various voltage steps. (Note different scales.)

The model of the T-type Ca^{++} current, which underlies the LTS, was that

---

[1]LGN cells are thought to meet the necessary criteria for equivalence to a cylinder [2].

developed by Wang *et. al.* [20], modified to incorporate GHK kinetics. Several models of the hyperpolarization-activated cation current, Ih, were developed from the data [12, 17], and one was chosen based on its ability to match current-clamp data from rat LGN cells.

The primary model of the fast sodium current was derived from data obtained in sympathetic neurons by Belluzi and Sacchi [1][2]. In some versions of the model, a delayed rectifier was used as the primary spike repolarizing current in place of the Ca^{++}-activated K current. The kinetics used were those of Lytton and Sejnowski [11][3] The model of the L-type Ca^{++} current was taken from Lytton and Sejnowski [11], and was used both with and without GHK kinetics.

Once derived from the voltage-clamp data, we considered each current's kinetics as fixed, adjusting only the peak conductances in current clamp simulation. These were adjusted largely by trial and error, but close attention was paid to the often widely varying values reported in the literature, and the final values obtained were usually within an order of magnitude.

3 RESULTS

The simple model incorporating a delayed rectifier replicates the basic behavior of thalamic neurons, including the switch between tonic and burst modes. This collection of currents is capable of replicating fairly detailed aspects of thalamic behavior. For instance, at an appropriate holding potential, the model responds to a current step stimulus with a greater than 100 msec delay to LTS, seen in real thalamic cells under a narrow range of stimulation conditions. This delay is due to the interaction between the membrane leak and the kinetics of the T and H currents, without a requirement for activation of an inactivating K^+ conductance. When depolarized to near threshold from a hyperpolarized potential, thalamic cells can show a delay to high-threshold firing of up to as many as 9 seconds, resulting from the slow inactivation of Ias [13]. The model is also capable of replicating this behavior. However, sufficient Ias to robustly cause this behavior is also sufficient to severely attenuate the tail of the LTS. This suggests that the activation processes of Ias and It, which have never been measured in the same cell type, may be carefully balanced against each other, or that some long-duration modulatory process is also involved in this firing delay.

A major inadequacy of this version of the model is that a large delayed recti-fier has not been seen in these cells, even in several extensive studies of their potassium currents [6, 7]. Simulations indicate that the current playing this role must be large and deactivate very rapidly to account for after-hyperpolarization

---

[2]Some preliminary experiments with a persistent sodium current have been performed, but it was not used in the simulations described here.

[3]In these cases, the fast sodium current also derived by Lytton and Sejnowski [11] was also used.

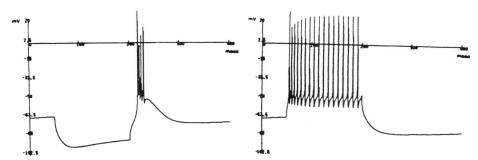

Figure 2   The response of the model to depolarizing and hyperpolarizing current steps, showing tonic and burst firing behavior.

shape in these cells. Such characteristics have been described for the fast Ca^{++}-dependent K^+ current termed "Ic".

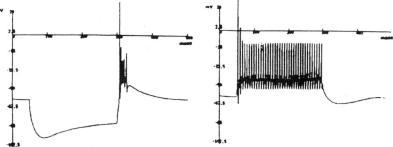

Figure 3   An Ic-like current can substitute for a delayed rectifier to generate gross aspects of thalamic cell behavior.

Huguenard and Prince [7] described in isolated thalamic cells a fast, Ca^{++}-dependent potassium current, Ik1, which shares many characteristics of Ic, but they did not extensively characterize it. Preliminary experiments in cultured rat thalamic neurons have shown the presence of high concentrations of "maxi-K" (BK) channels (Esguerra and Smetters, unpublished), the channel known to underly Ic. Replacing the delayed rectifier in the model with a K^+ current based on maxi-K channels [15], as in Figure 3, shows that such a current can indeed take over its functional role.

Though the model does function with an Ic in place of the delayed rectifier, to do so requires that the current is affected differently by Ca^{++} entering through the L and T-type Ca^{++} channels. The Ca-based LTS has been shown to allow entry of much more Ca^{++} than enters during a high-threshold spike. In the model, this causes excessive Ic activation during the LTS. If Ic is only activated by Ca^{++} entering through the L-type channel, as if the two channel types were colocalized, the model becomes much more stable. Such colocalization has been previously suggested by hippocampal modeling studies [3] and supported experimentally by recordings in cultured GH3 cells [10] and isolated hair cells [16].

The behavior of the other K^+ currents will also influence spike repolarization. Simulations indicate that the fastest K^+ current available (in order: Iaf, Ic, Ik2) will perform the bulk of repolarization, leaving the other currents largely unactivated. If the faster current is blocked, then the next fastest current will take over, completing the job of spike repolarization with perhaps only minimal increase in spike width. This suggests that current-clamp experiments attempting to derive the roles for various currents by blocking them may be difficult to interpret.

Finally, the model points out the significant role active currents may play in the "passive" properties of the cell. Even though the model without any currents has an input resistance of 60 $M\Omega$, the full model has an apparent input resistance much lower than this. These changes are apparent even with the smallest current steps, and depend significantly on holding potential.

4 SUMMARY AND CONCLUSIONS

The detailed model of a thalamic relay cell described in this study shows that the combination of currents recorded physiologically in these cells can indeed account for most of their behavior. However, the model suggests that the interactions between several of these currents, which may have significant effects on cell behavior, have not been adequately characterized. The model may be a useful tool in identifying areas most profitable for future experimental work, and for interpreting the results of such experiments.

This work supported by HHMI.

REFERENCES

[1] Beluzzi, O. and O. Sacchi. (1986) J. Physiol., 380:275–291.
[2] Bloomfield, S.A., J.E. Hamos and S.M. Sherman (1987) J. Physiol., 383:653–692.
[3] Borg-Graham, L. (1987) Master's Thesis. MIT.
[4] Crunelli, V., N. Leresche, and J.G. Parnavelas (1987) J. Physiol., 390:243–256.
[5] Halliwell, J. and P.R. Adams (1982) Brain Research, 250:71–92.
[6] Huguenard, J.R., D.A. Coulter and D.A. Prince (1991) J. Neurophysiol., 66:1304–1315.
[7] Huguenard, J.R., and D.A. Prince (1991) J. Neurophysiol., 66:1316–1328.
[8] Jahnsen, H. and R. Llinas (1984a) J. Physiol., 349:205–226.
[9] Jahnsen, H. and R. Llinas (1984b) J. Physiol., 349:227–247.
[10] Lang, D. and A. Ritchie (1988) Pflug. Arch., 410:610–622.
[11] Lytton, W.W. and T.H. Sejnowski (1991) J. Neurophysiol., 66:1059–1079.
[12] McCormick, D.A. and H-C. Pape (1990) J. Physiol., 431:291–318.
[13] McCormick, D.A. (1991) J. Neurophysiol., 66:1176–1189.
[14] McFarlane, S. and E. Cooper (1998) J. Neurophysiol., 66:1380–1391.
[15] Modczydlowski, E. and R. Latorre (1983) J. Gen. Physiol., 82:511–542.
[16] Roberts, W.M., R.A. Jacobs and A.J. Hudspeth (1990) J. Neurosci., 10:3664–3684.
[17] Soltesz, I., S. Lightowler, N. Leresche, D. Jassik-Gerschenfeld, C.E. Pollard and V. Crunelli (1991) J. Physiol., 441:175–197.
[18] Spruston, N. and D. Johnston (1992) J. Neurophysiol., 67:508–529.
[19] Storm, J.F. (1988) Nature, 336:379–381.
[20] Wang, X-J., J. Rinzel and M. Rogawski (1991) J. Neurophysiol., 66:839–850.

AN ENTROPY MEASURE FOR REVEALING DETERMINISTIC STRUCTURE IN SPIKE TRAIN DATA

Garrett T. Kenyon
David C. Tam*

*Department of Neurobiology and Anatomy, University of Texas Medical School, P.O. Box 20708, Houston, TX 77225, *Center for Network Neuroscience, Department of Biological Sciences, University of North Texas, P.O. Box 5218, Denton, TX. 76203*

ABSTRACT

The limiting behavior of the time-dependent Kolmogorov entropy, $K_{\Delta t}(T)$, obtained from computer generated spike trains digitized using rectilinear grids of length $T \to \infty$ and grid spacing $\Delta t \to 0$, is used to identify chaotic (i.e., deterministic) structure, which is a necessary (but not a sufficient) condition for significant information to be encoded in the precise temporal spike sequence.

7.1. INTRODUCTION

Statistical measures, such as the peristimulus time histogram (*PSTH*), have been successfully employed throughout the nervous system in order to analyze spike activity, both for characterizing receptive fields and for studying the information processing capabilities of neurons. The rationale underlying the use such measures depends strongly, however, on the validity of the *frequency code hypothesis*, which may be summarized as follows: 1) the information content of a spike train is encoded in the (time dependent) firing rate, $f(t)$, which may be measured by constructing a *PSTH*, and 2) individual spikes occur stochastically with a probability per unit time given by $f(t)$. Inherent in the frequency code hypothesis is the assumption that the precise temporal sequence of spikes does not encode significant information. This is because whenever the generation of firing events is governed by a stochastic process, the spike sequence will not be uniquely determined. There will be infinitely many temporal sequences which all represent consistent realizations for the same *PSTH*.

The *chaotic hypothesis* is an alternative to the frequency code hypothesis which asserts the following: 1) complex, multi-unit firing patterns encode significant information in the precise spatio-temporal spike sequence, 2) these patterns result from deterministic dynamical interactions between neurons and 3) the associated multiunit spike trains generate information at a (reasonably low) finite rate. Initial motivation for the chaotic hypothesis comes from the observation that even relatively simple dynamical systems can generate very complex looking signals, which, on the basis of standard correlation measures, may be indistinguishable from stochastically generated activity. This raises the possibility that the multi-unit spike trains generated by any given nerve cell network, although qualitatively 'noisy' in appearance, may actually be chaotic.

An important consequence of the chaotic hypothesis is that due to the underlying deterministic structure, a chaotic multi-unit spike train is capable of encoding significant information in the detailed spatio-temporal firing sequence. There exists both experimental[1-7] and theoretical[8-10] evidence which implies, although by no means confirms, that the precise spatio-temporal firing sequence does encode significant information. Since an event train cannot support a precise temporal code if the spikes are distributed stochastically, a necessary (though not sufficient) condition for the chaotic hypothesis to be valid is that the temporal firing sequence possess a deterministic structure. We describe here techniques for detecting deterministic structures in naturally occurring spike trains. A description of a parallel algorithm which implements these techniques and its application to general time series data will appear elsewhere[11].

7.2. THEORY AND METHODS

7.2.1. Digitized Firing Patterns

The firing patterns are digitized by superimposing on the spike trains a series of rectilinear grids, with total length T and temporal grid spacing Δt, such that the left edge of each grid is aligned along each spike, referred to here as a reference spike. Each grid square is then assigned a value 0 if it is empty, 1 if it is not. The occurrence of multiple spikes in the same grid square is not distinguished from the case of a single spike. Any temporal structure present on time scales outside this range will not be resolved by the foregoing analysis.

7.2.2. Time-Dependent Kolmogorov Entropy

Formally, we define the time-dependent Kolmogorov entropy, $K_{\Delta t}(T)$, as

$$K_{\Delta t}(T) = K_s + \left[H_{\Delta t}(T) - H_{\Delta t}(T - \Delta t)\right]/f\,\Delta t \qquad (1)$$

where $H_{\Delta t}(T)$ gives the entropy across all patterns with length T, and we have divided through by the time averaged firing rate, f, which provides a natural unit of time (the quantity K_s is defined below).

7.2.3. The Limiting Behavior of $K_{\Delta t}(T)$ for Discriminating Stochastic from Chaotic Event Sequences

It is straightforward to show that for a Poisson process, $K_{\Delta t}(T)$ is independent of T and in the infinitesimal limit is given by

$$K_{\Delta t}(T) \approx (1 - \ln f\Delta t)/\ln 2 + O(f\Delta t) \qquad (2)$$

Equation 2 shows that for a Poisson process, $K_{\Delta t}(T)$ scales as $\ln f \Delta t$, growing to arbitrarily large values as the temporal resolution is increased. We may then generalize to an arbitrary stochastic process by allowing f to be a function of the pattern of previous events. We assume that the stochastic process is sufficiently smooth such that there exists an expression of the same form as eq. 2 for each previous event pattern. It follows that there exists a constant, f, such that the average of these expressions, over all event patterns, will exhibit a characteristic $\ln f \Delta t$ dependence. Thus, for any stochastic process satisfying the above smoothness condition, we expect that $K_{\Delta t}(T)$ will scale as $\ln f \Delta t$.

For a deterministic process, the conditional probabilities collapse to singularities (delta-functions), and the assumption of Poisson like behavior over infinitesimal time intervals breaks down. Based on the analysis of nonlinear dynamical systems, we look for $K_{\Delta t}(T)$ for any deterministic process to asymptotically approach a finite value as $T \to \infty$, $\Delta t \to 0$[12,13]. The minimum pattern size for which $K_{\Delta t}(T)$ approaches its asymptotic value then estimates the number of previous spike events which are sufficient to predict the next event in the sequence. In theory, this quantity is related to the number of dynamical variables which would be necessary to model the relevant network interactions[14].

7.2.4. Shuffling to Control for Artifacts Arising from Insufficient Ensemble Size

When the sample size is too small, it will not be possible to accurately estimate the occurrence frequencies for those patterns characterized by small, but finite, probabilities, which may be incorrectly assigned a weight of zero. This effectively increases the order, or predictability, across the ensemble of patterns, thus reducing the measured value of $K_{\Delta t}(T)$. To estimate the size of this artifact, $K_{\Delta t}(T)$ is recomputed after shuffling the grid squares corresponding to the last time slice between all patterns in the ensemble. This defines the spurious Kolmogorov entropy, K_s, which tends to zero in the limit of an infinite sample size.

7.3. RESULTS

7.3.1. Comparison of Two Pseudo-Random Number Generators

Figure 1 demonstrates that an analysis of the scaling behavior of $K_{\Delta t}(T)$ can be used to detect subtle temporal structures not resolved by a standard correlation analysis. Auto-correlograms for event

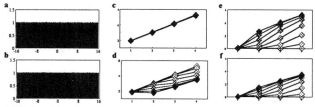

Figure 1. Comparison of two pseudo-random number generators.

sequences obtained from either a CM system supplied pseudo-random number generator (fig. 1a) or from a linear congruential routine (fig. 1b) are essentially flat, as required for Poisson distributed events. For the system supplied routine, a plot of $K_{\Delta t}(T)$ vs. $|\log_2 f\Delta t|$ displays a strict linear dependence (fig 1c), as predicted by eq. 2. In contrast, for the linear congruential method (fig 1d), plots of $K_{\Delta t}(T)$ vs. $|\log_2 f\Delta t|$ for eight different pattern sizes, $fT = n$, $n = \{1,...,8\}$, exhibit clear deviations from linearity. At larger pattern sizes, (darker points), the curves become progressively more nonlinear, revealing the existence of subtle temporal correlations. At the largest pattern size, $K_{\Delta t}(T)$ decreases initially with Δt, indicating deterministic structure at least down to this temporal resolution.

Figures 1e-f shows that K_s as a function of $|\log_2 f\Delta t|$ becomes significant as both $\Delta t \rightarrow 0$ and $T \rightarrow \infty$, which demonstrates that spurious correlations resulting from an insufficiently large ensemble size are dominating the calculation in this limit. Since the spurious correlations are systematically overestimated by our shuffling control, the calculated values of $K_{\Delta t}(T)$ will also be too large. This explains why $K_{\Delta t}(T)$ fails to approach a finite asymptotic limit even though both event trains are entirely deterministic.

7.3.2. Comparison Between a Highly Correlated Stochastic and a Chaotic Event Train

Figure 2 compares event trains derived from either a highly correlated stochastic or a chaotic process. The stochastic event train is constructed by modifying a Poisson process through the imposition of an absolute refractory period equal to 90% of the mean inter-event interval. This relatively large refractory period causes the firing probability to peak at nearly periodic intervals, as shown by the autocorrelogram in figure 2a. The autocorrelogram obtained from a chaotic event train (figure 2b), constructed from the natural logarithm of successive iterates of the logistic map ($r = 3.999$), reveals the existence of temporal correlations only at relatively short time scales.

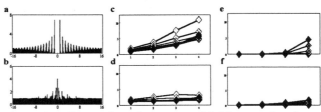

Figure 2.   Comparison between a highly correlated stochastic and a chaotic event train.

Figure 2c shows that in spite of the strong temporal correlations seen in figure 2a, $K_{\Delta t}(T)$ still diverges uniformly as $\Delta t \rightarrow 0$, correctly indicating that the events are stochastically distributed. In contrast, an analysis of $K_{\Delta t}(T)$ for the chaotic event train, shown in figure 2d, demonstrates that at intermediate pattern sizes, $K_{\Delta t}(T)$ approaches a finite asymptotic value for $|\log_2 f\Delta t| \leq 3$. This indicates that the event sequences possess a deterministic structure over this range of time scales. The minimum pattern size for which $K_{\Delta t}(T)$ approaches its asymptotic value implies that the minimum embedding dimension is approximately 3, and thus the number of relevant dynamical variables is between 1 and 3.

Figure 2e shows that K_s for the stochastic train is essentially negligible at all but the largest temporal resolutions examined, indicating that the divergence in $K_{\Delta t}(T)$ as $\Delta t{\to}0$ cannot be attributed to finite data artifacts. For the chaotic train (fig. 2f), the time course of K_s at larger pattern sizes follows closely the slight growth seen in $K_{\Delta t}(T)$ in figure 2d. The magnitude of K_s, however, exceeds the growth in $K_{\Delta t}(T)$, which is consistent with the proposition that spurious correlations account entirely for the observed divergence of $K_{\Delta t}(T)$ as $\Delta t{\to}0$, and that in the absence of finite data artifacts, $K_{\Delta t}(T)$ would approach a finite asymptotic value.

7.4. CONCLUSION

Spike trains possessing a chaotic distribution can be distinguished from qualitatively similar firing activity generated by a stochastic process. The criterion employed, the limiting behavior of the time dependent Kolmogorov entropy, can reveal a deterministic organization even when the autocorrelogram exhibits little or no detectable correlations. Furthermore, even highly correlated stochastic processes are not falsely identified as being deterministic. A reliable control is employed to prevent the false identification of deterministic order due to an insufficient ensemble size. The identification of a chaotic spike train organization is a necessary, but not a sufficient, condition to establish the existence of a precise temporal code.

REFERENCES.

1. Eckhorn R, Bauer R, Jordan W, Brosch M, Kruse W, Reitboeck HJ, *Biol. Cyber.*, 60 (1988) 121.
2. Gray CM, König P, Engel AK, Singer W, Nature, 388 (1989) 334.
3. Carr CE, Konishi M, *J. Neurosci.*, **10**(10) (1990) 3227.
4. Reinis S, Weiss DS, Landolt JP, *Biol. Cyber.*, **59** (1988) 41.
5. Abeles M, *IEEE Trans. Biomed. Eng.*, **30** (1983) 235.
6. Dayhoff JE, Gerstein GL, *J. Neurophys.*, **49** (1983) 1349.
7. Mpitsos GJ, Burton Jr RM, Creech HC, Soinila SO, *Brain Res. Bull.*, **21** (1988) 529
8. Kenyon GT, Fetz EE, Puff RD, in *Advances in Neural Information Processing Systems #2*. Touretzky DS (ed) (Morgan Kaufmann 1990) 141.
9. Kenyon GT, Puff RD, Fetz EE, *Biol. Cyber.*, **67** (1992) 133.
10. van der Malsburg, in *Brain Theory*, Palm G, Aertsen (eds), (Spinger-Verlag, Berlin, 1986) 161.
11. Kenyon GT, Tam DC, to appear in *Proceedings of the Second Annual Conference on EEG and Nonlinear Dynamics*, Jansen B (ed) (World Sci. Pub. Co.)
12. Kolmogorov AN, *JDokl. Akad. Nauk.*, **124** (1959) 754.
13. Sinai J, *JMath USSR Sb.*, **63** (1964) 23.
14. Takens F, in: *Lecture Notes in Mathematics*, Rand DA, Young LS, (eds.) (Springer, New York, 1981).

8

A MULTI-NEURONAL VECTORIAL PHASE-SPACE ANALYSIS FOR DETECTING DYNAMICAL INTERACTIONS IN FIRING PATTERNS OF BIOLOGICAL NEURAL NETWORKS

David C. Tam

Center for Network Neuroscience, University of North Texas, Denton, TX 76203
E-mail: dtam@sol.acs.unt.edu

ABSTRACT

A multi-neuronal vectorial phase-space analysis is introduced to detect spatio-temporally correlated firing patterns in a network of n neurons. The cross-interval analysis uses a vectorial approach to establish spatio-temporal correlation between firing intervals among multiple neurons. The resultant vectorial sum captures an n-wise correlation among all n neurons in the network. Applying this vectorial phase-space analysis on multiple spike trains recorded simultaneously, an n-tuple correlation among all n neurons in the network can be computed. The results show clusters of these vector trajectories clearly identified the spatio-temporal firing patterns. This vectorial phase-space analysis also provides quantitative descriptions of dynamical interactions from the trajectories and clusters of these vectors in the phase-plane.

8.1. INTRODUCTION

It is well-known that neurons in a network do not operate independently since there are synaptic connections and synaptic couplings among these neurons. Therefore, it is a fundamental question to address what kind of dependence or interactions are exhibited in a network of neurons. In particular, we want to reveal not only the temporal correlation among neurons, but also spatial correlation at the same time. The spatio-temporal patterns revealed from the interactions among neurons may provide us with better understanding of the dynamics underlying neural network processing.

Furthermore, we are able to record from a large number of neurons simultaneously using the multi-unit recording microelectrode technology. The firing activity of a network of neurons can be recorded extracellularly using these multi-electrodes. Time series of action potentials recorded from these neurons can be represented by *spike trains* when digitized and spike train analyses can be performed to detect and identify the signal processing properties of these neurons.

Traditional cross-correlation methods are often used as the standard analytical method for analyzing spike train data [4]. But, the cross-correlational analysis is limited to *pair-wise* correlation between *two* spike trains. To analyze a large number of neurons in a network, the number of correlograms (correlation graphs) required to analyze the whole set is prohibitively large, i.e., on the order of n^2, where n is the number of neurons in the network. There are other specialized spike train analytical techniques developed to reveal firing patterns in multiple neurons, such as [1], [2], [3], [5], [6] and [7].

We have developed yet another powerful statistical measure to identify specific spatio-temporal firing patterns in neurons and their correlated interactions. This descriptive statistic is represented by a *vectorial* measure of firing intervals among neurons. The statistic is used to establish not only temporal correlation but also spatio correlation among firing times of neurons. It provides not only the *scalar* measure of the statistic (the *magnitude* of a vector), but also the *vectorial* measure of the statistic (the *direction* of a vector). This new representation provides additional information for identifying the firing relationships.

8.2.   METHODS

A statistical measure we developed is called the "cross-interval vector" It is used to detect spatio-temporally correlated firing patterns in multiple spike train data. The cross-interval is defined as the time interval between adjacent spikes in two spike trains. There are two adjacent cross intervals for every given reference spike: the *pre-cross interval* and the *post-cross interval*. From these two cross-intervals, a vector can be constructed by representing the x- and y-components of this cross-interval vector by the post-cross interval and the pre-cross interval, respectively.

The cross-interval vector, $\mathbf{v}$, can be represented by:

$$\mathbf{v} = I\mathbf{x}' + J\mathbf{y}' \tag{8-1}$$

where I is the post-cross interval, J the pre-cross interval, and $\mathbf{x}'$ and $\mathbf{y}'$ are the unit vectors.

The angle of this vector is given by the arctangent of the ratio of the post-cross interval and the pre-cross interval. The scalar quantity of this cross-interval vector, $\mathbf{v}$, is analogous to the first-order "conditional cross-interval" measure introduced by Tam *et al* [5].

Given n spike trains, there are a total of $n - 1$ such cross-interval vectors for any given pair of neurons. A resultant vectorial sum, $\mathbf{V}$, of these $n - 1$ vectors can be obtained for each given reference spike:

$$\mathbf{V} = \sum_{i=1}^{n-1} \mathbf{v}_i = \sum_{i=1}^{n-1} (I_i\mathbf{x}' + J_i\mathbf{y}') \tag{8-2}$$

where $\mathbf{v}_i$ denotes the cross-interval vector with respect to the i-th spike train. This resultant cross-interval vector will provide a quantitative measure of a specific firing pattern for n neurons based on a reference spike in a given train. It also captures the correlation of firing intervals between all n neurons simultaneously, not just the pair-wise correlation between two neurons as in conventional cross-correlation methods.

The changes in firing patterns among n neurons can be described by time evolution of these resultant vectors quantitatively. When the trajectories of these vectors are mapped onto the phase-space plot, the evolution of how firing pattern changes over time can be shown. Clustering of these resultant vectors can be used to delineate repeated firing patterns. The Euclidean distance between consecutive resultant vectors can be used to quantify repeated firing patterns among different sets of neurons.

8.3. RESULTS

There are five different sets of cross-interval vectors from the combination of cross-intervals given a reference spike. The normalized resultant vectors for each of the five distinct vector sets are given below:

$$\mathbf{A} = \frac{1}{n}\sum_{i=1}^{n}\mathbf{a}_{ii} = \frac{1}{n}\sum_{i=1}^{n}(I_{ii}\vec{x} + J_{ii}\vec{y}) \tag{8-3}$$

$\mathbf{A}$ is the normalized resultant vector of all interspike-interval vectors, $\mathbf{a}_{ii}$. In general, I_{ij} represents the post-cross interval between the reference train i and an observed train j, and J_{ij} represents the pre-cross interval between the reference train i and an observed train k. Similarly, the other normalized resultant cross-interval vectors $\mathbf{B}$, $\mathbf{C}$, $\mathbf{D}$ and $\mathbf{E}$ are given by:

$$\mathbf{B} = \frac{1}{n(n-1)}\sum_{i=1}^{n}\sum_{\substack{j=1 \\ j\neq i}}^{n-1}\mathbf{b}_{ij} = \frac{1}{n(n-1)}\sum_{i=1}^{n}\sum_{\substack{j=1 \\ j\neq i}}^{n-1}(I_{ii}\vec{x} + J_{ij}\vec{y}) \tag{8-4}$$

$$\mathbf{C} = \frac{1}{n(n-1)}\sum_{i=1}^{n}\sum_{\substack{j=1 \\ j\neq i}}^{n-1}\mathbf{c}_{ij} = \frac{1}{n(n-1)}\sum_{i=1}^{n}\sum_{\substack{j=1 \\ j\neq i}}^{n-1}(I_{ij}\vec{x} + J_{ii}\vec{y}) \tag{8-5}$$

$$\mathbf{D} = \frac{1}{n(n-1)}\sum_{i=1}^{n}\sum_{\substack{j=1 \\ j\neq i}}^{n-1}\mathbf{d}_{ij} = \frac{1}{n(n-1)}\sum_{i=1}^{n}\sum_{\substack{j=1 \\ j\neq i}}^{n-1}(I_{ij}\vec{x} + J_{ij}\vec{y}) \tag{8-6}$$

$$\mathbf{E} = \frac{1}{n(n-1)(n-2)}\sum_{i=1}^{n}\sum_{\substack{j=1 \\ j\neq i\neq k}}^{n-1}\sum_{\substack{k=1 \\ k\neq i\neq j}}^{n-2}\mathbf{e}_{ij}$$

$$= \frac{1}{n(n-1)(n-2)}\sum_{i=1}^{n}\sum_{\substack{j=1 \\ j\neq i\neq k}}^{n-1}\sum_{\substack{k=1 \\ k\neq i\neq j}}^{n-2}(I_{ij}\vec{x} + J_{ik}\vec{y}) \tag{8-7}$$

It should be noted that these resultant vectors are a measure of the averaged cross-interval vectors. Moreover, it captures not only the n^2 pairs of correlation, but n^3 correlational cross-interval vectors. The total n^3 combinations can be obtained by considering the summation limits of the above five vectors $\mathbf{A}$, $\mathbf{B}$, $\mathbf{C}$, $\mathbf{D}$ and $\mathbf{E}$:

$$n + n(n-1) + n(n-1) + n(n-1) + n(n-1)(n-2) = n^3 \tag{8}$$

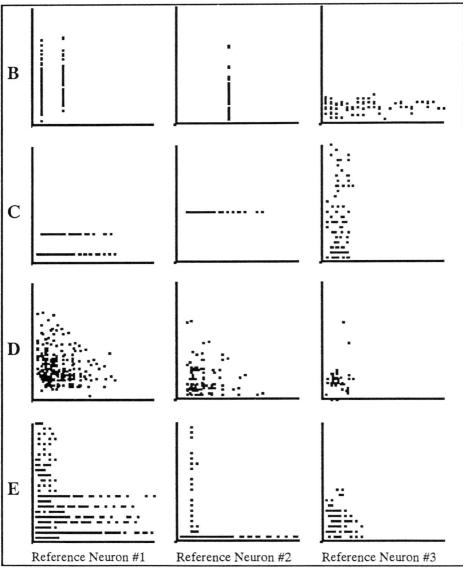

Fig. 1. Phase-space plots of three reference neurons showing the spatio-temporal patterns represented by the distribution of the cross-interval vectors. The y-axis represents the pre-cross interval, the x-axis the post-cross interval.

The major advantage of this n-tuples measure lies in its ability to collapse all n^3 combinations into five vector sets quantifying variations of firing intervals between spikes.

Figure 1 shows the phase-space plot of the trajectories of the cross-interval vectors for three reference neurons using simulated spike trains recorded simultaneously. The cluster plots are used to display the "tips" of the cross-interval vector as points on the x-y plot. The phase-space plots clearly show patterns in the location of these vectors with respect to each neuron. The cross-interval vectors **B**, **C**, **D** and **E** are displayed in rows denoted by B, C, D and E in Fig. 1. Distinct firing patterns can be observed in these plots, such as periodic firing with respect to neuron # 2 is seen as the single band of cross-intervals in the phase-space plot (column 2), and doublet firing (i.e., firing with short and long intervals consecutively) for neuron # 1 is seen as the double-band of alternating cross-interval (column 1). Clusters of vectors are also found, indicating preferred repeated firing patterns. These phase-space plots also show that a description of interactions among all n neurons can be accomplished without necessary restricted to using pair-wise correlation techniques. This method can also be used to analyze chaotic systems.

SUMMARY

The multi-unit spike train analysis techniques using cross-interval vectors are used to reveal the spatio-temporal firing patterns among multiple neurons in a network. The vectorial measures capture the correlated patterns among all n neurons providing a spatio-temporal description of the interactions among these neurons.

Acknowledgments

This research was supported by Office of Naval Research grant number ONR N00014-90-J-1353 and USPHS grant number RR05425-30.

REFERENCES

[1] G. L. Gerstein and A. Aertsen. Representation of Cooperative Firing Activity among Simultaneously Recorded Neurons. *Journal of Neurophysiology*, 54: 1513-1528, 1985.

[2] G. L. Gerstein, D. H. Perkel and J. E. Dayhoff. Cooperative Firing Activity in Simultaneously Recorded Populations of Neurons: Detection and Measurement. *J. of Neurosci.* 5: 881-889, 1985.

[3] B. G. Lindsey, R. Shannon, G. L. Gerstein. Gravitational representation of simultaneously recorded brainstem respiratory neuron spike trains. *Brain Res.* 483: 373-378, 1989.

[4] D. H. Perkel, G. L. Gerstein, and G. P. Moore. Neuronal Spike Trains and Stochastic Point Process. II. Simultaneous Spike Trains. *Biophysical Journal*, 7: 419-440, 1967.

[5] D. C., Tam, T. J. Ebner and C. K. Knox. Cross-Interval Histogram and Cross-Interspike Interval Histogram Correlation Analysis of Simultaneously Recorded Multiple Spike Train Data. *J. of Neurosci. Methods*, 23: 23-33, 1988.

[6] D. C. Tam and D. H. Perkel. A Model for Temporal Correlation of Biological Neuronal Spike Trains. *Proc. of the IEEE International Joint Conference on Neural Networks* 1: 781-786, 1989.

[7] D. C. Tam. A hybrid time-shifted neural network for analyzing biological neuronal spike trains. *Progress in Neural Networks* (O. Omidvar, ed.) Vol. 2, Ablex Publishing Corporation: Norwood, New Jersey. (*in press*)

THE 'IDEAL HOMUNCULUS': STATISTICAL INFERENCE FROM NEURAL POPULATION RESPONSES

Peter FÖLDIÁK

MRC Research Centre in Brain and Behaviour, University of Oxford
South Parks Road, Oxford OX1 3UD, U.K., e-mail: peter@psy.ox.ac.uk

What does the response of a neuron, or of a group of neurons mean? What does it say about the stimulus? How distributed and efficient is the encoding of information in population responses? It is suggested here that Bayesian statistical inference can help answer these questions by allowing us to 'read the neural code' not only in the time domain[2, 5] but also across a population of neurons. Based on repeated recordings of neural responses to a known set of stimuli, we can estimate the conditional probability distribution of the responses given the stimulus, $P(response|stimulus)$. The behaviourally relevant distribution, i.e. the conditional probability distribution of the stimuli given an observed response from a cell or a group of cells, $P(stimulus|response)$ can be derived using the Bayes rule. This distribution contains all the information present in the response about the stimulus, and gives an upper limit and a useful comparison to the performance of further neural processing stages receiving input from these neurons. As the notion of an 'ideal observer' makes the definition of psychophysical efficiency possible[1], this 'ideal homunculus' (looking at the neural response instead of the stimulus) can be used to test the efficiency of neural representation. The Bayes rule is: $P(s|r) = P(r|s)P(s)/P(r) = P(r|s)P(s)/\sum_S P(r|s)P(s)$, where in this case s stands for stimulus, r for response, and S is the set of possible stimuli. To calculate the distribution based on the responses of two neurons, we need to calculate $P(s|r_1, r_2) = P(r_1, r_2|s)P(s)/P(r_1, r_2)$. Assuming that the variability of the individual neurons is statistically independent $P(r_1, r_2|s) = P_1(r_1|s)P_2(r_2|s)$, so $P(s|r_1, r_2) = P_1(r_1|s)P_2(r_2|s)P(s)/\sum_S P_1(r_1|s)P_2(r_2|s)P(s)$. Note that the assumption is about the independence of the *variability* of the responses, and *not* about the independence of the responses themselves. The validity of this assumption will need to be tested when the method is later to be applied to data recorded from several cells simultaneously with multiple electrodes. The formula can similarly be derived for more than two cells. As a simple demonstration, the method is applied to the responses of neurons in the primary visual cortex to sine-wave gratings of different orientations. The electrophysiological experiments were carried out and all recording data was kindly provided by F Sengpiel[1]. Responses of 8 neurons in the primary visual cortex of an anaesthetised cat were recorded using standard extracellular techniques as described by Blakemore and Price[3]. Stimuli consisted of high contrast, drifting sine-wave gratings with parameters that elicited some response from most cells. All parameters except for orientation were kept constant during the experiment. The analysis shows that uncertainty about the stimulus orientation decreases quickly with the number of cells included in the calculation (Figure 1,2). Reliable estimation of stimulus orientation is possible using a small number of cells. A comparison is made between this and the 'population vector' method of Georgopoulos et al.[4] (Figure 3). The failure of that method suggests that the high number of cells necessary to derive reliable estimates using population vectors (see also [7]) is due to the sub-optimality of that method, and *not* to the highly distributed nature of the neural representation. The Bayesian method can also be applied in cases (e.g.[6]) where there is no simple and natural space in which to define the population vector.

Figure 2   A) The number of spikes recorded form 8 different neurons. The range of responses [min,max] are indicated in the upper right hand corner of each plot. B) The conditional probability estimates of the cell responses given the stimulus orientation (as in Figure 1c)). C) The conditional probability distributions of orientation given the response recorded when a stimulus of 30° was applied to the corresponding cell, calculated separately for each cell. The responses used in the calculation were those recorded on the first (single) trial. The true stimulus orientation (30°) is marked by an arrow on the plots. Note that cell 8, which did not show significant response to any orientation gives a nearly flat distribution, expressing its uncertainty. D) The combined conditional probability distributions of orientation given the responses recorded from 1, 2 and 3 cells when a stimulus of 30° was applied to the corresponding cells (single trial). Note that cell 5 alone gives an uncertain estimate of stimulus orientation, while additional information from cell 6 sharpens the distribution. Cells 5, 6 and 7 together give a sharp peak at the true stimulus orientation, reducing the probability of all false orientations.

[1]I am very grateful to Frank Sengpiel, University Laboratory of Physiology, University of Oxford for including stimuli in his experiment and kindly making the data available to me. I would also like to thank A Treves, H Barlow, M Young, S Nowlan and others for useful comments and suggestions. This work has been supported by the University of Oxford MRC Research Centre in Brain and Behaviour, and a travel grant has been provided by the University of Oxford McDonnell-Pew Centre for Cognitive Neuroscience and the NSF.

REFERENCES

[1] H B Barlow, *Vision Research*, 18, 637-650, 1978.

[2] W Bialek, F Rieke, R R Steveninck, D Warland, *Science*, 252, 1854-1857, 1991.

[3] C Blakemore, D J Price, *J. Physiology*, 384, 263-292, 1987.

[4] A P Georgopoulos, R E Kettner, A B Schwartz, *The Journal of Neuroscience*, 8(8), 2928-2937, 1988.

[5] L M Optican, B J Richmond, *J. Neurophysiology*, 57, 162-178, 1987.

[6] M P Young, S Yamane, *Science*, 256, 1327-1331, 1992.

[7] E Zohary, *Biological Cybernetics*, 66, 265-272, 1992.

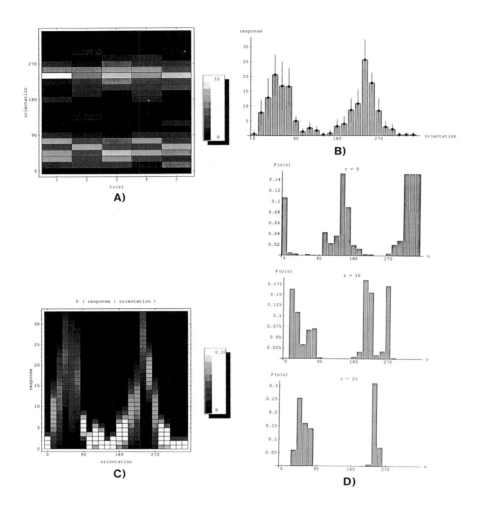

Figure 1   A) The total number of spikes recorded form a neuron in primary visual cortex during 1.25s presentations of drifting sine-wave gratings at 24 different orientations $(0 - 345^o)$ in random order, each presentation repeated 5 times (trials 1-5). The brightness levels represent the spike counts. B) The orientation tuning curve of the cell (mean and st. dev.). C) The conditional probability estimate of the cell response given the stimulus orientation. For each orientation, a Gaussian with the same mean and standard deviation as that of the cell response sampled at each response level, truncated at 0, and normalized to sum to 1 was used as an estimate of the conditional probability distribution. D) The conditional probability distribution of the stimulus orientation given the cell response, calculated using the Bayes rule and plotted for 3 response levels (r=0,10,25). A uniform prior distribution was assumed for the stimulus orientations.   This prior assigns 0 probability to all stimuli outside the set considered here.   Note that for r=25 (high response) the distribution peaks near where the tuning curve does. For r=0, high probability values are in the low response regions of the tuning curve while for r=10 they are on the sides of the tuning curve peaks.

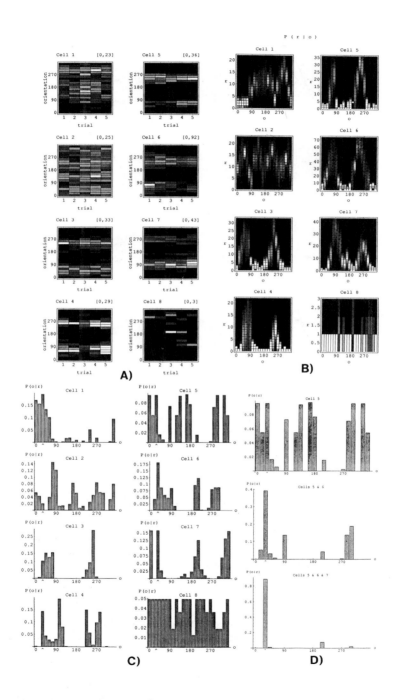

Figure 2 *(continued)* See legend above References section.

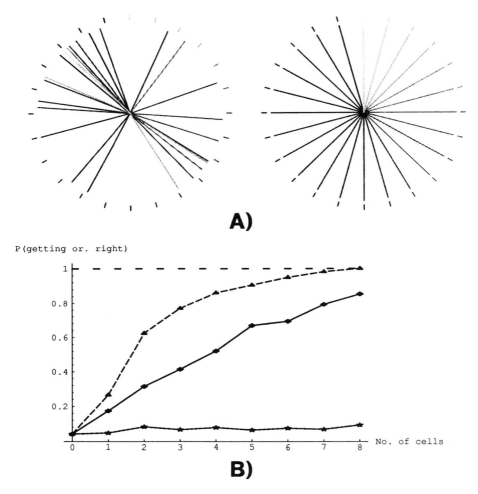

Figure 3   A) Orientation estimate using the responses of all 8 recorded
cells. For both plots the brightness of each line in the figure corresponds to
the true stimulus orientation (as mapped by the marks on the circumference
of the circles), the estimated orientation given by the orientations of the lines.
On the left, the population vector estimate (normalized weighting function 6 in
[4]) was used. On the right, the estimate of the 'ideal homunculus' is plotted,
taking the orientation with maximum posterior probability as the estimate.
Responses averaged over all 5 trials were used in this figure for both methods.
Note that while the population vector gives near chance level performance on
8 cells, the ideal homunculus estimate corresponds to the true orientation for
each of the 24 orientations. B) The relative frequency of correct estimates as a
function of the number of cells used in the calculation. For each cell number,
500 random combinations of cells were taken (with repetitions of combinations
but without repetition of cells in a combination). The top curve plots the
performance of the ideal homunculus using the responses averaged over all 5
trials, while the middle curve plots that using the response of each cell from a
single (randomly picked) trial. The lower curve plots the performance of the
population vector estimate (calculated using rounding to the nearest presented
stimulus orientation).

10

THE EFFECT OF SLOW SYNAPTIC COUPLING ON POPULATIONS OF SPIKING NEURONS

C.A. van Vreeswijk*, L.F. Abbott*, A. Treves**

*Physics Department and Center for Complex Systems
Brandeis University, Waltham, MA 02254-9110*

**Department of Experimental Psychology
Oxford University, Oxford OX 13UD, UK*

ABSTRACT

We examine the conditions under which a population of spiking neurons with all-to-all excitatory coupling can fire asynchronously. Synapses with time constants satisfying computed constraints assure the stability of an asynchronous firing state even in the absense of inhibition.

1 MODEL

We model a system of a large number N of tonically firing, identical integrate and fire neurons, with a global excitatory coupling. The timepath of the voltage x_i of neuron i is given by

$$\frac{dx_i}{dt} = F(x_i) + G(x_i)E(t)$$

with $F(x) > 0$ and $G(x) > 0$ for $0 \leq x \leq 1$. (The threshold potential is set at $x = 1$ and the reset at $x = 0$). $E(t)$ is given by

$$E(t) = \frac{1}{N} \sum_{j,\alpha} \frac{1}{\tau_1 - \tau_2} \left(e^{-\frac{t-t_{j,\alpha}}{\tau_1}} - e^{-\frac{t-t_{j,\alpha}}{\tau_2}} \right) \theta(t - t_{j,\alpha})$$

Here $t_{j,\alpha}$ is the time when neuron j fires for the αth time.

2 ASYNCHRONOUS STATE

For $\int_0^1 dx\ G(x)^{-1} > 1$ there is a solution with firing rate E^0. We replace x_i with a phase ϕ_i so that

$$\frac{d\phi_i}{dt} = E^0 + \Gamma(\phi_i)e(t)$$

with

$$\Gamma(\phi) = \frac{E^0 G(x)}{F(x) + E^0 G(x)} \quad \text{and} \quad e(t) = E(t) - E^0.$$

With the density $\rho(\phi, t) \equiv \frac{1}{N} \sum_i \delta(\phi - \phi_i(t))$ and the current $J(\phi, t) \equiv (E^0 + \Gamma(\phi)e(t))\rho(\phi, t)$ the system has to obey the equations [1-4]

$$\frac{\partial}{\partial t}\rho(\phi, t) = -\frac{\partial}{\partial \phi}J(\phi, t)$$

$$\left(\tau_1 \frac{\partial}{\partial t} + 1\right)\left(\tau_2 \frac{\partial}{\partial t} + 1\right)E(t) = J(1^-, t) \tag{1}$$

With boundary conditions $J(0^+, t) = J(1^-, t)$. These equations have a static solution $\rho^0(\phi) = 1$ and $J^0(\phi) = E^0$.

3 STABILITY

If one expands around the static solution $J(\phi, t) = J^0 + j(\phi, t)$, $E(t) = E^0 + e(t)$, etc. one has after linearisation

$$j(\phi, t) = j(\phi)e^{\lambda t} \quad \text{and} \quad e(t) = \varepsilon e^{\lambda t}.$$

Where λ has to obey

$$(\lambda\tau_1 + 1)(\lambda\tau_2 + 1)(e^{\lambda/E^0} - 1) = \frac{\lambda}{E^0}\int_0^1 d\phi\ \Gamma(\phi)e^{\lambda\phi/E^0}$$

The static solution is stable if this equation only has solutions with the real part of λ less than 0.

4 SMALL COUPLING

The eigenvalue equation has infinitely many solutions. In the uncoupled case, $G(x) = 0$, one has solutions $\lambda^0 = -1/\tau_1$, $\lambda^0 = -1/\tau_2$ and $\lambda_n^0 = 2\pi n i E^0$ with

$n \neq 0$. The system is marginally stable. For small $G(x)$, λ will be close to these values. For stability the real part of $\lambda_n = 2\pi n i E^0 + \delta_n$ must be less than 0. To the lowest order in δ_n one has

$$(1 + 2\pi n E^0 \tau_1 i)(1 + 2\pi n E^0 \tau_2 i)\frac{\delta_n}{E^0} = 2\pi n i \int_0^1 d\phi\, \Gamma(\phi)e^{2\pi n i \phi}$$

There is no solution with $\Re(\delta_n) < 0$ for large n if $\Gamma(0) > \Gamma(1)$. If $d\Gamma/d\phi > 0$ the real part of the right hand side of the eigenvalue equation is larger than 0, so that if τ_1 and τ_2 are large enough the asynchronous state will be stable.

5 EXAMPLE

We consider a system in which

$$\frac{dx_i}{dt} = k(x^0 - x_i) + g(x^E - x_i)$$

and $\tau_1 = \tau_2 \equiv \tau$. In this case the timedependence of the synaptic current is given by a sum of alpha-functions

$$E(t) = \frac{1}{N}\sum_{j,\alpha}\frac{t - t_{j,\alpha}}{\tau^2}e^{-\frac{t - t_{j,\alpha}}{\tau}}\theta(t - t_{j,\alpha})$$

and for the eigenvalues one has

$$(\lambda\tau+1)^2(e^{\lambda/E^0}-1) = \frac{g}{\ln(R)}\frac{\lambda}{E^0}\int_0^1 d\phi\left(1 + \frac{(R-1)x^E - R}{R}e^{\ln(R)\phi}\right)e^{\lambda\phi/E^0}.$$

E^0 and R are given by the two equations $E^0 = k(\ln(R) - g)^{-1}$ and $R = (kx^0 + gE^0 x^E)/[k(x^0 - 1) + gE^0(x^E - 1)]$. Stability of the asynchronous state is in this case possible only if $x^E > x^0$ and for small g this state is stable if

$$\left[k\tau - \left(\frac{\ln(\frac{x^0}{x^0-1})}{2\pi}\right)^2\right]^2 > \left(\frac{\ln(\frac{x^0}{x^0-1})}{2\pi}\right)^4 + \left(\frac{\ln(\frac{x^0}{x^0-1})}{2\pi}\right)^2.$$

The model only has static solutions for $g < \ln(\frac{x^0}{x^0-1})$. ¿From the eigenvalue equation we determined numerically the value of τ above which the asynchronous state is stable. The results are plotted in figure 1. In figure 1a $k\tau$ is plotted against g for $x^E = 2.0$ and different values of x^0, in figure 1b $k\tau$ is plotted x^0 for fixed x^E and g.

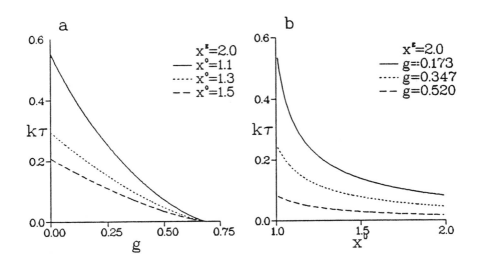

Figure 1   The value of τ above which the asynchronous state is stable is plotted in a against g for $x^E = 2.0$ and and different values of g.

Finally we did numerical simulations with $k = 1.0$, $x^E = 2.0$, $x^0 = 1.5$ and $g = 0.5\ln(x^E/[x^E - 1])$ for a system with 100 neurons, with the initial distribution close to $\rho^0(x)$. ¿From figure 1 one has that, for these parameters, the static state is stable if τ is larger than ± 0.8 and unstable if τ is smaller. Figure 2 shows the result for one simulation with $\tau = 0.15$ and one with $\tau = 0.05$ with the same initial voltages for the neurons. As expected the density stays close to the theoretical $\rho^0(x)$ for the system with $\tau = 0.15$, while for $\tau = 0.05$ the neurons start to synchronize.

6   REFERENCES

[1] D. Golomb, D. Hansel, B. Shraim and H. Somoplinsky, Phys. Rev.   A45, 3516(1992).

[2] S.H. Strogatz and R.E. Morollo, Physica D31, 143(1988).

[3] S.H. Strogatz and R.E. Mirollo, J. Stat. Phys. 63, 613(1991).

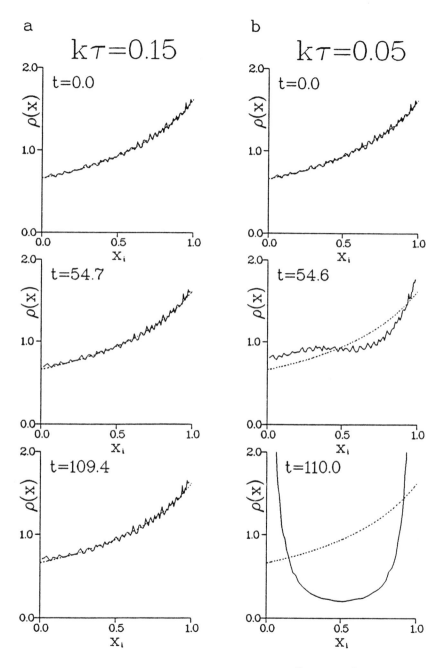

Figure 2 Numerical simulation with $k = 1.0$, $x^E = 2.0$, $x^0 = 1.5$ and $g = 0.5 \ln(x^E/[x^E - 1])$ in a system of 100 neurons. In *a* $\tau = 0.15$ the calculated density does stay close the theoretical density (dotted line). In *b* $\tau = 0.05$ the density changes and the neurons start to synchronize.

[4] A. Treves, Oxford University preprint (unpublished)(1992)

NEURAL GEOMETRY REVEALED BY NEUROCOMPUTER ANALYSIS OF MULTI-UNIT RECORDINGS

Andras J. Pellionisz
David L. Tomko
*James R. Bloedel

NASA Ames Research Center, 242-3, Moffett Field, CA 94035
* Barrow Neurological Institute, 350 West Thomas Rd, Phoenix, AZ 85013

ABSTRACT

A basic goal of neurocomputing is to identify, from biological neural nets, the mathematics intrinsic to neural net function. This is not only a basic research goal - but fundamental to electronic implementation to neurocomputers. To experimentally measure neural geometry, from up to ten cerebellar Purkinje cells of locomotory cats, multi-electrode recordings were obtained and the metric tensors of the functional manifold were calculated. The covariant metric tensor was calculated by cross-correlation analysis and the contravariant metric tensor was obtained by its Moore-Penrose inverse. For analysis of massively parallel data, a transputer-based neurocomputer was built. Results demonstrate the non-Euclidean geometry intrinsic to neural nets and its modifiability with learning.

11.1. NEURAL GEOMETRY

A decade ago a geometrically framed hypothesis was formulated portraying sensorimotor function as a process requiring generalized coordinate transformations that utilize non-Cartesian vectors expressed in frames of reference that are intrinsic to biological systems [10]. This geometrical approach to neural networks was experimentally confirmed [8], [11], and was developed to a sensorimotor neural net theory with implications to an array of fields [9]. Tensor network theory reformed vestibulo-cerebellar gaze research, from system-analysis [15] to analysis of parallel distributed neural networks [3]. Within a decade of its inception, a geometrical approach is taken also by leading European and Japanese neural net researchers, [6], [2] including some of its former critiques [4].

11.2. NEUROCOMPUTER ANALYSIS OF NEURAL GEOMETRY

The goal of this experimental study was to determine the geometric, non-Euclidean properties characterizing the population responses of neurons in specific CNS nuclei and

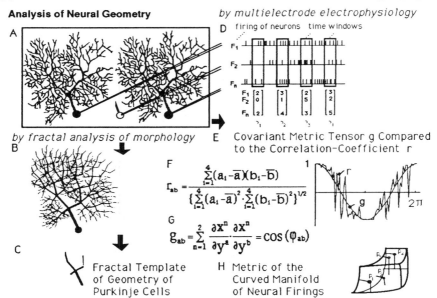

Fig. 11.1. Mathematical basis of fractal (B-C) and metrical neural geometry (D-H); cf. [9]. Correlation of firings (D-E-F) yields covariant metric (G) of curved geometry

to relate those properties to the geometrical theory. For this purpose, responses of up to 10 simultaneously recorded Purkinje cells were investigated by one of us (J.R.B.) during perturbed and unperturbed locomotion in acutely decerebrated cats. Findings were related to a computer model of the skeletomotor system of the cat forelimb [5].

11.3. NEURAL NET ALGORITHM: TENSOR NET THEORY

Population responses were analyzed based on the method [9] of calculating the non-Cartesian axes from the cross-correlogram of the neural activity of n neurons, a procedure that also yields the matrix characterizing the covariant metric tensor, from which the Moore-Penrose generalized inverse (the contravariant metric tensor) is calculated, see Fig. 1. The dual metric tensors fully characterize the geometry of population response.

11.4. ANALYSIS BY TRANSPUTER-BASED NEUROCOMPUTER REVEALS INTRINSIC COORDINATES OF MUSCLES

Calculation of cross-correlation-matrix and its Moore-Penrose generalized inverse was performed by a transputer based neurocomputer (developed by A.J.P and D.L.T. by a NASA "Director's Discretionary Fund", awarded by "Best Projects"). Basic features of the neurocomputer are shown in Fig. 2. Responses were recorded to intermittent perturbations of the forelimb as well as during the acquisition and performance of the movement conditioned to avoid an obstacle. The results indicate (shown in Fig. 3.) that the geometry describing the population responses is expressed in non-Cartesian coordinates and that the characteristics of this geometry are comparable to characteristics of the locomotor skeletomuscular system. Specifically, the multi-unit analysis revealed, to a

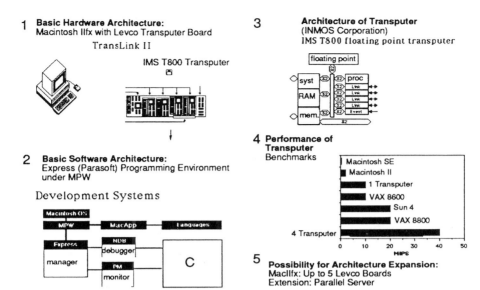

Fig.11.2. Basic features of transputer-architecture of the neurocomputer developed for experimental analysis of neural geometry. Macintosh II host boosted to twice VAX

remarkable degree, those non-Cartesian axes of pulling-directions of the muscles that are predicted by the model of skeletomuscular system of the limb (see Fig. 3).

11.5. NEURAL ADAPTATION CONCEIVED AS ALTERATION OF FUNCTIONAL GEOMETRY

Results presented in a preliminary account further showed [12], see Fig. 4, that the geometry of the responses is modified when the characteristics of the movement are changed either by perturbing the swing phase of the ipsilateral forelimb. These findings suggest that the derived geometry of the population responses may be the basis for reconstructing physical invariant such as movement direction and distance. If so, this theoretical-experimental method is a candidate for deciphering the neural code internally representing invariants of the external world, and the actually measured alterations in the non-Euclidean functional geometry could be a novel basis of cerebellar adaptation.

11.6. MATHEMATICAL LANGUAGE OF NEUROCOMPUTERS: NON-EUCLIDEAN GEOMETRY

By discerning from existing biological neural networks the functional geometry intrinsic to neural net computation, neurobiological vestibulo-cerebellar gaze research can progress towards a vectorial reconceptualization of reflexes [14], and construction of electronic neurocomputers will become possible that mimic the function of the biological brain, resulting (by surpassing pre-neurocomputing attempts [1],[7]) in the development of an artificial electronic (vestibulo)cerebellar sensorimotor neurocontroller [12],[13],[14],[16].

Supported by NASA DDF T4967 (AJP & DLT), NASA-NRC (AJP) and NS21958 (JRB)

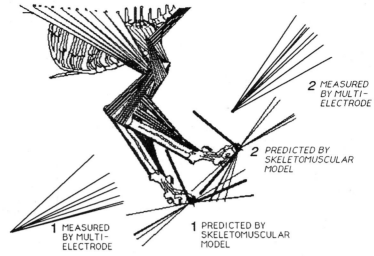

Fig. 11.3. Non-Cartesian axes of limb-muscle pulling directions as predicted from skeletomuscular model [5], compared to axis-directions computed from population-responses of cerebellar Purkinje cells in primary- (1) and obstructed positions (2). Neurocomputer analysis of population responses of Purkinje cells can reveal, to a remarkable degree, the generalized coordinate systems intrinsic to neural nets.

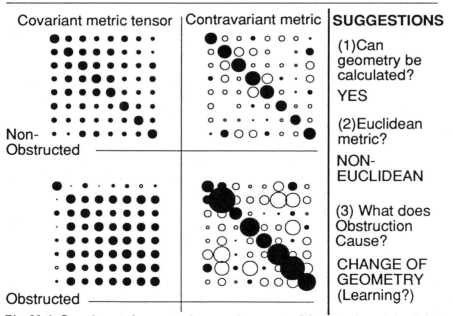

Fig. 11.4. Covariant and contravariant metric tensors of the neural geometry inherent in cerebellar coordination as calculated from 8 Purkinje cell responses. Positive and negative matrix-components are shown by filled and empty circles. Since co- and contravariant tensors are greatly different, neural geometry is non-Euclidean. Also, cerebellar functional geometry is altered when the cat learns to step over obstacle.

REFERENCES

[1]   J. R. Albus. *"Brains, behavior and robotics."* McGraw-Hill. New York. 1981.

[2]   S. Amari. Dualistic Geometry of the Manifold of Higher-Order Neurons. *Neural Networks.* 4(4): 443-451, 1991.

[3]   T. J. Anastasio and D. A. Robinson. Distributed Parallel Processing in the Vestibulo-Oculomotor System. *Neural Comp.* 1(2): 230-241, 1989.

[4]   M. Arbib and S. I. Amari. Sensorimotor Transformations in the Brain with a Critique of Tensor Network Theory of the Cerebellum. *J. Theor. Biol.* 112: 123-155, 1985.

[5]   J. R. Bloedel, S. Tillery and A. J. Pellionisz. Experimental-Theoretical Analysis of the Intrinsic Geometry of Limb Movements. *Soc. Neurosci. Absts.* 14/2: 953, 1988.

[6]   R. Eckmiller. "Concerning the Emerging Role of Geometry in Neuroinformatics." *Parallel Processing in Neural Systems and Computers.* Eckmiller, Hartmann and Hauske ed. Elsevier Science Publishers B.V (North Holland). 5-8, Amsterdam. 1990.

[7]   J. Fiala and R. Lumia. *Adaptive Inertia Compensation Using a Cerebellar Model Algorithm.* IJCNN Singapore: 1-8, 1991.

[8]   C. C. A. M. Gielen and E. J. van Zuylen. Coordination of Arm Muscles During Flexion and Supination: Application of the Tensor Analysis Approach. *Neuroscience.* 17: 527-539, 1986.

[9]   A. Pellionisz. "Vistas from Tensor Network Theory: A Horizon from Reductionalist Neurophilosophy to the Geometry of Multi-Unit Recordings." *Computer Simulation in Brain Science.* Cotterill ed. Cambridge Univ. Press. 44-73, Cambridge. 1988.

[10]   A. Pellionisz and R. Llinás. Tensorial Approach to the Geometry of Brain Function: Cerebellar Coordination via Metric Tensor. *Neurosci.* 5: 1125-1136, 1980.

[11]   A. Pellionisz and B. W. Peterson. "A Tensorial Model of Neck Motor Activation." *Control of Head Movement.* Peterson and Richmond ed. Oxford University Press. 178-186, Oxford. 1988.

[12]   A. J. Pellionisz and J. R. Bloedel. *Functional Geometry of Purkinje Cell Population Responses as Revealed by Neurocomputer Analysis of Multi-Unit Recordings.* Soc. Neurosci Absts. 21: 920,1991.

[13]   A. J. Pellionisz, C. C. Jorgensen and P. J. Werbos. *Cerebellar Neurocontroller Project, for Aerospace Applications, in a Civilian Neurocomputing Initiative in the "Decade of the Brain".* IJCNN 92. III: 379-384, 1992.

[14]   A. J. Pellionisz, B. W. Peterson and D. L. Tomko. "Vestibular Head-Eye Coordination: A Geometrical Sensorimotor Neurocomputer Paradigm." *Advanced Neurocomputing.* R. Eckmiller ed. Elsevier, North-Holland. 126-145, Amsterdam. 1990.

[15]   D. Robinson. The Use of Control Systems Analysis in the Neurophysiology of Eye Movements. *Ann Rev Neurosci.* 4: 463-503, 1981.

[16]   P. J. Werbos and A. J. Pellionisz. *Neurocontrol and Neurobiology: New Developments and Connections.* IJCNN 1992. III: 373-378, 1992.

***

SIGNAL DELAY IN PASSIVE
DENDRITIC TREES

Hagai Agmon-Snir
Idan Segev

Department of Neurobiology, Hebrew University, Jerusalem 91904, Israel

ABSTRACT

A novel approach for analyzing transients in passive structures is introduced. It provides, as a special case, an analytical method for calculating the signal delay in any passive dendritic tree. Total dendritic delay (TD) between two points (y,x) is defined as the difference between the centroid (the center of gravity) of the transient current input at point y and the centroid of the voltage response at point x. The TD for x = y is the local delay (LD) and the propagation delay, PD(y,x), is TD (y,x) - LD (y). With these definitions, the delay between any two points in a given tree is independent of the properties of the transient current input. Also, TD(y,x) = TD(x,y) for any two points, x,y. The local delay (also TD) in an isopotential isolated soma is τ, the time constant of the membrane whereas the LD in an infinite cylinder is $\tau/2$. In finite cylinders coupled to a soma the TD from end-to-soma is always larger than τ. In dendritic trees equivalent to a single cylinder (Rall [1]), the TD from a given input site ($X = x/\lambda$) at an individual branch to the soma is identical to the total delay in the equivalent cylinder for current injected at the same X. The LD(X), however, is shorter in the full tree for any $X \neq 0$. In real dendritic trees the total delay between the synaptic input and the voltage response at the soma is on the order of τ. However, electrical communication between adjacent synapses in distal arbors can be more than 10-times faster. Consequently, exact timing between inputs is critical for local dendritic computations and less important for the input-output function of the neuron. These results have important functional significance for both the input-output characteristics of neurons and for processes underlying learning and memory.

12.1 INTRODUCTION

An important property of single neurons is that they behave as delay lines. Indeed, time elapses from the arrival of an action potential to the presynaptic (input) site, to the triggering of transmitter release by an action potential at the output site of the postsynaptic cell. This delay in transmitting information plays an important role both for the input-output function of the nervous system as well as for processes involved in learning and memory [2-8]. Time delays in single neurons arise from three distinct sources: the *synapse*, the *dendrites*, and the *axon*. Synaptic delay was thoroughly explored [9, 10]; in normal conditions this delay is less than 1 msec. Axonal propagation also introduces a delay which varies with axon lengths

74

and diameter, axonal delay may range from 1 - 20 msec [11-13]. Passive propagation in dendritic trees is also a source of delay [14-16]. First, because of the RC properties of the membrane, the synaptic potential lags in time behind the synaptic input. Second, synaptic potentials are delayed when propagating between the input site and other locations, including the cell body and the initial segment of the axon. Detailed exploration of dendritic delay was somewhat neglected both because experimentally it is difficult to measure voltage transients in dendrites and because it is mathematically a difficult problem. Consequently, analytical solution for the signal delay in electrically passive structures was obtained only for simple cases such as infinite cylinders [16]. Exploration of signal delay in dendritic trees relied primarily on numerical computations [16-19].

Here we present briefly a new theoretical approach for analyzing transients in passive dendrites. It provides a simple analytical solution for the delay in passive dendritic trees with arbitrary branching. A complete formal treatment will be given elsewhere.

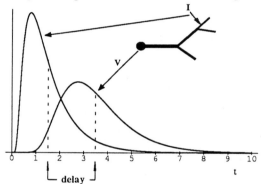

Figure 12.1 Delay is defined as the difference between the centroid of current input (I) and centroid of voltage response (V).

12.2 DEFINITIONS AND THEOREMS

Signal's time $(\hat{t}_f)$. Let $f(t)$ be some transient function of time. The time of $f(t)$, $\hat{t}_f$, is the centroid of $f(t)$ defined as,

$$\hat{t}_f \equiv \frac{\displaystyle\int_{-\infty}^{\infty} t \cdot f(t)dt}{\displaystyle\int_{-\infty}^{\infty} f(t)dt} \tag{12.1}$$

Total Delay (TD) - the difference between the time of the transient current input at some point (y) in the tree and the time of voltage response at another location (x) (Fig. 12.1). Formally,

$$TD(y,x) \equiv \hat{t}_V(x) - \hat{t}_I(y) \tag{12.2}$$

where $TD(y,x)$ is the dendritic total delay, $\hat{t}_I(y)$ is the time of current at the input point y, and $t_V(x)$ is the time of the voltage response at point x. The *Local Delay* (LD) is the TD when x = y, and the *Propagation Delay* (PD) is TD (y,x) - LD (y). With these definitions several general theorems can be proven for any passive structures:

Theorem I (shape invariance). The total delay between any two points (y,x) in a given passive structure is independent of the shape of the transient input current at point y. Clearly, this is true also for LD(y) and PD(y,x).

Theorem II (reciprocity). Given two points (y,x) in an arbitrary passive structure, $TD(y,x) = TD(x,y)$. Note that this is not necessarily true for the propagation delay.

Theorem III (equivalence). When analyzing delays in a tree, one can compute delays in any segment, replacing the subtrees at its boundaries by isopotential "somas", each has the same input resistance and input (local) delay as the corresponding original subtree. This theorem is very useful for computing delays in arbitrary trees by means of a recursive procedure.

12.3 THE ANALYTICAL METHOD

For computing delays in passive structures, we introduce a novel approach called "*the method of moments*". In this approach the i-th moment of a transient signal, $f(t)$, is defined as,

$$M_{f,i} = \int_{-\infty}^{\infty} t^i \cdot f(t) dt \tag{12.3}$$

Note that the centroid, $\hat{t}_f$, in eq. (12.1) is exactly the ratio between the 1-st and 0-th moment. In general, it is possible to define other properties of $f(t)$, such as width, skewness, etc., using moments and explore analytically how these properties change as the signal spreads in passive structures. To compute the delay in passive trees utilizing the method of moments, Rall's one dimensional cable equation was employed. The cable equation in dimensionless units $(X = x/\lambda, T = t/\tau)$ is,

$$\frac{\partial^2}{\partial X^2}V(X,T) - \frac{\partial}{\partial T}V(X,T) - V(X,T) = - R_\infty \cdot I(X,T) \tag{12.4}$$

where R_∞ is the input resistance of a corresponding semi-infinite cylinder and $I(X,T)$ is the input current. Multiplying both sides of eq. (12.4) by T^i and integrating over T, we get a linear ODE with constant coefficients for the i-th moment, i=1,2,3...,

$$\frac{d^2}{dX^2}M_{V,i}(X) - M_{V,i}(X) = - R_\infty \cdot M_{I,i}(X) - i \cdot M_{V,i-1}(X) \tag{12.5}$$

where $M_{I,i}$ is the i-th moment of the current input. Here the moments are defined over T. Using these ODE's the moments can be analyzed by the same methods used in steady-state analysis of passive trees [1]. Note also that the equation for the i-th moment depends only on moments of smaller- or- equal- order.

The equations for $M_{V,i}$ in a given tree and a given transient input, $I(t)$, can be solved analytically either directly or by using Laplace transforms. For the purpose of computing delays, only the 0-th and 1-st moment should be derived (eq. (12.1)). Because for computing delay any tree can be lumped into an equivalent "soma" (*Theorem III*), it is sufficient to find the delay in the case of a cylinder coupled to a soma. Delays in arbitrary trees can be computed in each cylindrical segment separately and a recursive method for calculating delays over the tree can then be used [20].

12.4 RESULTS AND DISCUSSION

The total delay between an injection point, Y, and the sealed end at the origin (X = 0) is shown in Fig. 12.2 for three cylinders with L = 0.5, 1, 2. The total delay from one end of a cylinder to the other is always larger than τ. Namely, compared to an isopotential structure where the TD (which is also LD) is τ, the total delay increases in the cylinder when the signal propagates between its two ends. Nonetheless, cylinders allow for faster communication (smaller delays than τ) for some values of X < L. This communication can be much faster in dendritic trees.

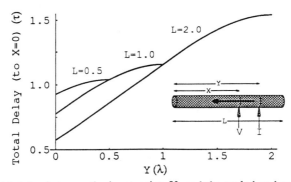

Figure 12.2 Total delay between the input point, Y, and the sealed-end origin (X = 0) of a cylinder. Three cylinders are examined with L = 0.5, 1.0 and 2.0.

In Fig. 12.3 an idealized tree (equivalent to a cylinder) is modeled. Continuous lines show TD along the tree for input at terminal, T_1, and the dashed line depicts TD in the equivalent cylinder. Numbers denote PD over individual branches. Along the path from T_1 to the soma, PD over the first three branches increased (0.21, 0.30, 034) then it decreases to 0.22 in the stem dendrite. The TD from T_1 to origin is ~1.3; this is also the TD for the equivalent cylinder. The LD at T_1, however, is significantly smaller than the LD at the corresponding point in the equivalent cylinder. Hence, the delay to various possible "target" points in the tree may differ significantly. This delay (and the communication speed) is not only a function of the distance from the input point to the target point but also a function of the morphology between the two points (e.g., compare the TD to the two points that are 0.5λ from T_1). Hence, electrical communication in dendritic trees operates in multiple time-scales. Delay to soma is on the order of τ whereas local interactions may be 10-times faster.

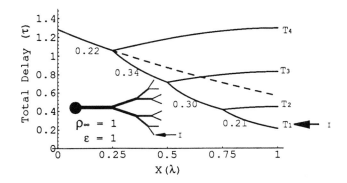

Figure 12.3 Delay in an idealized tree equivalent to a cylinder. Each branch is 0.25 λ long. Input is to the lower terminal tip (T_1). The total delay from T_1 to all other dendritic sites is depicted by the continuous line. Numbers denote PD over individual branches. Dashed line is the total delay when the input is at the end of the equivalent cylinder.

12.5 ACKNOWLEDGMENTS

We thank Wilfrid Rall for many insightful suggestions and to La Netta Dietrichson for her skillful assistance in formatting the layout of this manuscript. This work was supported by the ONR and by a grant from the US-Israel Bionational foundation.

12.6 REFERENCES

[1]   W. Rall. *Exp. Neurol.* 1, 491 1959.
[2]   W. Reichardt, W. *J. Comp. Physiol.* A 161, 533 1987.
[3]   C. E. Carr., M. Konishi. *Proc. Natl. Acad. Sci. USA.* 85, 8311 1988.
[4]   H. B. Barlow., W. R. Levick. *J. Physiol. (Lond).* 178, 477 1965.
[5]   C. Koch., V. Torre., T. Poggio. *Proc. Natl. Acad. Sci. USA* 80, 2799 1983.
[6]   E. R. Kandel. *Cellular Basis of Behavior.* Freeman, San Francisco, 1976.
[7]   T. Brown., P. Chapman., E. Kairiss., C. Keenan. *Science.* 242, 724 1988.
[8]   W. Rall. In: *Neural Theory and Modeling*, R. F. Reiss, editor. Stanford University Press, pp. 74-94 1964.
[9]   J. C. Ecceles., B. Katz., S. W. Kuffler. *J. Neurophysiol.* 4, 362 1941.
[10] B. Katz., R. Miledi. *Proc. R. Soc. B* 167, 8 1967.
[11] S. Waxman., W. M. L. Bennett. *Nature (Lond.),* 238, 217 1972.
[12] B. I. Khodorov., E. N. Timin. *Prog. Biophys. Mol. Biol.* 30, 145 1975.
[13] Y. Manor., C. Koch., I. Segev. *Biophys, J.* 60, 1424 1991.
[14] W. Rall. *Ann. NY Acad. Sci.* 96, 107 1962.
[15] R. Rinzel., W. Rall. *Biophys. J.* 14, 759 1974.
[16] J. Jack., D. Noble, R. Tsien. *Electric Current Flow in Excitable Cells.* Oxford 1975.
[17] J. Jack., S. Redman. *J. Physiol. (London).* 215, 283 1971.
[18] W. Rall. In: *Neural Theory and Modeling*, R. F. Reiss, ed. Stanford University Press, pp. 73-94, 1974.
[19] J. F. Iles. *Proc. R. Soc. Lond. B.* 197, 225 1977.
[20] H. Agmon-Snir., I. Segev. *J. Neurophys.* (in press).

SECTION 2

MODELING

This section brings together research that focusses on modeling techniques, modeling tools, and model validation. A variety of approaches and techniques, from software to hardware and from single cell compartmental modeling to whole systems modeling, is represented here. Related work can be found in chapter 6 (section 1), chapter 23 (section 3), chapter 37 (section 4), chapter 43 (section 5), chapter 45 (section 6), and chapter 51 (section 7).

13

MATCHING NEURAL MODELS TO EXPERIMENT

William R. Foster[1,2]
Julian F.R. Paton[2]
John J. Hopfield[3]
Lyle H. Ungar[1]
James S. Schwaber[2]

[1]Che. Eng. Dep.; U. of Penn.; 330 S. 33rd; Philadelpha, PA 19104
[2]Neural Computation Group; DuPont Co.; Wilmington, DE 19880-0323
[3]Division of Biology; Caltech; Pasadena, CA 91125, U.S.A.

ABSTRACT

Methods of quantitatively fitting single compartment neuron models to experimentally observed neuron behavior in current clamp are investigated. Standard minimization techniques relying on gradient information are not useful for matching Hodgkin and Huxley style models to current clamp data. It is, however, possible to fit current clamp data with minimal experimental information by making use of channel kinetics descriptions taken from the literature. The fitting procedure uses random search and an appropriately defined metric. This approach eases the task of creating neuron models for use in network simulations. The resulting fits to experiment are not unique and therefore the utility of the models must be judged in terms of their ability to represent important aspects of observed neuronal behavior for relevant inputs.

13.1   INTRODUCTION

This paper presents a method for creating Hodgkin and Huxley single compartment neural models. This method, like all neural modeling methods, arises from the interplay of different sources of knowledge. These sources include experiment, the existing biological literature, computation, and the modeler's labor and intuition. The demands placed on these knowledge sources are determined by the purpose of the neural model. Each of these knowledge sources has, however, a cost of exploitation and these costs may be traded against one another in order to achieve an acceptable model. The work presented here gives a particular example of the interplay among knowledge sources which emphasizes the use of the biological literature, computation and purpose of the model in order to ease the requirements placed on experiment and the modeler.

The purpose of our neural models is to represent neurons in the brain stem which participate in the baroreceptor vagal reflex. Specifically we are interested in neurons which receive inputs from baroreceptors. These neurons receive synaptic volleys which are a consequence of variations in blood pressure. The neurons must then process these inputs depending on their morphology and the types, amounts, and distribution of ion channels present. The result of this processing is a train of action potentials. We are interested in models of relevant neurons which reproduce the timing of these action potentials in response to physiological inputs. As a first step, however, we seek models which reproduce the timing of action potentials in response to intracellular current injection at the recording site, acknowledging that this overlooks the role that the spatial distribution of inputs may play in information processing. Nontheless, this simple approach may provide insights into the role of membrane non-linearities in input processing.

Whole cell patch was used to stimulate relevant neurons *in vitro*. The stimuli used were current clamp step inputs, sinusoidal inputs, and synaptic currents recorded from similar neurons previously. These various stimuli allowed us to match model response to a stimulus and then test model generalization on very different and even somewhat physiological stimuli. In addition, pharmacological experiments gave reasons to include TTX, 4AP, TEA, cobalt, and noradrenaline sensitive currents in the models. (See 1 for a description of relevant methods and findings.)

13.2   PROBLEM STATEMENT

Our modeling goal is to approximate the input-output mappings of neurons in current clamp based on minimal experimental data and minimal modeling labor. The approach used is to assume that some classes of ionic channel models, particularly ones derived from a full series of voltage clamp experiments on other neural systems, are useful approximations to the channels which actually exist in the neurons we wish to model. The problem then becomes one of adjusting given channel descriptions and the degree to which they contribute to the behavior of a neuronal model in order to achieve a match with observed experimental data. (See 2 for a fuller description of this philosophy.) The resulting fit to experiment is then subject to the concerns of fitting techniques in general, that is: do the resulting parameters have any physical significance, is the fit unique, and does the fit extrapolate well. In the present context the last concern is of primary importance.

To achieve the modeling goal, we started with a simple model structure and measure of success. The models used are single compartment neural models based on the formalism of Hodgkin and Huxley (3), termed HH. To reproduce observed qualitative behaviors and pharmacological results, the presence of channels other than fast sodium and delayed rectifier were made available to models. The specific kinetics of these channels were taken from the literature and viewed as useful approximations which would need adjustments, certainly given the difference in the source preparations and that under study, neurons in the rat cardiorespiratory NTS (1). The number and kinetics of the channels could, therefore, be adjusted to reproduce desired behaviors.

These ideas can be placed into a quantitative measure of goodness of fit and the whole problem cast as a minimization. Firstly, the mathematical description of an HH single compartment model can be abstracted as a system of non-linear differential equations:

$$\frac{dy}{dt} = \underline{f}(\underline{y}, i(t), \underline{\alpha}) \qquad (13.1)$$

Where $\underline{y}$ are the variables representing voltage, calcium concentration, and the activation and inactivation parameters. Perhaps $y(1)$ is voltage, $y(2)$ calcium concentration, $y(3)$ sodium activation parameter and so forth. $i(t)$ is the stimulating current, and the symbol $\underline{\alpha}$ is a vector of model parameters which can be used to alter model behavior. These parameters may be such things as voltage shifts in steady state activation curves or the maximal conductance of a channel type.

The initial conditions used by us are determined by assuming steady state at a holding current, with the corresponding voltage measured.

$$\frac{dy}{dt} = \underline{f}(\underline{y}, i_h, \underline{\alpha}) = \underline{0}; \ y(1) = v_h \qquad (13.2)$$

This condition creates a non-linear system of equations which determines the values at time zero of calcium concentration as well as of the activation and inactivation parameters and of the leakage reversal potential. Note it may be that no solution exists or that the solution is unstable.

These equations are then integrated from the initial conditions and in response to a current stimulus. The time that action potentials occur is then determined, since as a first approximation we view this as the output of the neuron. If the number of action potentials differs between experiment and model then this difference squared represents an error, while if the number is the same, then the differences in action potential timing can represent an error. The minimization problem to be solved is, therefore:

$$\min_{\underline{\alpha}} \ \sum_{stimulus j} [(\tilde{n}_j - n_j)^2 + \sum_1^{n_j} (\tilde{t}_{i,j} - t_{i,j})^2] \qquad (13.3)$$

where the tilde denotes model values. The second sum which is the error in the timing of action potentials between the model and experiment is only

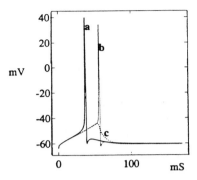

Figure 13.1   Model sensitivity to parameter changes.

computed if both model and experiment produce the same number of action
potentials for a given stimuli, i.e. $\tilde{n}_j = n_j$.

To solve this problem several general approaches exist. The first is to
adjust model parameters by hand to reproduce desired model behavior (2,4,5).
That is, a set of model parameters is chosen, and the model's response to
current injection is determined by numerical integration. The results are compared, often by eye, to the experimentally observed behavior. What are deemed
appropriate adjustments are then made, and the process is repeated until the
modeler is satisfied with the match. This process has the advantage of making use of the modeler's intuition regarding the role each ionic channel type
is likely to play in the observed neural behavior. Unfortunately, the process
is extremely time consuming and tedious. Another approach is to use common minimization techniques which rely on gradient descent. These methods
such as conjugate gradient take derivatives of model behavior in order to find
changes in parameters which reduce error. Due to the character of HH equations, these methods are generally not useful. The reason for this becomes
apparent by the example shown in figure 13.1. In the figure a model is shown
responding to a step input current injection. A is the control model, B is the
model with an increase of 3.96 mmho/cm2 of Kdr, and C shows the model behavior after an additional 1.1E-14 mmho/cm2 of Kdr are added. If the 1.1E-14
mmho/cm2 change were made to model A, the behavior would have changed
by an insignificant amount; while the same change made at B produces a very
large change in model behavior. This shows that derivatives of HH model behavior with respect to parameter changes can lead to an unpredictable mix of
near zero values and extremely large ones. This is simply an expression of the
thresholding behavior of neurons with which everyone is familiar. Thus, methods relying on derivative information will fail. This failure was seen in practice
by us and only later the reasons for the failure were investigated. Interestingly,
almost identical curves B and C may be arrived at by shifting sodium activation to depolarized levels by 0.63 mV and and then an additional 2.4E-15
mV. This shows that models having different parameter values may behave

the same. We will see later that this non-uniqueness can appear in solutions to the minimization.

Since manual parameter adjustment is tedious and time consuming and standard minimization techniques do not work, another approach is to use stochastic methods. These methods use pseudo-random changes in parameter values and repeated evaluations to search for a solution. We have successfully used random search (6) for solving the stated problem. The method used is very simple:

1) Pick an initial guess for model parameters and give them bounds. These bounds would ideally be tightly determined by good experimental information. However, if this is not possible then looser bounds based on plausible guesses and physical constraints, such as non-negativity of conductances, are used.

2) Evaluate the error measure at the initial parameter values. In general this error measure may involve the combination of multiple criteria and of responses to multiple stimuli. Here we will confine our examples to single stimulus training and an error measure composed of an exclusive combination of action potential number and timing. It is to be noted that whenever combinations are done in creating an error measure a minimization can result in undesirable compromise solutions.

3) For each parameter α_i which is to be varied:

$$\alpha_{i,new} = \alpha_{i,old} + N(0,\beta)(\alpha_{i,max} - \alpha_{i,min}); \beta_{new} = 0.95\beta_{old} \qquad (13.4)$$

If $\beta < 0.01$ then $\beta = 0.20$. $N(0,\beta)$ is a pseudorandom number drawn from a gaussian distribution having zero mean and β variance. Note that β shrinks such that small steps are favored, and then re-expands. This is common in random search methods. If a chosen parameter value lies outside of its accepted bounds, then it is given the boundry value.

4) Solve for the initial conditions and then integrate the model as driven by the selected stimuli. Note that it is possible that the parameters chosen will lead to unacceptable initial conditions or to a failed numerical integration due to some pathology. In such cases parameters are rechosen.

5) Compute the error measure. If $Error(\alpha_{new}) < Error(\alpha_{old})$ then accept α_{new}, otherwise keep α_{old}.

6) Go to 3 and repeat an arbitrary number of times or until some stopping criteria is met. In this report an arbitrary repetition of 1000 trials was used. With integrations spanning a few hundred milliseconds and several action potentials, the 1000 trials took less than one hour on a workstation.

13.3   RESULTS

The utility of the random search method depends on its ability to converge and the uniqueness and generalizeability of the resulting solutions. A short exploration of these issues is presented here.

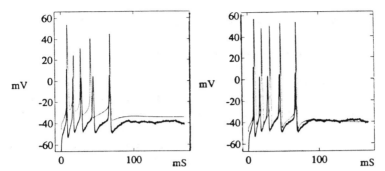

Figure 13.2 Two model solutions to the minimization.

Computer simulations showed that matches to current clamp data can be found using random search even when starting from very distant starting guesses. For instance, a single compartment model with an initial guess of no active channels present was able to find a solution matching action potential timing within 1000 trials (Figure 13.2, left panel). The model response is shown dotted while the data is solid. The solution involved adjusting the amounts of Na, Kdr, A (5), T (7), KCa, and M (8) currents as well as five voltage shifts and two uniform scalings of time constants of inactivation, 12 parameters in all. Note that only action potential timing is fit, not any measure of voltage. The right panel shows another solution which was found starting with a model having only Na and Kdr initially. A comparison between the parameter values of these two models revealed that some parameters differed by over 100%. A test of the sensitivity of model behavior of the second model to 10% changes in three of the model parameters was done, and it was found that spike timing changed very little (not shown). Furthermore, computer simulations were done in which the two models were presented with 2000 stimuli which were either steps with pseudorandomly drawn amplitudes of 0 to 1.0 nA or sine waves of amplitude 1.0 nA and pseudorandomly chosen frequencies of from 0 to 65 hz. On average the two models differed by 0.5 action potentials for steps and 1.4 action potentials for sines. Their mean difference in action potential times was 4.6 ms for steps and 1.0 ms for sines. Thus these models showed similar behavior arising from very different parameter values. The ability of model fits to generalize was explored by fitting two structurally different models to a neuron's response to a current step. One of the models was allowed only to use Na, Kdr, and A current, while the other was in addition allowed to contain a T current and a KCa current (same sources as above). The resulting fits were identical in terms of action potential timing, however it was found that the behavior of the simpler model was much more sensitive to changes in parameter values. When the generalization of the two models was tested by driving them with previously recorded synaptic currents (increased by a

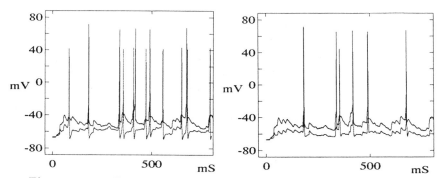

Figure 13.3 Two different models' responses to simulated synaptic input. Each model was trained on a single step response only.

factor of four upon presentation so that action potentials could be achieved), the behavior of the five conductance model (Figure 13.3, right panel) was very similar to that of the actual neuron, while the three conductance model was not (Figure 13.3, left panel. Models are shown dashed, neuron solid. Neuron had higher action potentials and baseline than the models.) Also, when the five conductance model was driven by sine wave stimuli, it reproduced many of the features of the actual neural response (not shown). These observations suggest that though random search may be able to match a stimulus using a variety of models, an inadequate model may result in behavior which is very sensitive to parameter values and does not generalize well. On the other hand the increased robustness of the five conductance model and its generalization to a complex and physiological stimulus after being fit only to the response from a single current step is encouraging. Such an outcome would be very improbable with a completely black box approach such as an artificial neural network.

13.4 CONCLUSIONS

We have shown that the creation of Hodgkin and Huxley single compartment models which match experimental behaviors can be achieved by stochastic minimization. This approach helps overcome the tedium of manual trial and error and the limitations of traditional minimization techniques. Furthermore, the method eases the burden on the experimenter and modeler by computationally exploiting the knowledge of ion channel kinetics contained in the existing biological literature. Such an approach helps avoid treating the biological literature on neurons as merely a catalogue of special cases, with every neuron model requiring a complete program of voltage clamp and pharmacological studies which then presumably result in "the definitive model." Of course, the use of random search to adapt kinetics from the literature may lead to models which reproduce desired behaviors to acceptable accuracy but are demonstrably incorrect. Again the model purpose then determines whether an incorrect model is actually inadequate.

The fits done here typically involved on the order of ten parameters and single compartment models. In principle there is no reason why multiple compartment models and models with many more parameters may not be fit using random search. Of course with many parameters to search over, very bad initial guesses may make finding acceptable solutions nearly impossible in finite time. Also, it may be that a model is asked to reproduce behaviors for a group of stimuli. In general, we can expect the fitting of multiple stimuli to have a smaller set of acceptable solutions than that for the fitting of the individual stimuli. Such solutions will be harder to find.

The neural models which result from random search are capable, though certainly not guaranteed, of generalizing to novel and physiological stimuli, but these matches may not be unique. However, this non-uniqueness can be highly desirable. For instance it is unlikely that nature is able to control the maximal conductance of a channel to arbitrary precision. Therefore, if nature desires a particular behavior from a neuron, it would be best if this behavior, which ultimately arises from parameter values, is robust within the precision that these values can be maintained. Furthermore, to the extent that behavior is what is desired by nature, then exact parameter values for neural models are not needed, and in some cases may be inappropriate.

Acknowledgements. We wish to acknowledge the support of the DuPont Company and of NSF grant BCS-9109246. We also would like to thank M. Piovoso for helpful discussions during the course of this work.

REFERENCES

[1] Paton, J.F.R., Foster, W.R. & Schwaber, J.S.(1992). *Brain Res.* , in press.

[2] Schwaber, J.S., Graves, E.B. & Paton, J.F.R.(1992). *Bra in Res.* , in press.

[3] Hodgkin, A.L. & Huxley, A.F.(1952). *J. Physiol.* **117**, 500-544.

[4] Yuen, G.L.F. & Durand, D.(1991). *Neuroscience* **41**, 411-423.

[5] Connor, J.A., Walter, D. & McKown, R.(1977). *Biophys. J.* **18**, 81-102.

[6] Salcedo, R., Concalves, M.J. & Azevedo S.F.(1990). *Computers chem. Engng.* **14**, 1111-1126.

[7] Coulter, D.A., Huguenard, J.R. & Prince, D.A.(1989). *J. Physiol.* **414**, 587-604.

[8] Yamada, W.M., Koch,C. & Adams, P.R.(1989). *Methods in Neuronal Modeling*, ed. Koch, C. & Segev, I., pp. 97-133. Cambridge: MIT Press.

14

NEURONAL MODEL PREDICTS REPONSES OF THE RAT BAROREFLEX

James S. Schwaber
Julian F.R. Paton
Robert F. Rogers
Eliza B. Graves

*Neural Computation Program, DuPont Company, Wilmington, DE 19880-0323 &
Dept. of Neuroscience, University of Pennsylvania., Philadelphia. PA 19104*

ABSTRACT

Computer models were constructed for neurons recorded *in vitro* in the cardiorespiratory NTS of the rat. Experimental results for many of these neurons showed time- and voltage-dependent transient responses to sustained inputs. Neuron models following the Hodgkin-Huxley formalism were constructed and used to produce and match this dynamic behavior. Both experimental and modeling results point to the importance of large transient potassium conductances in many of these neurons. Network models composed of one class of these neurons were synaptically driven with patterns of baroreceptor input to explore systems level computation by these neurons. Highly nonlinear integration of inputs was observed in simulations. Experiments inspired by these simulation results have subsequently shown partially similar nonlinear processing of baroreceptor inputs.

14.1 INTRODUCTION

Model neurons that capture the dynamic voltage behavior of neurons within the specific dorsal region of the nucleus tractus solitarii (NTS) known to receive inputs predominantly from cardiovascular and respiratory afferents (cardiorespiratory subdivision of the NTS; crNTS; Bradd et al., 1989; Escardo et al., 1991) were used to explore systems-level computation in baroreceptive neurons, and resulting predictions have been explored experimentally. It is well known that blood pressure and heart rate are regulated on a heartbeat-to-heartbeat time scale by the circuitry of the baroreceptor vagal reflex in the lower brainstem (reviewed in Schwaber, 1987; Spyer, 1990; Fig. 1). We have been interested in the reflex as a relatively simple multiple-input, multiple-output (MIMO) adaptive biological control system. Although fundamental circuit information is partially known (see Fig. 1 legend), the system is poorly

BAROREFLEX

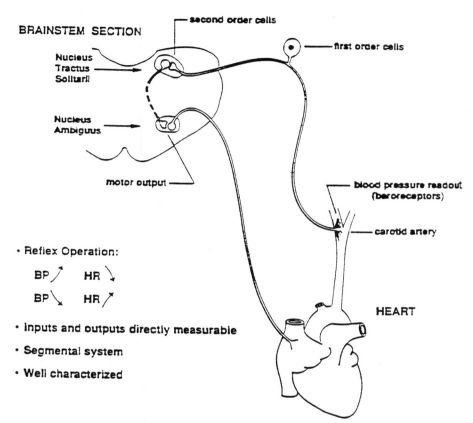

Fig. 1. Schematic of the baroreceptor vagal reflex circuit. On the afferent arm of this reflex, cardiovascular data is read by stretch receptors at the major blood vessels. These first order neurons project their input onto second order neurons in restricted regions of the nucleus tractus solitarii, here termed "cardiorespiratory" (crNTS).. There are connections between crNTS neurons and neurons located at the same anterior-posterior level of the brainstem in the caudal ventro-lateral medulla (cVLM; Paton, Rogers and Schwaber, 1991). Much of the output of the reflex is from nucleus ambiguus (NA) neurons, partially intermixed within the cVLM, and this output regulates the cardiac rate and rhythm (Escardo et al., 1991). In operation, if the blood pressure rises above desired set point, for example, then the heart is slowed, reducing cardiac output and thus blood pressure

understood, for example in terms of the representation of sensory space by first order neurons, or its mapping by second order neurons. Thus we have begun our analysis of the system with the first and second order neurons.

14.2 RESULTS

First order (sensory) model neurons

The response properties of A-fiber or Type I first order neurons are well described as normally silent below their pressure threshold, highly sensitive (with a narrow dynamic pressure range), rapidly adapting neurons that transduce and encode each pressure pulse with a train of spikes on the rising phase of pressure, in which the amount of activity is sensitive to dP/dt (Abboud and Chapleau, 1988; Seagard et al., 1990). To capture this function our first order (sensory) neuron model has two terms for the transducer current, one proportional to P(t) and the other to dP/dt, and encodes the transducer current into spike trains by an active membrane with gNa(fast), gK(DR) and gK(A) ionic currents. There are fewer than 100 of these fibers in a nerve in rat, and variations in the pressure thresholds of these fibers cover a range from well below (approx. 35 mm/Hg) to well above (170 mm/Hg) the normal range of pressure at rest. We analyzed raw data from the laboratories of F. Abboud (U. Iowa), M. Andresen (U. Texas, Galveston) and J. Seagard (U. Wisconsin, Milwaukee) to determine the integrated probability distribution function of pressure thresholds and matched this distribution in an array of model first order neurons by varying the transducer current. Fig.3B shows the simulated activity of a sample of 5 neurons from this array driven by recorded arterial pressure (Fig. 3A) before, during and after a rise in pressure due to clamping of the descending aorta. The spatio-temporal spike distribution in the plot shows the prominent encoding of the dP/dt transient and the resulting encoding of rate, the recruitment of activity as pressure rises, and encoding of mean pressure and the shape of the pressure wave form, e.g. the pulse height and the dicrotic notch.

crNTS neurons

It is not clear how the activity pattern seen in the afferent array is represented in the second-order population. Recorded second order neurons very rarely demonstrate the decrescendo activity pattern or the pulse-rhythmical activity of their sensory inputs, although they show strong excitatory responses to shock of the afferent nerves or to activation of the pressure transducers by stretch (see Schwaber, 1987; Spyer, 1990 for review). As a result it was not possible to specify *a priori* a transfer function or integral by which the observed second order responses were produced.

In order to determine whether dynamic response properties of second order neurons might contribute to their processing of afferent inputs, we recorded intracellularly with sharp electrodes from crNTS neurons *in vitro* (Paton et al., 1991; Paton et al., in press). The response of these cells proved to have large non-linearities dependent on time and voltage. Four distinct cell classes were described based on differences in response dynamics. We hypothesized that these dynamics would greatly

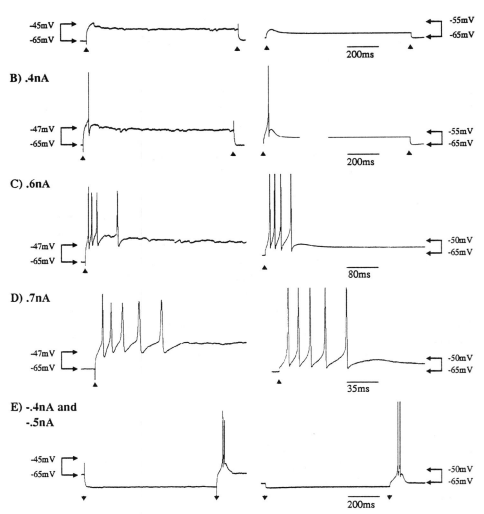

Fig. 2. The second order neuron model (right panel) as compared to the *in vitro* recordings at different current injections; a) 0.3nA, b) 0.4nA, c) 0.6 nA, d) 0.7nA, and e) - 0.4nA in the experimental case and in order to see the same rebound response -0.5nA in the simulated case. This neuron was modeled using gNa(fast), gK(DR), gK(A), gK(Ca), gK(M), and gCa(T) channels. The second order neuron is complex due to its broad range of features: initial burst, subsequent quiescence, accommodation of spikes, and rebound depolarization upon release from a hyperpolarizing pulse. The figure shows how similar the model is to the intracellular data in terms of number of spikes, interspike interval, and subthreshold behavior. Time scales for simulation and intracellular recordings at the same stimuli are equal

influence the response of the cells to the afferent input and that capturing these dynamics in neuron models might allow further analysis of the function of second order cells.

Dynamic model neurons

Computer models were constructed using membrane ion channels to produce and match the observed neuronal voltage behavior. The description of channel kinetics follows the Hodgkin-Huxley form. Different neuronal sources from the literature were investigated in order to assemble kinetic equations. The channels needed to create crNTS neuron models include gNa(fast), gK(DR), gK(A), gK(AHP), gK(M), gCa(T), gCa(L), and gK(Ca). Neuron models were tuned to achieve qualitative fits to the recorded voltage behavior of the crNTS cell types, primarily by adjusting the maximum channel densities and time-dependence of kinetics. *The model neurons demonstrate the contribution of an interaction of various voltage- and calcium-dependent membrane conductances to neuronal response dynamics* (Schwaber et al., 1992, in press).

The transient outward potassium conductances proved to be necessary to simulate the observed behavior of crNTS neurons, particularly the gK(A) but also the gK(Ca). For example, the close correspondence of simulation with experimental results for one of the crNTS neuron cell types, the S3 cell, can be seen in Fig.2. The simulation and recorded responses have similarly graded responses to successively larger depolarizing currents and similar rebound discharge following withdrawal of hyperpolarizing current. The shape of the rebound discharge and afterhyperpolarization in simulation primarily depends on the gK(A), while the accommodation of neuron firing-frequency reflects influences of both potassium conductances.

Experimental predictions and results

We have now taken these simulation results as a prediction, and further studied the crNTS neurons types in current and voltage clamp using the whole-cell patch to define the presence and role of certain ionic currents in the firing response patterns of the four cell types. In three of the four types, including the S3 type simulated here, the key role of a 4-aminopyridine (4-AP) sensitive gK(A)-like conductance to the discharge pattern, including firing rate and the shape of the afterhyperpolarization was shown. In the same cell types an additional, cobalt-and NE-sensitive gK(Ca) or gK(AHP)-like conductance was demonstrated and made important contributions to the accommodation of spike-frequency (Paton et al., 1992, in press).

Network simulation, predicted experimental results.

In order to explore the possible implications of crNTS neuron dynamics for the processing of afferent inputs, the S3 neuron model was driven synaptically by the first

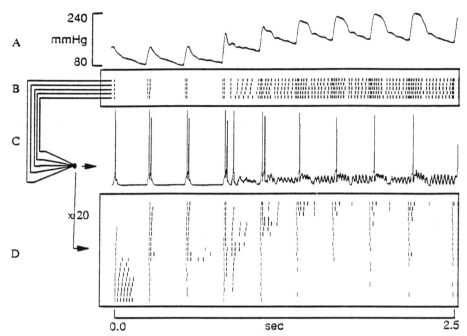

Fig. 3. A is a recording of carotid arterial pressure before, during ,and after a rise in pressure due to clamping of the descending aorta. In B neuronal activity is shown of 5 first order afferent models between 100-105mm/Hg. Neuronal activity is shown as a "raster plot" in which each mark is a spike, each horizontal row of marks is the response of a single neuron. These afferents drive the neuron in C, which is the model neuron S3 from Fig. 2. D is 20 of these model neurons all driven by 5 inputs taken sequentially from 50-150mm/Hg so that the bottom trace is 50-55mm/Hg and the top trace is 145-150m/Hg.

order neuron model in simulation. Synapses were modeled as ligand-gated ($t•e^{-\alpha t}$) conductances, with reversal potentials and time constants matched to *in vitro* measured epsp's. Panels A and B in Fig.3 show the simulated response of five baroreceptor afferents with pressure thresholds between just above the mean arterial pressure at the beginning of the trace. Panel C shows the response of the putative second order model neuron from Fig. 3 to the convergent input of the five afferents in panel B. Panel D shows an array of second order model neurons having an organization such that the bottom trace is receiving input from low pressure inputs, and the top trace from high pressure-threshold inputs. *The model second order neuron response in Panels C&D is not linearly related to pressure or afferent input.* Its response never reflects the bursting pattern of the inputs, shows nonlinear responses to increases in the input, and paradoxically actually decreases activity with increases in afferent input. This unexpected result is the product of the membrane-dependent dynamics of the model neuron. The cardiac period is clearly represented in the population response in Panel D, but not by all cells. Many cells have bursts of action potentials during the change in pressure. This suggests that this second order neuron model can detect and respond to pressure transients.

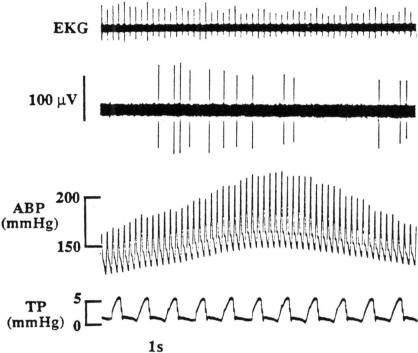

Fig. 4. Response of a crNTS second order cell that responded with short latency to aortic nerve stimulation to a rise in blood pressure. Phenylephrine (2µg i.v.) was used to produce the rise in arterial blood pressure seen in the lower trace. The extracellularly recorded neuronal response in the upper trace changed from almost silent to a burst of spikes, and then returning to a lower rate as pressure continues to rise. TP-tracheal pressure; EKG-electrocardiogram; ABP-arterial blood pressure.

14.3 CONCLUSIONS

We have no way of knowing whether our S3 neuron model, derived from *in vitro* experiments, is baroreceptive, and we recognize that many influences such as inhibitory feedback and drive from other brain regions would influence the response of baroreceptive cells *in vivo*. Nonetheless, taking the network simulation results as a hypothesis, we have recorded second order cell responses *in vivo* during experimentally induced increases in blood pressure. Fig.4 shows a putative second order cell response, representative of several hundred recorded to date, to increases in blood pressure. That its maximal rate is on the slope of the pressure rise is completely unexpected, but resembles the response pattern of the model neuron to inputs reflecting a pressure increase in that (1) the cell increases rate as pressure rises to a point, but (2) then slows such that a lower rate is seen at peak pressures. The nonlinear responses of both the recorded second order neuron and the S3 neuron model to primary afferent input may indicate the significant role neuronal dynamics play in the computation in this network.

Other classes of crNTS neurons may differently process the same inputs, dividing sensory space into "channels". However, the present S3 neuron type does not appear to integrate input patterns it would plausibly receive from A-fiber or Type I baroreceptor afferents so as to encode mean pressure, but does appear to encode pulse rate and changes in pressure. Experimental results confirm that this appears to be the case for at least some crNTS neurons. We are continuing to interact simulation and experimental work to determine the basis of these and related results thus improving this neuron model, and to construct a model reflex network with additional neuron models.

Acknowledgments. This work was supported by the Dupont Company, the ONR, the NIH and the NSF. Several colleagues have contributed to the views expressed here, among whom are KM Spyer, JJ Hopfield, B Ogunnaike, M Piovoso, W Rogers and J Lazzaro. Thanks to J Seagard, M Chapleau, F Abboud and M Andresen for generously sharing raw data for analysis.

REFERENCES

Abboud, F.M. & Chapleau, M. W. Effects of pulse frequency on single-unit baroreceptor activity during sine-wave and natural pulses in dogs. *J. Physiol.* **401**: 295-308, 1988.

Bradd, J., Dubin, J., Due, B., Miselis, R.R., Montor, S., Rogers, W.T., Spyer, K.M. & Schwaber, J.S. Mapping of carotid sinus inputs and vagal cardiac outputs in the rat. *Soc. Neuroscience Abstr.* **15**: 593, 1989.

Escardo, J.A., Schwaber, J.S., Paton, J.F.R. & Miselis, R.R. Rostro-caudal topography of cardiac vagal innervation in the rat. *Soc.Neuroscience Abstr.* **17**: 993, 1991.

Paton, J.F.R., Rogers, W.T. & Schwaber, J.S. The ventrolateral medulla as a source of synaptic drive to rhythmically firing neurons in the cardiovascular nucleus tractus solitarius. *Brain Res.* **561**: 217-229, 1991b.

Paton, J.F.R., Foster, W.R. & Schwaber, J.S.Characteristic firing behavior of cell types in the cardiorespiratory region of the nucleus tractus solitarii of the rat. Brain Res, in press.

Schwaber, J. S. Neuroanatomical substrates of cardiovascular and emotional-autonomic regulation. In *Central and Peripheral Mechanisms in Cardiovascular Regulation*, ed. A. Magro, W. Osswald, D. Reis & P. Vanhoutte. pp. 353-384. New York: Plenum Press. 1986.

Schwaber, J.S., Graves, E.B. & Paton, J.F.R. Computational modeling of neuronal dynamics for systems analysis: application to neurons of the cardiorespiratory NTS in the rat. Brain Res., in press.

Seagard, J. L., van Brederode, J. F. M., Dean, C., Hopp, F. A., Gallenburg, L. A. & Kampine, J. P. Firing characteristics of single-fiber carotid sinus baroreceptors. *Circ. Res.* **66**: 1499-1509, 1990.

Spyer, K.M. The central nervous organization of reflex control. In *Central Regulation of Autonomic Functions*, edited by A.D. Loewy, and K.M. Spyer. New York: Oxford University Press. 1990.

15

SIMULATIONS OF SYNAPTIC INTEGRATION IN NEOCORTICAL PYRAMIDAL CELLS

Paul C. Bush
and Terrence J. Sejnowski

Computational Neurobiology Laboratory
Salk Institute and Howard Hughes Medical Institute
La Jolla, CA 92037

ABSTRACT

Despite their electrotonic compactness, neocortical pyramidal cells cannot be considered as point neurons because of nonlinear interactions between inputs on the same dendritic branch. Using compartmental simulations, we have shown that dendritic saturation is significant for physiological levels of synaptic activation. We also show that the firing of about 10% of the total number of inhibitory synapses on a cortical pyramidal cell is sufficient to reduce and even completely suppress the firing of neurons receiving strong excitatory input. Finally, we present a reduced pyramidal cell model (9 compartments) that runs significantly faster yet faithfully reproduces the behavior of the full 400 compartment model. The reduced model will be used for future physiological network simulations.

15.1 DENDRITIC SATURATION

The basal and oblique dendrites of neocortical pyramidal cells are electrotonically compact [1]. A single EPSP produces virtually the same effect at the soma where ever it originates from on the basal/oblique tree. It may seem that such a neuron would resemble the point neurons used in most connectionist-type networks, but this ignores the fact that dendritic terminal segments are quite isolated from each other and have high input impedances relative to the soma [2]. These properties introduce the potential for nonlinear interactions of inputs arriving on the same dendritic branch;

the large depolarization produced in a dendrite by a single EPSP reduces the driving force on simultaneous and subsequent EPSPs on the same branch.

Ferster and Jagadeesh [3] have recently demonstrated that the size of an EPSP evoked in visual cortical cells by electrical stimulation of the LGN is reduced during depolarizations caused by visual stimulation compared to its size at the resting potential. The reduction in EPSP size was proportional to the somatic depolarization caused by the visual stimulation. Thus an EPSP that peaked at about 6mV at rest was reduced to less than 1mV when the cell was depolarized from a resting potential of -60mV to -40mV by visual stimulation (fig 9a of [3]). Ferster and Jagadeesh interpret their results as indicating that the synaptic sites (dendrites) are significantly more depolarized than the soma during synaptic activation of the cell, thus producing saturation by the process described above.

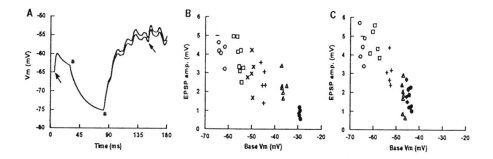

Figure 1. Dendritic saturation during simulated physiological synaptic activation of a layer 2 pyramid. (A) Somatic membrane potential (V_m) during simulation. A constant current of -0.1 nA is injected at $t = 30$ ms to prevent firing (first asterisk), then 70 excitatory synapses are activated at a mean frequency of 50 Hz to simulate visual stimulation (second asterisk). 35 additional synapses are given a simultaneous stimulus to produce a control (first arrow) and test (second arrow) EPSP. The amplitude of the test EPSP is significantly reduced with respect to the control. Upper trace is result of simulation with all excitatory synapses directly on dendritic shafts, lower trace is result of simulation with all synapses on the heads of dendritic spines. (B) Peak amplitude of the test EPSP plotted against V_m just before the EPSP, for a variety of firing frequencies of the 70 excitatory synapses; $* = 0$ Hz, E = 25 Hz, G = 50 Hz, I = 75 Hz, D = 100 Hz, C = 200 Hz, F = 300 Hz, J = 400 Hz. The amplitude of the EPSP decreased linearly with V_m. (C) Concurrent inhibition is included in the simulation (33 inhibitory synapses at twice frequency of excitatory synapses).

We have simulated Ferster's experiment to test this hypothesis using a digitised model layer 2 pyramid, simulated as a 400 compartment model using CABLE [4]. 70 excitatory synapses (each 0.5nS peak conductance) were placed randomly on the basal/oblique dendrites, and were activated at a mean frequency (eg. of 200Hz) at 80ms. A constant current of -0.1nA was injected starting at 30ms to prevent firing during the synaptic activation. 35 additional excitatory synapses were placed on the same dendritic segments as the initial 70, and were given a single simultaneous stimulation at 5ms and again at 150ms. The firing of these 35 synapses represented the effects of electrical stimulation of the LGN - the first stimulation gave a control EPSP, the second, occuring during 'visual stimulation', gave a test EPSP. Different trials could be produced by using a different seed for the random number generator that produced the Poisson-distributed trains of EPSPs. We found that the test EPSP was significantly reduced with respect to the control EPSP. Figure 1 shows the peak height of the test EPSP plotted against the somatic membrane potential just before the EPSP occurs, for a variety of input frequencies.

In order to produce a steep enough slope on this graph, 33 inhibitory synapses (12 somatic, others on preterminal dendrites) were active concurrently with the 70 excitatory ones. This is consistent with physiological results showing that inhibitory input to excitatory cortical cells is weak during null responses and is in fact strongly correlated with the degree of synaptic activation of the excitatory cells [5]. This is also consistent with microanatomical data indicating that excitatory cells make direct contacts with inhibitory cells, which then make direct contacts back onto the same excitatory population [6].

15.2 EFFECTIVENESS OF INHIBITION

When we added active conductances to the above model to produce adapting trains of action potentials, we found that the inhibition produced by 33 synapses significantly reduced the firing rate when the inhibitory synapses were discharging at a high rate (800Hz). Cortical pyramidal cells receive hundreds of inhibitory synaptic contacts on their somata and proximal dendrites [7]. Therefore, we increased the number of active inhibitory synapses in our simulation to determine if the concerted activity of a larger fraction of the cell's inhibitory input was sufficient to suppress firing. We found that the activity of about 200 somatic inhibitory synapses is sufficient to prevent a cell that is receiving strong excitation from firing. Consequently, we conclude that strong cortical inhibition is able to prevent the firing of even strongly driven pyramidal cells, contrary to previous conclusions [8].

The input resistance (R_{in}) of cortical neurons shows no significant reduction during the response to nonpreferred stimuli or even during sustained hyperpolarizations that are part of an optimal response to a visual stimulus [9],[3]. Tests of our model show that R_{in} decreases by at least 50% during the simulations using only 33 inhibitory synapses. This indicates that the level of inhibition used in our simulations was as least as great as the level of inhibition occuring during null or hyperpolarizing visual responses. This level of inhibition was not strong enough to suppress significant synaptic excitation. Consequently, the reason that the cell is not firing at these times must be because of a lack of synaptic excitation.

15.3 REDUCED COMPARTMENTAL MODELS

Further research into the synaptic activation of cortical cells will require the simulation of the network which is providing that activation. At present the speed of available computers is such that 400 compartment models are too large to be used in network simulations. Most model neurons composed of just a few compartments have been assigned a somewhat arbitrary geometry, with no systematic testing of the reduced model against the real cell or a more complete model. Such models may have the same R_{in} and time constant as the real cell, but typically will not accurately simulate the integration of inputs from the dendritic compartments into the soma. We have produced a 9 compartment model that faithfully reproduces the performance of the full 400 compartment model.

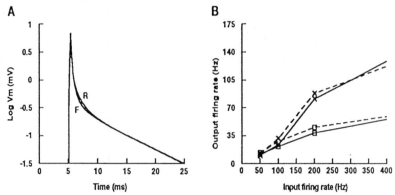

Figure 2. A) Semi-log plot of voltage response of reduced (R) and full (F) model layer 2 pyramidal cells to a 0.44 ms 0.3 nA somatic current injection at t = 5 ms. B) Comparison of firing rates of reduced (dashed traces) and full (solid traces) model layer 5 cells as a function of the firing rate of their 140 excitatory inputs. The somata of both models contain active conductances with exactly the same kinetics and densities. Each model also receives 45 inhibitory synapses, active at twice the rate of the excitatory ones. I = Initial, peak firing rate. G = Steady, adapted firing rate. The close fit of the two models demonstrates that the reduced model integrates excitatory and inhibitory synaptic input in the same manner as the full model.

The 'equivalent dendritic profile' [1] conserves surface area but at the expense of axial resistance - the equivalent dendrite is too wide and does not produce significant attenuation of synaptic input. Our approach is to conserve axial resistance by setting the diameter of the equivalent cylinder to a value such that the reciprocal of the axial resistance of the equivalent cylinder is equal to the sum of the reciprocals of the axial resistances of all the dendrites represented by that equivalent cylinder. The length of the equivalent cylinder is just the average length of all the dendrites represented by the equivalent cylinder. The membrane resistance and capacitance are then scaled for the loss in membrane area by matching the R_{in} and time constant with that of the full model. The resulting reduced model not only has the same R_{in} and time constant as the full cell, but also gives the same voltage response to a brief somatic current pulse [1], which depends on the rate of flow of current from soma to dendrites (Figure 2). When active conductances are taken from the full model and put directly into the reduced model, both models show almost exactly the same firing response to identical dendritic synaptic input. This means that the reduced model has the same E-S coupling characteristics as the full model.

REFERENCES

1.   Stratford, K., A. Mason, A. Larkman, G. Major and J. Jack. "The modelling of pyramidal neurons in the cat visual cortex." The Computing Neuron. Durbin, Miall and Mitchison ed. 1989 Addison-Wesley. Wokingham, England.

2.   Rinzel, J. and W. Rall. Transient response in a dendritic neuron model for current injected at one branch. Biophys. J. **14**: 759-789, 1974.

3.   Ferster, D. and B. Jagadeesh. EPSP-IPSP interactions in cat visual cortex studied with *in vivo* whole-cell patch recording. J. Neurosci. **12**: 1262-1274, 1992.

4.   Hines, M. L. A program for simulation of nerve equations with branching geometries. Int. J. Biomed. Comp. **24**: 55-68, 1989.

5.   Ferster, D. Orientation selectivity of synaptic potentials in neurons of cat primary visual cortex. J. Neurosci. **6**: 1284-1301, 1986.

6.   Douglas, R. J., K. A. C. Martin and D. Whitteridge. A canonical microcircuit for neocortex. Neural Comp. **1**: 480-488, 1989.

7.   Farinas, I. and J. DeFelipe. Patterns of synaptic input on corticocortical and corticothalamic cells in the cat visual cortex. I. The cell body. J. Comp. Neurol. **304**: 53-69, 1991.

8.   Douglas, R. J. and K. A. C. Martin. Control of neuronal output by inhibition at the axon initial segment. Neural Comp. **2**: 283-292, 1990.

9.   Douglas, R. J., K. A. C. Martin and D. Whitteridge. Selective responses of visual cortical cells do not depend on shunting inhibition. Nature. **332**: 642-644, 1988.

MODELING STIMULUS SPECIFIC HABITUATION: THE ROLE OF THE PRIMORDIAL HIPPOCAMPUS*

DeLiang Wang

Department of Computer and Information Science and Center for Cognitive Science
The Ohio State University, Columbus, OH 43210-1277, USA

ABSTRACT

We present a neural model for the organization and neural dynamics of the medial pallium, the toad's homolog of mammalian hippocampus. A neural mechanism, called cumulative shrinking, is proposed for mapping temporal responses into a form of population coding referenced by spatial positions. Synaptic plasticity is modeled as an interaction of two dynamic processes which simulates both short-term and long-term memory. Successful modeling allows us to provide an account of the neural mechanisms of stimulus specific habituation.

16.1 INTRODUCTION

After repeated presentation of the same prey dummy in their visual field, toads and frogs reduce the strength of orienting responses toward the moving stimulus. This visual habituation exhibits stimulus specificity. Another stimulus given at the same locus may restore the response habituated by a previous stimulus. Only certain stimuli can dishabituate a previously habituated response, forming a dishabituation hierarchy (Fig.16.1, [1]).

Anatomically, the medial pallium (MP) has been thought to be the homolog of the mammalian hippocampus ("primordium hippocampi" by Herrick [4]). Finkenstädt [2] identified three types of visual sensitive neurons in MP which exhibit spontaneous firing activities. MP1 neurons strongly increase, MP3 neurons decrease, and MP2 neurons do not alter their discharge rates in response to 30 minutes of repetitive visual stimulation. After bilateral lesions of MP, toads showed no visual habituation [3]. In addition, both effects of conditioning and associative learning ability in naive toads are abolished due to MP-lesions. All these data strongly suggest that MP is the neural structure where learning occurs.

* The research described in this paper was supported in part by grant no. 1RO1 NS 24926 from NIH (M.A. Arbib, Principal Investigator).

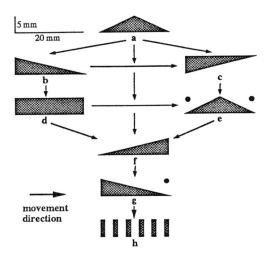

Figure 16.1. Dishabituation hierarchy for worm stimuli used in stimulus-specific habituation. One stimulus can dishabituate all the stimuli below it. On the same level the left stimulus can slightly dishabituate the right one.

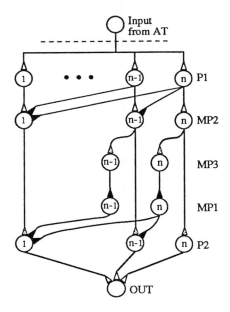

Figure 16.2 Diagram of an MP column model. Each cell type is a layer of cells numbered from left to right as 1, 2, ..., n. Empty triangles indicate excitatory, black ones inhibitory, and filled ones habituation synapses.

16.2 A MP COLUMN MODEL

Anatomical studies of the medial pallium indicate that this structure is organized in an orientation vertical to the telencephalic ventricle, with cells mainly projecting in this direction [5]. This leads us to suggest that MP processes information by means of functional units, in a form of vertical columns. We will model one such functional unit to demonstrate stimulus-specific habituation. For visual habituation, MP receives inputs from the anterior thalamus (AT), a cell type of which has been modeled by Wang and Arbib [7]. In the model, an MP column receives input from an AT neuron, representing a specific visual location. The MP column model includes arrays of five types of neurons, MP1, MP2, and MP3 [2] and hypothetical cell types P1 and P2. The organization of a MP column is shown in Fig. 16.2. The connections from

MP2 to P2 and MP3 are the only modifiable connections.

P1 layer: The layers P1 and MP2 together transform different temporal activities of AT cells into spatial activity distribution centered at different positions. All P1 cells receive the same input from AT, but their thresholds are chosen so that a higher intensity stimulus triggers more P1 neurons than does a lower intensity stimulus.

MP2 layer: In order to map temporal activity into different spatial locations, we propose the membrane potential of the *i*th MP2 cell

$$\frac{dm_{mp2}(i,t)}{dt} = -A_{mp2}\, m_{mp2}(i,t) + (B_{mp2} - m_{mp2}(i,t))I_{ii}(t) + m_{mp2}(i,t)\sum_{j>i} I_{ij}(t) \qquad (1)$$

A_{mp2} and B_{mp2} are parameters, and $I_{ij}(t)$ represents the input from neuron $P1_j$. With the unilateral shunting inhibition of (1) and the P1 layer, the network exhibits a phenomenon called *cumulative shrinking*, which normalizes and shrinks the activity in the MP2 layer along one direction. Because of cumulative shrinking, a stimulus pattern is represented by a distribution of normalized cell activities maximized the so-called *representative* cell of the stimulus. In this cell group, the more a cell is to the left (see Fig.16.2) the less it contributes to the representation due to shrunk activity.

The projections from MP2 to MP3 and P2 are one-to-one correspondence, and the weight of the projection from cell $MP2_i$ is denoted by $y_i(t)$ that is habituatable with initial value $y_i(0) = y_0$. The MP1 layer receives its inhibitory input from MP3, and the P2 layer receives excitatory input from layer MP2 through habituatable projections, and unilateral inhibition from layer MP1. Finally, as the sole efferent neuron of the MP column, OUT simply integrates activities from the P2 layer.

Wang and Hsu [8] used an S-shaped curve to model the build-up of habituation. Adapting their scheme, we provide the dynamics for the synaptic weights $y_i(t)$ as

$$\tau \frac{dy_i(t)}{dt} = \alpha\, z_i(t)\, (y_0 - y_i(t)) - \beta\, y_i(t)\, N_{mp2}(i,t) \qquad (2)$$

$$\frac{dz_i(t)}{dt} = \gamma\, z_i(t)\, (z_i(t) - 1)\, N_{mp2}(i,t) \qquad (3)$$

where τ is the time constant for controlling the rate of habituation. The first term in (4) regulates recovery towards the initial value y_0. The product $\alpha\, z_i(t)$ has an activity dependent control on the rate of forgetting. The second term regulates habituation, and parameter β controls its speed. The factor $N_{mp2}(i,t)$ is the input from layer MP2. Equation (3), which exhibits an S-shaped curve, achieves both short-term and long-term conditions. Before the inflection point $z_i(t)$ holds a relatively large value, and forgetting as manifested by the first term of (2) is relatively fast; after the inflection point, $z_i(t)$ holds a relatively small value, and forgetting is relatively slow. These two phases are used to model two phases of memory: short-term memory (STM) and long-term memory (LTM). The time course of transition from STM to LTM can be fully controlled by the model parameters γ and $z_i(0)$.

16.3 COMPUTER SIMULATION AND MODEL PREDICTIONS

The stimuli were originally presented to the model of the toad retina, and all the calculations from the retina to AT have been carried out by the Wang and Arbib model [7]. For each array of an MP cell type, 50 cells have been simulated ($n = 50$). Time was measured like this: a basic discretization step 0.05 corresponded to 1 sec. We present a typical set of simulations, where two pairs of stimuli (Fig.16.1) were studied. In Fig.16.3A, stimulus f was presented first. The response of the model was habituated after continuous presentation of the stimulus. After 60 min simulation time, stimulus d was presented, and it triggered a new response. The reverse order of presentation was studied in Fig.16.3B, and apparently no dishabituation was demonstrated. We can see from Fig.16.3C and D that stimulus b was able to dishabituate the habituated response to stimulus d, but d was unable to dishabituate the habituated response to b. Figure 16.3E shows the corresponding experimental results. From the comparison between Fig.16.3A–D and Fig.16.3E, it can be concluded that our model clearly reproduces the experimental data. Similarly good results were found in other simulations corresponding to the Ewert and Kehl experiments and some further behavioral studies (see [6] for more simulations).

The theory we have developed yields a number of neurobiological predictions. In particular, the model allows us to predict that dishabituation is nothing but the release of a normal prey-catching behavior. The reason that presentation of a second stimulus fails to release a new response is because of inhibition caused by habituation to the first stimulus. Also, we predict that, due to the buildup of long-term memory traces, toads will saturate (see Fig.16.3 for the saturation effect) at lower and lower levels of activity as a series of training proceeds, and eventually reach the zero level.

REFERENCES

[1] J.-P. Ewert, and W. Kehl. *J Comp Physiol A* 126: 105-114, 1978.
[2] T. Finkenstädt. In: J.-P. Ewert and M.A. Arbib (eds.) *Visuomotor coordination: amphibians, comparisons, models, and robots*. Plenum, New York, pp 767-797, 1989
[3] T. Finkenstädt, and J.-P. Ewert. *J Comp Physiol A* 163: 1-11, 1988.
[4] C.J. Herrick. *J Comp Neurol* 58: 737-759, 1933.
[5] E. Kicliter, and S.O.E. Ebbesson. In R. Llinás and W. Precht (eds) *Frog Neurobiology*. Springer, Berlin Heidelberg New York, pp 946-972, 1976.
[6] D.L. Wang. Neural networks for temporal order learning and stimulus specific habituation. TR 91-06, Center for Neural Engineering, University of Southern California, 1991.
[7] D.L. Wang, and M.A. Arbib. *Biol Cybern* 64: 251-261, 1991.
[8] D.L. Wang, and C.C. Hsu. *Simulation* 55: 69-83, 1990.

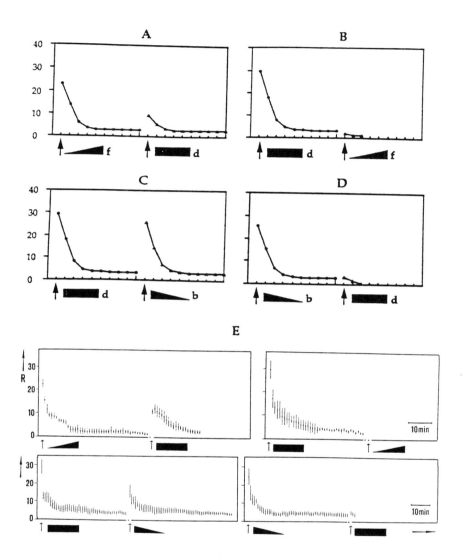

Figure 16.3 Computer simulation of habituation and dishabituation of prey orienting response in toads. The response was taken from the OUT cell of the MP column, and it was measured as the relative value of the initial response of each frame that is scaled to the same value measured experimentally. In a frame, the response to the first stimulus is indicated by little circles and that to the second one is indicated by little triangles. A. Stimulus f was first presented and habituated, and then d was tested. B. The reverse order of presentation. C. Stimulus d was first presented and habituated, and then b was tested. D. The reverse order of presentation. E. The experimental results obtained by Ewert and Kehl [1] are shown using the same combination of the stimuli.

EFFECTS OF SINGLE CHANNEL KINETICS UPON TRANSMEMBRANE VOLTAGE DYNAMICS

Adam F. Strassberg

Computation and Neural Systems Program
California Institute of Technology
Pasadena, California 91125

Louis J. DeFelice

Department of Anatomy and Cell Biology
Emory University School of Medicine
Atlanta, Georgia 30322

ABSTRACT
A standard membrane model, based upon the continuous deterministic Hodgkin-Huxley equations, is compared to an alternative membrane model, based upon discrete stochastic ion channel populations represented through Markov processes. Simulations explore the relationship between these two levels of activity: the behavior of the macroscopic membrane currents versus the behavior of their microcopic ion channels.

Introduction
Action potentials within the neuron arise from the time-variant and voltage-dependent changes in the conductance of the neural membrane to specific ions. Hodgkin and Huxley produced their famous model of active membrane, based upon an assumption that, whatever ion permeation processes exist within the membrane, that these processes can be approximated as both continuous and deterministic [1]. However, the permeation processes existing within active membrane are known to be neither continuous nor deterministic. Active membrane is studded with discrete ion channels undergoing random fluctuations between open and closed stable states [2]. This paper investigates the relationship between these two regimes of membrane activity through a comparison of the standard membrane model, based upon the continuous Hodgkin-Huxley equations, to an alternative membrane model, based upon discrete ion channel populations represented through Markov processes. When both models are used to simulate the active membrane of the squid giant axon, the explicit con-

vergence of the alternative model to the standard model can be observed. However, under certain conditions, the behavior predicted by this alternative model will diverge from the behavior predicted by the standard model. Under these conditions, the simulation suggests that random microscopic behavior, such as single channel fluctuations, becomes capable of producing random macroscopic behavior, such as entire action potentials.

Methods

The neural membrane of a space-clamped squid giant axon is modeled with an equivalent electric circuit. The space-clamp technique removes the spatial dependence of the membrane voltage and so the axon becomes effectively equivalent to an isopotential patch of membrane. A simple lumped circuit model thus can interpret the electrical characteristics of the membrane. The macroscopic membrane conductances are represented by the conductive elements g_{Na}, g_K, and g_L and the transmembrane voltage V_m behaves according to the equation

$$\frac{dV_m}{dt} = -\frac{1}{C_m}[(V_m - E_L)g_L + (V_m - E_K)g_K + (V_m - E_{Na})g_{Na} - I_{\text{inject}}]$$

Simulations of the membrane behavior are performed using GENESIS, an object-oriented general purpose neural simulator for the UNIX/X-windows environment. All biophysical parameters for squid axonal membrane are chosen to be identical to those values used by Hodgkin and Huxley in their seminal paper [1]. Behavior predicted by the standard Hodgkin-Huxley equations for the membrane conductances g_{Na} and g_K [1] is compared to behavior predicted by alternative descriptions for these conductances based upon their underlying ion channel populations. These voltage-gated ion channels are modeled well by Markov processes [3, 4, 5, 6, 2]. Observed microscopic potassium channel behavior is reproduced by the Markov kinetic scheme

$$[n_0] \quad \underset{\beta_n}{\overset{4\alpha_n}{\rightleftharpoons}} \quad [n_1] \quad \underset{2\beta_n}{\overset{3\alpha_n}{\rightleftharpoons}} \quad [n_2] \quad \underset{3\beta_n}{\overset{2\alpha_n}{\rightleftharpoons}} \quad [n_3] \quad \underset{4\beta_n}{\overset{\alpha_n}{\rightleftharpoons}} \quad [n_4]$$

where $[n_4]$ = Open State, and α_n and β_n are the voltage-dependent rate constants from the Hodgkin-Huxley formalism [7, 8, 9, 2]. Observed microscopic sodium channel behavior is reproduced by the Markov kinetic scheme

$$[m_0 h_1] \quad \underset{\beta_m}{\overset{3\alpha_m}{\rightleftharpoons}} \quad [m_1 h_1] \quad \underset{2\beta_m}{\overset{2\alpha_m}{\rightleftharpoons}} \quad [m_2 h_1] \quad \underset{3\beta_m}{\overset{\alpha_m}{\rightleftharpoons}} \quad [m_3 h_1]$$
$$\alpha_h \uparrow \downarrow \beta_h \qquad\qquad \alpha_h \uparrow \downarrow \beta_h \qquad\qquad \alpha_h \uparrow \downarrow \beta_h \qquad\qquad \alpha_h \uparrow \downarrow \beta_h$$
$$[m_0 h_0] \quad \underset{\beta_m}{\overset{3\alpha_m}{\rightleftharpoons}} \quad [m_1 h_0] \quad \underset{2\beta_m}{\overset{2\alpha_m}{\rightleftharpoons}} \quad [m_2 h_0] \quad \underset{3\beta_m}{\overset{\alpha_m}{\rightleftharpoons}} \quad [m_3 h_0]$$

where $[m_3 h_1]$ = Open State, and α_m, β_m, α_h, and β_h are the voltage-dependent rate constants from the Hodgkin-Huxley formalism [10, 11, 2, 9].

Results

Figure 1 shows the voltage-clamp step response of g_K and figure 2 shows the voltage-clamp step response of g_{Na}. Either the continuous Hodgkin-Huxley equations or the discrete channel population Markov models are used alternatively to represent the membrane conductances. Note that as the size of the channel populations increase, the responses from the discrete channels converge to the responses from their continuous currents.

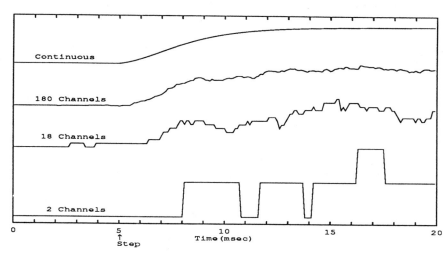

Figure 1: Potassium Conductance Voltage-Clamped Step Response

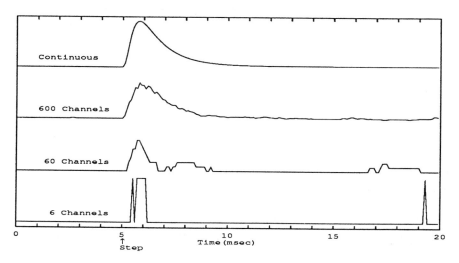

Figure 2: Sodium Conductance Voltage-Clamped Step Response

Figure 3 shows the response of the membrane model to a constant current injection and figure 4 shows the resting response with *no* current injection. As the membrane surface area is increased, the response from the simulation of a constant density of ion channels converges to the response from the standard Hodgkin-Huxley model. Both models predict that a train of action potentials occurs when constant current is injected into the axon and that no activity occurs when no current is injected into the axon. Note that, as the membrane surface area is decreased, the behavior predicted by the channel model diverges dramatically from the behavior predicted by the Hodgkin-Huxley model. These simulations suggest that, as the area of an isopotential membrane patch is decreased, the fluctuations of single channels become capable of eliciting entire action potentials.

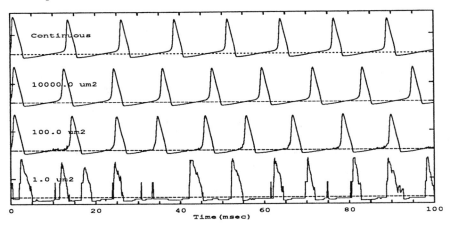

Figure 3: Membrane Response to Constant Current Injection

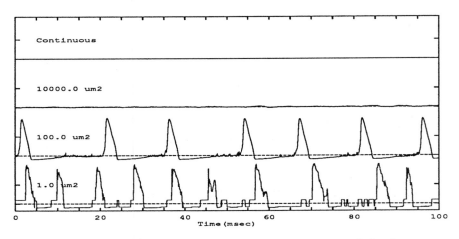

Figure 4: Membrane Response without Current Injection

Discussion

The divergent behavior can be explained through an analysis of the voltage perturbations due to single channel fluctuations. Both the open state conductance and the voltage-dependence of any individual ion channel are *independent* of the surface area of the surrounding membrane. However, the voltage perturbations to this surrounding membrane produced by brief openings of these channels are *dependent* upon this membrane surface area. As the surface area is decreased, both the total effective membrane capacitance and the total effective membrane resistance will decrease. This will cause a decrease in the rise-time and an increase the asymptotic value of any membrane voltage perturbations, so brief open fluctuations from any embedded channels will effect larger voltage changes within smaller time periods. For sufficiently small membranes, these voltage perturbations will be large enough to alter the the transition state probabilities of other voltage-dependent ion channels within the membrane. If a sufficient number of these channels enter their open conductive states, then the brief fluctuation of a single ion channel will have initiated an entire action potential cascade.

Acknowledgments

This material is based upon work supported under a National Science Foundation Graduate Fellowship and an NIH HL-27385. We would like to express our deep appreciation to Dr. Christof Koch for his comments and suggestions throughout the preparation of this manuscript. We also would like to thank Hsiaolan Hsu and Dr. Henry Lester for helpful insights.

References

[1] Hodgkin, A.L., and A.F. Huxley. 1952. A quantitative description of membrane current and its application to conduction and excitation in nerve. *J. Physiol. (Lond.)* 117, 500-544.

[2] Hille, B. 1992. *Ionic Channels of Excitable Membrane*, 2nd Ed. Sinauer Associates, Inc., Sunderland, Mass., 607 pp.

[3] Strassberg, A.F., and L.J. DeFelice. 1992. Limitations of the Hodgkin-Huxley Formalism. *Computation and Neural Systems Program - Memo 14*, California Institute of Technology.

[4] Kienker, P. 1989. Equivalence of aggregated Markov models of ion-channel gating. *Proc. R. Soc. Lond. B* 236, 269-309.

[5] Clay, J., and L. DeFelice. 1983. Relationship between Membrane Excitability and Single Channel Open-Close Kinetics. *Biophys. J.* 42, 151-157.

[6] Conti, F., L. J. DeFelice and E. Wanke. 1975. Potassium and sodium ion current noise in the membrane of the squid giant axon. *J. Physiol. (Lond.)* 248, 45-82.

[7] Armstrong, C.M. 1969. Inactivation of the potassium conductance and related phenomena caused by quaternary ammonium ion injected in squid axons. *J. Gen. Physiol.* 54, 553-575.

[8] Llano, I., C.K. Webb and F. Bezanilla. 1988. Potassium conductance of the squid giant axon. *J. Gen. Physiol.* 92, 179-196.

[9] Fitzhugh, R. 1965. A kinetic model of the conductance changes in nerve membrane. *J. Cell. Comp. Physiol.* 66, Suppl., 111-117.

[10] Bezanilla, F. 1987. Single sodium channels from the squid giant axon. *Biophys. J.* 52, 1087-1090.

[11] Vandenberg, C.A. and F. Bezanilla. 1988. Single-channel, macroscopic and gating currents from Na channels in squid giant axon. *Biophy. J.* 53, 226a.

18

AN OBJECT-ORIENTED PARADIGM FOR THE DESIGN OF REALISTIC NEURAL SIMULATORS

David C. Tam[1]
R. Kent Hutson[2]

[1]*Center for Network Neuroscience, University of North Texas, Denton, TX 76203*
[2]*Keck Center for Computational Biology, Rice University, Houston, TX 77251*
E-mail: dtam@sol.acs.unt.edu, khustson@rice.edu

ABSTRACT

An object-oriented programming approach is used to implement a generic neural simulator to simulate the electrophysiological properties of neurons and interconnected neuronal networks. The neural simulator is designed with a generalizable principle to encourage users to make additions and modifications to the existing neurophysiological models. The object-oriented programming paradigm is used extensively to encapsulate the similarities and differences between different neurons and their components by various "objects". A neuron is constructed by a compartmental model where electrical properties of the membrane of a neuron are compartmentalized and then linked together to form a whole neuron. Similarly, an equivalent of a biological neural network can constructed by connecting model neurons together to form a network. The object-oriented approach is used not only in constructing the neuronal structural hierarchy but also in the methods for solving mathematical equations governing the electrical properties of the neurons. Using these principles, a truly generic and generalizable model for simulating neuronal properties is accomplished.

18.1. INTRODUCTION

There are a number of neural simulators currently available for investigating properties of the neural system (*e.g.*, MANUEL [4], GENESIS [5], CABLE [1,6], SABER [3] and MacGregor's [2]). Most of these biological neuronal simulators involve some complex representations of the inherently interrelated internal structures of the neural systems. As a result of this interdependence of the complex structures, modification or addition of new code to the existing simulators becomes extremely difficult since most of these neural simulators are implemented in traditional procedural programming languages. Furthermore, neuroscientists often want to develop his/her own specialized code for simulating some extra features that the original designers may not have foreseen. To make such changes, he/she may need to completely comprehend the complex details of the program before such venture can be attempted. If the program is not fully understood, undesirable results may emerge due to side-effects. But in reality, to comprehend programs that contain thousands of lines of code is not only time consuming but also impractical. With objective programming approach, such complexity can be reduced so that most of the details is encapsulated by "objects", leaving the core structure of the simulator comprehensible by any non-author of the original program design team. This makes modifications of the complex program possible without explicit knowledge of the entire program in detail.

There are many different levels of complexity of interactions in neuronal simulations: from ionic channel kinetics in membrane to synaptic interactions and signal processing in neurons. Although there are multiplicities of similar neuronal elements, such as different ionic channels, each neuron can be described by a similar set of biophysical and biochemical equations in principle. Therefore, a canonical neuron with generalizable properties can form the basis for reconstructing the functional equivalence of a biological neuron in a computer, whereby individual parameters of the model can be varied and the outcomes observed as if the same experiment were done on an animal. Since such model requires generalizable descriptions of individual elements of neurons from the levels of ionic channels in neurons to the levels of synaptic connections among neurons and network properties, well-orchestrated methods for describing the different levels of details are needed.

18.2. METHODS
18.2.1. CLASS STRUCTURE FOR NEURAL OBJECTS

We use an object-oriented approach to implement a biologically realistic neuronal network by reconstructing the functional properties of neurons based on known biophysics and biochemistry principles. This neural simulator (*MacNeuron*) is implemented on the Macintosh computer.

Central to the object-oriented paradigm is the concept of "class" memberships in each object where properties of the "superclass" are inherited by default. Generalization and specialization of objects can easily be incorporated using this class structure of inheritance. Based on these design principles, we represent the neuronal elements using different classes, such as classes of electrical characteristics of ionic channels in modeling neural circuits. Using a generic class at the highest level of organization, we are able to construct subclasses of neuronal elements for modeling specific differences in neurons.

For instance, with a generic ionic channel conductance-class, we are able to construct specific subclasses for voltage-dependent conductances such as Hodgkin-Huxley conductances of Na^+ and K^+, different types of Ca^{++} conductances and ligand-gated conductances at the synapses. Extending this membrane compartmental design, chemical compartments can be constructed to model the chemical reactions of synaptic neurotransmitter release, neurotransmitter-receptor binding and secondary messenger systems. In such case, each individual specialized membrane compartment (or chemical compartment) is an "instantiation" of an object in its class. This provides a generic description of a canonical neuron capturing the necessary specialization of the specific properties unique to that type of neuronal element.

18.2.2. NEURONAL CONNECTIVITIES

Extending the same design principles, individual neurons are constructed by connecting appropriate "patches" of membrane to form its morphology. An interconnecting network of neurons can be constructed by connecting individual neurons to form synaptic interconnections. At a higher level of description, interconnectivities between neurons can be described by the links between them. At a lower level of description, connectivities are actually associated with specific compartments of these individual neurons. These levels of description can be interchanged so long as they are conforming to one another. Such approach provides a generalizable functional equivalent description of a biological neuronal network.

18.2.3. SIMULATION PROCEDURES

Given that a physiological model is described by the class structures as discussed above, the procedures for simulating the model involve not only in solving the equations governing the principles of operations of the neuronal elements, but also involves in solving the equations semi-simultaneously with the couplings among all the neural elements under consideration. It is the interactions among the basic elements of the model that make the simulation environment complicated in a large-scale simulator. These interrelationships, if not managed well, will often create side-effects affecting the simulation results. They could potentially attribute to the unmanageability of simulating such complex models. Using a hierarchical structure for describing the elements of the model, most basic elements of the model can be accessed hierarchically and/or recursively. Thus, access to individual components of the system can be readily done without creating unnecessary complexity. This provides a simple, straight-forward approach to manage a large scale model with seemingly large quantities of dissimilar elements.

Generating the simulation results becomes straight-forward, once the natural hierarchy for accessing the simulation elements is established. The method for "kick-starting" the simulation is by sending messages to every element in the model. This message passing procedure is central to object-oriented programming where appropriate "methods" are invoked at the appropriate moment. ("Methods" in objective programming are similar to traditional "subroutines" or "procedures" or "functions", but are considered very different conceptually.) The key to this approach is to consider these methods as implementations of *what* need to be done, rather than *how* they are implemented. How the methods are implemented is deferred to the subsequent implementations, so long as the methods specify what needs to be done. Therefore, the actual implementation of the methods can be changed (or even overridden) by some other implementations subsequently. This enables modification of the program easily by changing the implementations alone without changing the procedures for invoking these methods. Thus the details of the methods are encapsulated in the actual implementation.

The execution of a simulation is started by sending appropriate messages to invoke the appropriate methods responsible for computing the results in each neural element. This message passing is accomplished in a hierarchical order via the "chain of command". The sequence of actions can be considered as "bootstrapping", resulting in completely accomplishing the desired goal by a simple "kick-start" message which invokes subsequent messages to be propagated appropriately. With this approach, a complex neural model with different levels of description can be simulated without the inherent complexity. Each level of description is responsible for receiving the appropriate messages from the level above, and then propagate another set of appropriate messages to the elements of the level below. By cascading the messages from top-down, all elements in the simulations can potentially be accessed regardless of the differences in the levels of description.

18.2.4. POLYMORPHISM OF NEURAL OBJECTS

With polymorphism of objects and methods, they can change their "identities" (or implementations) at run-time. In other words, the original implementation of the methods may be overridden by some other methods added to the simulation later. When the simulator is designed with generic objects, the original generic methods can be substituted by some other implementations of the same method later when there are needs for doing so, either by the original author or by someone else. The advantage is that these objects can be interchanged without *a priori* knowledge of the implementation of the original code, since they replace the original objects (code) of the same class while maintaining the original access methods. Contrary to the intuition that the original author may lose his/her control of the program using this object-oriented approach, it is this polymorphism that provides greater, more flexible functionalities to the simulator.

With object-oriented programming, the original generic code need not be modified if the functionalities of the simulator is extended. The only modifications are the overriding of the original objects. In fact, these new objects need not be completely replacing the original ones. Rather they can reuse the existing code by inheritance. Examples for using polymorphic objects in our simulator are the methods for solving the numerical equations, methods for plotting graphics and methods for displaying the user-interface with windows and icons. As a result, our neural simulator can solve the numerical simulation using different numerical algorithms without extensive modifications of the original core code, generate different ways to plot graphics without much difficulty and change the user-interface environments without necessarily interfering with the simulation core driver of the program. This provides a more robust program which is much more tolerant of minor changes, avoiding side-effect errors resulting from minor changes to code of typical large-scale, complex simulations.

18.2.5. INHERITANCE IN NEURAL OBJECTS

Another major benefit of object-oriented design in neural simulation is that new type of neural elements can be incorporated into the existing neural model without much difficulty. For instance, new types of ionic conductance channels are periodically discovered experimentally. When these new neural elements are discovered, the possibility of simulating the dynamics of the systems using these newly discovered channels are very appealing to neuroscientists. But if the neural simulator is implemented with traditional procedural programming approaches, major changes to the simulation program may be needed. The advantage offered by object-oriented programming in this respect is the inheritance of the existing objects. For instance, if a new type of potassium channel is found to be similar to the existing ones, they can be classified as belonging to the same class of channels but with its own unique characteristics. The properties of inheritance can be taken advantage of by inheriting the common properties of the generic superclass potassium channel. Thus, only the specialization of specifics in this new channel subclass needs to be implemented in the additional code. Thus, it reduces additional time and code needed to implement the new channel type. Since it reuses the existing code by inheritance automatically, it also reduces the chance of introducing side-effect errors since the original code of the superclass is not modified. Using this object-oriented design, the life-cycle of a neural simulator can be extended with ease, adding new functionalities to the existing program without detailed knowledge of the whole program.

SUMMARY

We have implemented a generalizable neural simulator using object-oriented design to allow quantitative experimentation on various physiological parameters of the neurobiological system. This design allows flexibility and ease of implementation of new features to the existing model. Future updates and modifications of the existing program can be done without requiring complete detailed knowledge of the entire complex large-scale simulation model. It also reduces the complexity of the interaction required for interfacing various levels of description in the complex neural model. Provided with such computational tool, many of the biophysical and biochemical principles underlying the mechanisms for neuronal functions can be explored and experimented on without constraint by the inflexibility of the simulator.

Acknowledgments

This research was supported by Office of Naval Research grant number ONR N00014-90-J-1353 and USPHS grant number RR05425-30. We also thank Garrett Kenyon, Laurie Feinswog and David Boney for the implementation of the simulator.

REFERENCES

[1] M. Hines. A program for simulation of nerve equation with branching geometries. *Int. J. Biomed. Comput.*, 24:55-68, 1989.

[2] R. J. MacGregor. *Neural and Brain Modeling.* Academic Press: San Diego, 1987.

[3] G. M. Shepherd and T. B. Woolf. Comparisons between active properties of distal dendritic branches and spines: implications for neuronal computations. *Journal of Cognitive Neuroscience*, 1: 273-286, 1989.

[4] D. C. Tam and D. H. Perkel. Quantitative modeling of synaptic plasticity. In: *The Psychology of Learning and Motivation: Computational Models of Learning in Simple Neural Systems,* (R. D. Hawkins and G. H. Bower, eds.) 23: 1 - 30. Academic Press: San Diego, 1989.

[5] M. A. Wilson, U. S. Bhalla, J. D. Uhley and J. M. Bower. GENESIS: A system for simulating neural networks. In: *Advances in Neural Information Processing Systems.* (D. S. Touretzky, ed.), Morgan Kaufmann Publishers, San Mateo, California. pp. 485-492, 1989.

[6] M. A. Wilson and J. M. Bower. The simulation of large-scale neural networks. In: *Methods in Neuronal Modeling: From Synapses to Network.* (C. Koch and I. Segev, eds.), The MIT Press: Cambridge, MA. pp. 291-333, 1989.

NEURAL NETWORK SIMULATION IN A CSP LANGUAGE FOR MULTICOMPUTERS

F. Bini, M. Mastroianni, S. Russo, and G. Ventre[1]

Dipartimento di Informatica e Sistemistica
Università degli Studi di Napoli "Federico II"
Via Claudio 21, Napoli, I-80125 Italy
[1] *International Computer Science Institute*
1947 Center Street, Suite 600
Berkeley, California, 94704

ABSTRACT

We show the simulation of neural networks on distributed-memory, message-passing multiprocessor systems, by means of a CSP language and related programming environment, that has been implemented at the University of Napoli. Among the features of the system, that make it suitable for simulation, and particularly for fast prototyping, of neural network models, are: simplicity, generality, portability, scalability and ease of use, at a good level of efficiency.

1   INTRODUCTION

In the recent years there have been several research efforts of simulating neural networks on parallel computers [1, 2, 3, 4]. Reason for the exploitation of parallel architectures is that simulation of large neural networks to solve real problems, such as pattern and speech recognition, usually requires huge amounts of computing time on conventional computers. Moreover, neural networks exhibit an intrinsic massive concurrency, and are therefore more adequate to be simulated on highly-parallel systems than on traditional sequential computers. However, several problems arise in implementing neural networks on distributed-memory message-passing multiprocessors systems, known as multicomputers [5]. In fact, in many concurrent programming languages for multicomputers, like Occam[6], it is difficult to write programs that are independent of the number of processing elements and of the topology of interconnection among them. This means that the programmer is not free to concentrate on his neural simulation program, and this results in programs that are neither portable nor scalable. This paper describes how neural networks can be simulated in the *DISC* system [7, 8, 9], an implementation of the Communicating Sequential Processes (CSP) model of computation [10] for multicomputer architectures.

2   NEURAL NETWORK SIMULATION

Basically, what is required to a general purpose parallel programming language for simulating a neural network is [11]:

- to allow the definition of the functionalities of each neuron;

- to allow the definition of inputs and outputs of each neuron;

- to allow the association of weights to the inputs;

- to allow the linking of the inputs and outputs of nodes to form a network;

- to allow communication among neurons;

- to support the interaction with the environment;

Many concurrent programming languages for multicomputers extend the sequential programming paradigm introducing parallel control structures based on processes, plus synchronization and communication mechanisms. A *process* is an independent program unit consisting of private data and sequential code. Processes can execute concurrently, each one operating on its local data, and interacting with each other by means of (explicit or implicit) *synchronization* and *communication* mechanisms. These are used respectively to enforce sequencing restrictions and to allow data exchange among processes. The basic concept of interacting processes seems to be appropriate to develop neural simulation programs. Each process simulates the behaviour of a neuron, with the sequential code describing the activity of the neuron. Common language abstractions like channels, ports, etc., are used to build the structure of the net, while communication primitives based on these abstractions express neuron interactions. It has been pointed out that the implementation of highly concurrent simulators would require at the language level the possibility of non-deterministic reception of messages, as long as ways to send messages in a on-blocking (asynchronous) manner, and to receive them in a blocking manner [4].

Actually, currently available concurrent programming languages differ chiefly for the way the programmer is allowed to define program units, to express interactions among them (*synchronous, asynchronous* mechanisms), or to define process creation (*static, dynamic*) and allocation. Many of them, however, do not offer to the programmer the possibility to keep separate the logical and the physical level in the design of distributed simulation programs. Thus, researchers interested in experimenting neural models are usually faced since from the early steps of the parallel software development cycle with a lot of software engineering problems, and cannot concentrate themselves on the properties of the neural system under simulation. Among these problems there are:

- process allocation: once the user has identified logical program units, he has to decide their allocation in the parallel architecture; unfortunately, in languages like Occam [6], changing the mapping policy usually causes the need for re-coding the simulation program;

- data routing: processor networks can have different sizes and topologies, and message exchange between non adjacent nodes have to be explicitly managed by the programmer; moreover, communication often occurs at the language level between physical processors, instead of logical processes; this makes more difficult the debugging task.

3   NEURAL SIMULATION IN DISC

DISC is a general-purpose parallel programming environment based on the CSP model. The current prototype runs on two different classes of multicomputer systems: a network of Inmos Transputer processors, and a local area network of UNIX workstations [9]. A neural network can be simulated in the DISC language [7] by activating an arbitrary number of concurrent *instances* of neuron processes by means of the *parallel* command (*par*), provided that the number of instances is known at compile time. Any type of neuron can be represented by a separate *process definition*, with its local constants, variables and functions. The language provides synchronous communication mechanisms, based on many-to-one monodirectional communication channels. Communication occurs among user processes, independently of their location in the processor network. The problem of mapping processes to processors must be solved just before execution, by providing allocation informations to a proper tool. A simple mechanism is also available, to allow a parent process to move variables to (from) the local memory of each neuron instance when it is activated (when it terminates) (*in/out* variables). Finally, non-deterministic reception of messages is accomplished through the *alternative* command (*alt*). Differently from Occam, DISC allows the use of output guards in guarded commands. Thus, with respect to the requirements discussed in section 2, DISC suffers for the lack of a non-blocking sending mechanism. This problem, however, can be overcome just as suggested in [4] for Occam, by introducing proper buffer processes. Implementing a neural network in DISC provides several advantages. First of all, any kind of internal neuron activity can be coded in the process corresponding to a neuron type, and since the language implementation guarantees that neuron instances actually run in parallel, this makes DISC adequate to simulate any type of neural network model. Moreover, the network is scalable in the number of neuron instances. DISC programs are portable and scalable too, in the sense that they are independent of the topology and of the number of nodes (workstations or Transputers) on which the application is executed. Combined with the availability of several tools (compiler, monitor, post-mortem debugger, makefile generator, user-friendly interface) that helps and guide the programmer in the development cycle, this makes the DISC environment particularly suitable for *fast prototyping* of neural networks, that can then be scaled-up both in the size and in the number of processors. Finally, full compatibility of the sequential part of the language with the ANSI C language [12] is another important feature, since it releaves the user from the need to learn a completely new programming language, and it results in reuse of existing software modules.

4   SIMULATION OF A HOPFIELD NET

As an example of design and coding of a distributed simulation program for a neural network, we consider the Hopfield's model. A possible DISC implementation of a distributed simulation program for the Hopfield network consists of *n* instances of a *neuron* process, emulating the neurons of the model, plus a *master* process. The master is responsible for interacting with the user, interpreting the commands; collecting the patterns to be learned, managing the I/O from a database; sending the patterns to the neuron network; collecting the state of the neurons, and checking wether: the network has learned all patterns, during the training phase, or stability has been reached, during the pattern retrieval phase. Communication between the master and the neurons makes use of two (monodirectional) one-to-n channels: channel *DataChan* conveys patterns sent by the master to the neurons; channel *StateChan* is used by neurons to send their state values to the master. We point out explicitly that each channel connects a single *owner* (the master) to n *users* (the neurons); however, communication is still synchronous, and no broadcast of messages is provided in DISC, according to the CSP model. The code for the master process is sketched out below.

```
#include "defs.h"
process master(DataChan, StateChan)
owned MESSAGE DataChan;
owned STATE StateChan;
::
{ . . . . . /* Declaration of local variables and external functions */
 quit = FALSE;
 while (!quit) {
     printf("1 - Training 2 - Retrieve 3 - Quit");
     switch(c = getchar()) {
       case TRAINING:
             . . . . .           /* Load patterns from database */
         trained = FALSE;
         while (!trained) {
            for (k=0; k < NPATTERN; k++) {
                for(i=0; i<N; i++) DataChan !! Pattern[k];
                for(i=0; i<N; i++) StateChan ?? state[i];
                if(check_convergence()) trained = TRUE;
                } } printf("O.K., trained");  break;
         case RETRIEVE:
         . . . . .         /* Get the pattern to be retrieved */
         stable = FALSE;
         while (!stable) {
            for(i=0; i<N; i++)  DataChan !! pattern;
            for(i=0; i<N; i++)  StateChan ?? state[i];
            if(check_stability()) stable = TRUE;
            else update_pattern();
         } break;
         case EXIT: quit = TRUE;  break;  } }
} endprocess
```

The *neuron* process has a simple indefinite loop, receiving the pattern from the master. Each neuron computes its state, updates the weights associated to the incoming links (during the training phase), and sends the state back to the master, according to the algorithm presented below.

```
while (TRUE) {
     DataChan ?? pattern
     switch(pattern.mode) {
   case  TRAINING:
       . . . . . . .      /* compute state and update weights */
   case RETRIEVE:
       . . . . . . .      /* compute state */
     }
```

```
StateChan !! mystate   /* send state to the master */
```

Thus, in the simulation program the weight matrix is actually stored in a distributed manner in the network. Both the learning and the retrieving phase happen in a truly parallel mode, provided that neurons are allocated on different nodes of the target multicomputer system.

REFERENCES

[1] B.P. Lester, *A Neural Network Simulation Language Based on Multipascal*, 1st IEEE Int. Conf. on Neural Networks, M. Caudill, C. Butler eds., 347-364, 1987.

[2] E. Deprit, *Implementing Recurrent Back-Propagation on the Connection Machine* , Neural Network, vol. II, pp. 143-150, 1988.

[3] J. Hicklin, H. Demuth, *Modeling Neural Networks on the MPP*, Proc. of the 2nd Symp. on the Frontiers of Massively Parallel Computation, 39-42, 1988.

[4] V.C. Barbosa, P.M.V. Lima, *On the Distributed Parallel Simulation of Hopfield's Neural Networks* , Software - Practice and experience, **20**(10), 967-983, Oct 1990.

[5] W. C. Athas, C. L. Seitz, *Multicomputers: Message-Passing Concurrent Computers*, IEEE Computer, Jan. 1988, 9-24.

[6] INMOS Ltd., *Occam 2 Reference Manual*, Prentice Hall, Cambridge, UK, 1988.

[7] G. Iannello, A. Mazzeo, G. Ventre, *Definition of the DISC Concurrent Language*, SIGPLAN Notices, Vol. 24 (6), June 1989.

[8] G. Iannello, A. Mazzeo, C. Savy, G. Ventre, *Parallel Software Development in the DISC Programming Environment*, Future Generation Computer Systems, North-Holland, Vol. 5, N. 4.

[9] A. Mazzeo, G. Ventre, S. Russo, *Using CSP Languages to program Parallel Workstation Systems*, Future Generation Computer Systems, North-Holland.

[10] C.A.R. Hoare, *Communicating Sequential Processes*, Comm. of the ACM, **21**(8), Aug. 1978.

[11] Treleaven, M. Recce, *Programming Languages for Neurocomputers*, Conf. on Neural Computing for Industry, 1988.

[12] B.W. Kernighan, D.N. Ritchie, *The C Programming Language*, 2nd edition, Prentice Hall, New York, 1988.

20

DESIGNING NETWORKS OF SPIKING SILICON NEURONS AND SYNAPSES

Lloyd Watts

Mail Code 116-81
California Institute of Technology
Pasadena California 91125
`lloyd@hobiecat.pcmp.caltech.edu`

1 INTRODUCTION

Biological neurons communicate with each other via short, fixed-amplitude pulses called *action potentials* or *spikes*. This discrete-amplitude continuous-time representation encodes intensity and timing information with high noise immunity, and thus is ideally suited for processing real-world sensory stimuli.

Since the properties of transistors are well matched to the properties of ionic channels in nerve membranes, Mead has advocated the use of analog VLSI for the construction of artificial neural systems [1]. The current state of the art in realistic single-neuron designs is the silicon neuron of Mahowald and Douglas [2], which exhibits the spiking, refractory, and adaptation characteristics of cortical neurons. The next step in this emerging engineering discipline is the construction of networks of spiking silicon neurons and synapses. This task will require the design and characterization of simple neuron and synapse circuit primitives, and the integrated development of suitable design tools, particularly for schematic capture and simulation.

This paper describes simple and compact silicon neurons and synapses, a neural schematic capture package, and a fast event-driven simulation program optimized for networks of these circuit primitives. The circuits and design tools have been used in the successful design of several small silicon networks.

2   CIRCUIT MODELS

The circuit models are shown in Figure 1. A refractory integrate-and-fire model
is the basis for the neuron circuit. The neuron circuit models the behavior
of voltage-gated sodium and potassium channels in a simple neuron [3]. An
optional tonic current may be injected into the neuron circuit. A simplified
synaptic cleft model is the basis for the synapse circuit. A spike from the
presynaptic cell causes the release of neurotransmitter into the synaptic cleft,
where it remains for a controlled duration. While neurotransmitter is in the
cleft, current is injected into the postsynaptic cell. Any number of synapses
may be connected to a neuron circuit. Symbols for the neuron, synapse, and
tonic input circuits are shown in Figure 2.

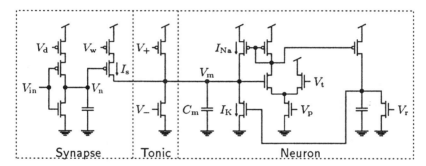

Figure 1   Circuit primitives for the neuron, tonic input, and excitatory
synapse. Currents may be injected onto the neuron membrane capacitance C_m
by the tonic input and synapse circuits. Within the neuron circuit, the currents
I_{Na} and I_K cause action potentials when the membrane voltage V_m reaches
the threshold voltage set by V_t. The pulse width and refractory period are
controlled by the bias voltages V_p and V_r. The output of the neuron circuit is
the membrane voltage V_m. The tonic current is controlled by the bias voltages
V_+ and V_-. The input to the synapse circuit is a spike on V_{in}, which results
in the appearance of "neurotransmitter", as represented by a low voltage on
V_n, for a duration controlled by the bias voltage V_d. The synaptic current I_s
is controlled by the weight voltage V_w and gated by V_n. An inhibitory synapse
circuit may be made by inverting the direction of current injection with a
current mirror.

3   DESIGN TOOLS

The circuits in Figure 1 may be interconnected to form a network of spiking
silicon neurons and synapses. However, it is desirable to simulate the operation

Figure 2   Symbols. The two parameters on the neuron symbol are an identifier label and the refractory period in ms. Default values for the membrane capacitance and threshold voltage are 1 pF and 1 V respectively. The two parameters on the synapse symbol are the synaptic weight current in nA and neurotransmitter duration in ms. The parameter on the tonic input symbol is the tonic input current in nA.

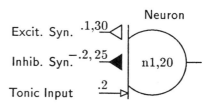

of a network before committing the design to silicon fabrication. A schematic capture program is used to describe the network in a machine-readable format; the correct operation of the network may then be verified via a simulation program.

Neural Schematic Capture

The prototype neural schematic capture package is based on the public-domain program **Analog**, by John Lazzaro and Dave Gillespie. **Analog** allows convenient placement and connection of user-definable symbols, specification of component parameters, and creation of a network description file in ntk format. The combination of the program **Analog** and the custom symbol definitions shown in Figure 2 constitutes the neural schematic capture package **Neuralog**.

Neural Simulator

The synapses act like switched constant-current sources onto the membrane capacitance, leading to piecewise-linear membrane voltage trajectories, as shown in Figure 3. A fast event-driven simulation program, called **Spike**, has been written to exploit this behavior. A similar program was written by Pratt [4], but without the schematic capture front-end or the direct link to VLSI circuit primitives. **Spike** maintains a queue of scheduled events; the occurrence of one event usually triggers the scheduling of new events, either immediately or at some later time. The simulation runs until there are no more events in the queue, or until the desired run time has elapsed.

Spike reads the ntk file generated by **Neuralog**, computes the network response, and outputs the results for plotting by the program **Cview** by Dave Gillespie, or by *Mathematica*.

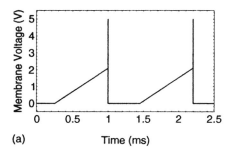 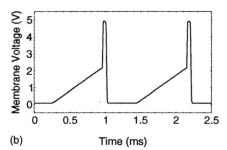

(a)   Time (ms)   (b)   Time (ms)

Figure 3   Response of a neuron to an excitatory tonic input. (a) Simulated response. (b) Measured chip response. The neuron integrates the tonic input current, fires an action potential, and remains refractory for a short time before the cycle repeats.

4 EXAMPLE

Central Pattern Generators (CPGs) are groups of neurons that generate rhythmic firing patterns during repetitive motor tasks [5]. A very simple CPG circuit, called a tonic burster, is shown in Figure 4, as an example of a simple network of spiking neurons. Many other networks of spiking neurons have been built and tested, including neural postprocessors for silicon cochleas, and an adaptive spiking network that learns a time delay.

5 CONCLUSIONS

The contribution of the present work is in developing a unified framework of circuit primitives and integrated software tools for neural schematic capture and simulation, to facilitate the investigation of new network designs before committing to silicon fabrication.

6 ACKNOWLEDGMENTS

The author gratefully acknowledges many helpful discussions with Carver Mead, Misha Mahowald, Sylvie Ryckebusch, Rahul Sarpeshkar, and Brad Minch. Special thanks go to John Lazzaro and Dave Gillespie for writing and supporting the Caltech design tools `Analog` and `Cview`. Chip fabrication was provided by DARPA and MOSIS. *Mathematica* is a registered trademark of Wolfram Research, Inc.

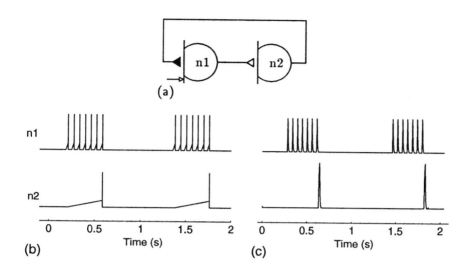

Figure 4   The tonic burster. (a) Schematic diagram. (b) Simulated response. (c) Measured chip response (spikes only). Neuron n1 has an excitatory tonic input that causes it to fire repeatedly. A weakly excitatory synapse couples it to neuron n2, which fires after several spikes from neuron n1. The firing of n2 inhibits n1 for a long time; when the inhibition subsides, n1 begins firing again.

REFERENCES

[1] C. A. Mead, *Analog VLSI and Neural Systems*, Addison-Wesley, 1989.

[2] M. A. Mahowald and R. J. Douglas, "A silicon neuron", Nature:354, 1991, pp. 515-518.

[3] R. Sarpeshkar, L. Watts and C. A. Mead, "Refractory neuron circuits", Internal Memorandum, Physics of Computation Laboratory, California Institute of Technology, 1992.

[4] G. A. Pratt, *Pulse Computation*, Ph.D. Thesis, Massachusetts Institute of Technology, 1989.

[5] S. Ryckebusch, J. M. Bower and C. A. Mead, "Modeling small oscillating biological networks in analog VLSI". In D. Touretzky (Ed.), *Advances in Neural Information Processing Systems*, San Mateo, CA: M. Kaufmann, 1989.

21

RALLPACKS : A SET OF BENCHMARKS FOR NEURONAL SIMULATORS

Upinder S. Bhalla
David H. Bilitch
James M. Bower

California Institute of Technology, Pasadena - CA 91125

ABSTRACT

The field of computational neurobiology has advanced to the point where there are several general purpose simulators to choose from. These cater to various niches of the world of realistic neuronal models, ranging from the molecular level to descriptions of entire sensory modalities. In addition, there are numerous custom designed simulations, adaptations of electrical circuit simulators, and other specific implementations of neurobiological models. As a first step towards evaluating this disparate set of simulators and simulations, and towards establishing standards for comparisons of speed and accuracy, we describe a set of benchmarks. These have been given the name 'Rallpacks', in honor of Wilfrid Rall who pioneered the study of neuronal systems through analytical and numerical techniques.

INTRODUCTION

The Rallpacks are primarily intended to objectively evaluate neuronal simulator performance on measures such as speed and accuracy. The performance information generated in this way also provides a means for making informed decisions on model parameters such as size and timestep which are constrained by accuracy requirements and computer time and memory limitations.

Our evaluation of the basic numerical capabilities of simulators will be confined to passive properties[1] and the computation of the Hodgkin-Huxley equations[2]. Most simulators provide additional capabilities which are not addressed by the present set of Rallpacks. Typical numerical capabilities of neuronal simulators include calcium dynamics, Calcium-dependent ion channels, Nernst potentials, synaptic input using alpha functions, and synaptic learning. From the user's perspective, the non-numerical capabilities such as interfaces and flexibility are often an even more important component of a simulator.

The components of the Rallpack benchmarks include :

• The establishment of standard computational tasks

- The provision of utilities and references for the model output.
- The definition of performance measures and a set of conditions for which measurements are to be taken.
- Specification of the format of the report on the simulator.

COMPUTATIONAL TASKS

The computational tasks set by the Rallpacks are simulations which address different aspects of modeling calculations. The three models currently comprising the Rallpacks are:

- Rallpack 1 : Linear cable model
- Rallpack 2 : Branched cable model
- Rallpack 3 : Linear axon model.

These models are specified in detail in Box 1.

The choice of this set of basic simulations for the benchmarks is determined by their relevance to the classes of model most commonly studied at the present time and by the availability of analytical and/or thoroughly investigated solutions. This set will be extended in the future as simulator capabilities broaden.

REFERENCE CURVES AND UTILITIES

The Rallpack suite provides the reference output data for each model, programs for generating this output, and programs for comparing the correct output and the output from the simulator being tested. These programs are written in the C programming language.

Reference output. The output for the two passive models can be analytically determined. These curves are calculated based on a Laplace transform solution for the time-varying behaviour of the cable[3], numerically calculated as the sum of a convergent series. The correct solution for the axon model in Rallpack 3 cannot be obtained analytically. We therefore provide the simulator setup code used to generate the results in NEURON[4] and GENESIS.[5] These models were run at a timestep of 1 μsec using full calculation of all exponential terms. The correspondence between the two, as calculated by the comparison programs (below) is better than 1%.

Comparison programs. These programs compare simulator output to the reference curves. A simple root mean squared difference in voltage is used for the exponential charging curves for the models in Rallpacks 1 and 2. For

Rallpack 3 spike peaks are aligned for the voltage calculations, and interspike interval differences are separately added to the root mean squared total.

PERFORMANCE MEASURES

Two main performance measures are defined for the benchmarks: *raw speed* and *accuracy*.

Raw speed is the number of compartmental calculations per second:

Raw speed = compartments * steps/simulation run time.

Accuracy is defined as the average normalized root mean squared difference between the simulator output and the reference curve, for each of the two output points on the model. These calculations are performed by the utility routines provided with the benchmark suite to ensure compatibility. Values are reported as percentage errors.

Each of these measures is dependent on specific model parameters such as model size and timestep. The following tables and/or curves in the detailed report for each model illustrate different aspects of simulator performance.

Accuracy vs. Timestep. This is a measure of the intrinsic numerical accuracy of the simulator algorithms. It provides one with information on the best expected accuracy, and the appropriate simulation timestep needed to obtain results with a desired accuracy.

Accuracy vs. Simulation Speed. The simulation speed is the product of timestep and raw speed. In most cases the raw speed does not depend on timestep, so the simulation speed is directly proportional to timestep for the purposes of this plot. This measure is most relevant to choosing between simulators. It directly displays which simulator will be fastest in a desired accuracy range, and assists in making the tradeoff between desired accuracy and speed.

Raw Speed vs. Model size. This curve shows how the speed performance of the simulator scales with increasing model size and gives an estimate of the computational overhead associated with the simulator.

Memory use per compartment vs. Model size. This auxiliary measure indicates the memory overhead associated with different kinds of model. It is valuable for estimating the memory that will be required to run a particular model.

FORMAT OF THE RALLPACK REPORT

The Rallpack report consists of a general section, followed by a detailed report for each of the three Rallpack models. The benchmark report is of the form displayed in Box 2, which contains the reports for NEURON and GENESIS. A complete report includes a detailed report for each of the three rallpacks.

The general section concisely (and sometimes misleadingly) summarizes the performance figures for the simulator. The Peak speed and compartment equivalent values are the basis for evaluating the sheer number crunching capabilities of a simulator. The peak speed corresponds to the best raw speed obtained for Rallpack 1, the linear cable. The compartment equivalents are the ratios of computer processing time required for compartments, branched compartments, and Hodgkin-Huxley-channels. Together these enable one to estimate the total time it would take to run a model involving any combination of these components.

A partial measure of the simulator accuracy is provided by the asymptotic accuracy (the best accuracy obtained with Rallpack 3) and semi-accurate timestep (the longest timestep which will give an error less than twice the asymptotic error). Rallpack 3 is the most demanding model of the three from the accuracy viewpoint, and is therefore used as the criterion. In most cases the semi-accurate timestep will be an adequate compromise between run time and accuracy.

The detailed report for each model includes data for speed, accuracy and memory use over a range of model sizes and timesteps. These curves are described above and illustrated in Box 2. The detailed reports are the basis for any but the most cursory evaluation of a simulator. They also provide a tool for making informed trade-offs between speed and accuracy for a particular simulation.

CONCLUDING REMARKS

The three models presented here as Rallpacks test only the very basic, universally accepted features of neuronal simulators. Existing simulators already support a much wider range of numerical capabilities[4,5,6,7] It is hoped that it will take less time than the decades since the introduction of the model of the squid axon, before some of these capabilities become universally accepted and embodied as benchmarks. Two classes of models already deserve consideration for future Rallpack tests : a complex single cell model with calcium dynamics, to test simulator capabilities on realistic neuronal models; and a model of a network with extensive synaptic interactions to evaluate performance of synapses. A performance related aspect of simulators that may be worth quantifying in future benchmarks is scaling of simulator speed with increasing numbers of nodes on parallel computer architectures.

Benchmarks as a whole offer a very narrow view of the real usability of simulators. Speed and accuracy are important, but less quantifiable features such as interfaces and documentation may have a much greater impact on the time taken to generate a simulation, and the validity of the results can never be any better than the model parameters. What these benchmarks do seek to offer are standards. It is a sign of the growing importance of neuronal simulations to neurobiology that there are so many disparate simulators available to the community. This is a valuable resource, in that each has its own strengths and is developing in different directions. However, it would be unfortunate if this profusion led to doubts about the generality of the results, since each simulator is far too complex for users to analyze in great detail. The Rallpacks are intended to be a common reference for the community, to lend a measure of confidence that all these sophisticated packages produce the same answers, and the *right* answers.

Acknowledgments

The development of the Rallpacks could not have occurred without the insights and suggestions of many people. The original idea was discussed at the symposium on Analysis and Modeling of Neural Systems in 1991 in San Francisco. Among those involved were Frank Eeckman, Michael Hines and John Miller. Feedback and assistance during the development of the Rallpacks was provided among others by Erik De Schutter, Michael Hines, Mark Nelson and Tony Zador.

Obtaining the Rallpacks

The Rallpack suite may be obtained by anonymous ftp from mordor.cns.caltech.edu. We would appreciate feedback on the suite and the results for simulators tested with it.

REFERENCES

[1] Rall, W., (1989), in *Methods in Neuronal Modeling:* From Synapses to Networks (Koch , C. and Segev, I., eds), pp. 9-62, MIT Press
[2] Hodgkin, A.L. and Huxley, A.F. (1952) *J. Physiol. Lond.* 117, 500-544
[3] Jack J.J.B., Noble, D. and Tsien R.W. (1983) in *Electrical current flow in excitable cells* , Oxford University Press
[4] Hines, M. (1989) *Int. J. Bio-Med. Comput.* 24, 55-68
[5] Wilson, M.A., Bhalla, U.S., Uhley, J.D., and Bower, J.M. (1989) *Advances in Neural Information Processing Systems* 1, 485-492
[6] Miller, J. (1990) *Nature* 347, 783-7847
[7] De Schutter, E. (1992) *Trends Neurosci. 15* (in press)

BOX 1: MODEL SPECIFICATIONS

Rallpack 1. Uniform unbranched cable with a length constant of 1, a diameter of 1 micron, and a total length of 1 mm, and sealed ends. The cable is divided into 1000 compartments. The membrane properties are:

RA $= 1.0\ \Omega m = 100\ \Omega\ cm$
RM $= 4.0\ \Omega m^2 = 40000\ \Omega cm^2$
CM $= 0.01\ F/m^2 = 1.0\ mF/cm^2$
EM $= -0.065\ V = -65\ mV$
$E_{Initial}$ $=EM$

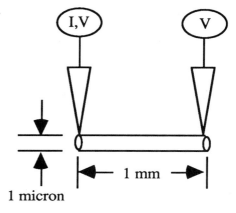

A current of 0.1 nA is injected in the first compartment. Membrane voltages are recorded for the first and last compartments.

Rallpack 2. Binary branched structure obeying Rall's 3/2 power law. It consists of 1023 compartments corresponding to 9 levels of branching. Each branch has an electrotonic length of 0.008 . At each branch away from the root the diameter decreases by a factor of $2^{2/3}$, in accordance with the power law, and the length by a factor of $2^{1/3}$. The detailed geometry of the compartments is

Depth	N	L	Dia
0	1	32.0	16.0
1	2	25.4	10.08
2	4	20.16	6.35
3	8	16.0	4.0
4	16	12.7	2.52
5	32	10.08	1.587
6	64	8.0	1.0
7	128	6.35	0.63
8	256	5.04	0.397
9	512	4.0	0.25

The membrane properties are as for Rallpack 1. A current of 0.1 nA is injected into the root compartment. Membrane voltages are recorded for the root and a terminal branch compartment.

BOX 1: MODEL SPECIFICATIONS (CONT)

Rallpack 3. This model consists of the cable from Rallpack 1 with the addition of squid sodium and potassium channels from Hodgkin and Huxley. The channel properties are as described by Hodgkin and Huxley, as follows : (We have reversed their sign convention, and take the resting potential to be -65 mV)

ENa	= 0.050 V	= 50 mV	
EK	= -0.077 V	= -77 mV	
GNa	= 1200 Siemens/m^2	= 120 mmho/cm^2	
GK	= 360 Siemens/m^2	= 36 mmho/cm^2	

(For each compartment :
  Gbar_Na   = 3.77e-9 Siemens   = 3.77e-6 mmho
  Gbar_K   = 1.131e-9 Siemens   = 1.131e-6 mmho)

The conductance of each channel is given by :

$$gNa = Gbar_{Na} m^3 h$$
$$gK = Gbar_K n^4$$

where m, h, and n satisfy equations of the type

$$dm/dt = a_m - (a_m + b_m)m.$$

a and b in units of ms^{-1} and mV are given by

$$a_m = 0.1(-V+25)/(e^{(-V+25)/10} -1) ; \quad b_m = 4e^{-V/18}$$
$$a_h = 0.07e^{-V/20}; \quad b_h = 1/(e^{(V+30)/10}+1)$$
$$a_n = 0.01(-V+10)/(e^{(-V+10)/10} -1); \quad b_n = 0.125e^{-V/80}$$

A current of 0.1 nA is injected in the first compartment. Membrane voltages are recorded for the first and last compartments.

BOX 2: RALLPACK REPORTS FOR GENESIS AND NEURON

General report.	**GENESIS 1.4**	**NEURON 2.43**
Peak speed	44,000	42,000
Compartment equivalents	1 : 1 : 0.5	1 : 1.8 : 0.75
Asymptotic accuracy	0.9%	0.9%
Semi-accurate timestep	50 μsec	50 μsec
Hardware information	Sun sparc 2; Sun OS 4.1; 26 MIPS	
Simulation setup time	13 Sec	1 Sec
Base memory	3.42 Megabytes	1.6 Megabytes
Integration method	Hines/Crank-Nicolson	

Specific report for Rallpack 3		
Peak speed/model size	22,000 >= 1000 comps	17,400 >= 100
Asymptotic accuracy	0.9%	0.94%
Semi-accurate timestep	50 usec	50 usec
Setup time	13 sec	1 sec

o GENESIS   ■ NEURON

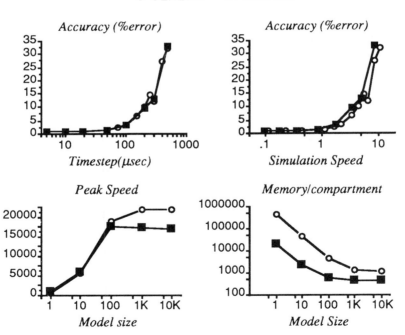

SECTION 3

VISUAL SYSTEMS

In this section on visual systems both "realistic" neural models incorporating explicit circuitry and more abstract artificial network models emphasizing psychophysical phenomena are presented. The individual chapters are organized by subject from the periphery to the central nervous system and from realistic to abstract models. The section concludes with two papers on attentional mechanisms. The reader may also want to consult chapters 5 and 6 in section 1.

INVARIANT CONTRAST ADAPTATION IN THE PRIMATE OUTER PLEXIFORM LAYER

Josef Skrzypek and George Wu

Machine Perception Laboratory
University of California
Los Angeles, CA 90024
email: skrzypek@cs.ucla.edu

ABSTRACT

Luminance contrast adaptation in the primate retina is localized to synaptic interactions of the outer plexiform layer (OPL). Experiments , with a computer model of the OPL derived from known anatomy and physiology, suggest that simple network interactions can implement invariant contrast adaptation while maintaining high luminance contrast sensitivity over 7 log units of retinal illuminance. Furthermore, simulation results predict behavior consistent with known psychophysics of primate foveal photopic vision.

Introduction

Primate photopic vision operates over 7 logarithmic units (LU) of light intensity without significant degradation of luminance contrast sensitivity [6]. The light intensity domain in individual cone photoreceptor is less than 4 LU [16]. The mechanism underlying invariant contrast adaptation remains unknown.

The invariance of luminance contrast is critical to lightness and color constancy [7, 5]. The invariance process begins with luminance contrast processing the retina. Although these phenomena have been extensively investigated in the past, the neural mechanism underlying the interpretation of surface reflectance as white, grey or black remains a matter of controversy. The most recent anatomical and physiological experimental results suggest that the site of luminance contrast processing in the primate retina is localized to the OPL [2, 3]. In this paper, we examine in depth the compatibility between computational theories based on psychophysics and physiological data through experiments with a physiologically and anatomically plausible computational model of the primate OPL. Our simulation results demonstrate that simple network interactions can transform absolute retinal illuminance into relative luminance contrast; center-surround antagonism in cone receptive field (RF), verified physiolgically in vertebrates, can explain contrast processing properties in primates observed psychophysically [18, 4].

Organization of the Primate OPL

The OPL of the primate retina plays a major role in luminance contrast processing. Light hyperpolarizes photoreceptors which drive horizontal cells (HC's) via non-inverting chemical synapses. Both types of horizontal cells, narrow (HC1) and broad receptive fields (HC2), contact all cone pedicles in their dendritic fields. In the fovea, each HC1 contacts 6-7 cones and each HC2 contacts 13-14 pedicles regardless of cone chromatic type [2, 3]. Both HC types have axon terminal (AT) that makes synaptic contacts with spherules (for HC2) and pedicles (for HC1) exclusively [2]. HC cell bodies and their AT's are thought to be electrically isolated and thus functionally independent [3]. Although HC1 and HC2 both appear to be equally sensitive to all light wavelengths, we hypothesize HC1's are more important than HC2's for foveal photopic vision because of HC2's lower density in the fovea [11]. In the present investigation only HC1's are modelled.

Fidelity of the model is critically dependent on the accuracy of the anatomical and physiological parameters such as the cone to HC1 density ratio (i.e. coverage factor). The cone to HC1 coverage factor has been estimated to be 3-4 [2, 17]. Pedicle size is critical since there is a receptoral areal magnification factor (more than 10 in the fovea [12]) going from the outer segment to the pedicle of a cone. Knowing the HC1 to pedicle coverage factor and the pedicle to HC1 RF size allows for the calculation of the HC1 RF overlap which is approximately 50%. To determine the extent of cone-cone and HC1-HC1 coupling one needs to estimate the gap junction conductance for cones and HC1's. Unfortunately, there is no known physiological data on primate gap junction conductance. Our estimate for cone-cone and HC1-HC1 RF sizes is obtained by interpreting psychophysical data. From [18], the cone-cone RF diameter is estimated to be 7 cones and HC1-HC1 RF diameter is 27 HC1's. Although no functional feedback has been observed in primate cones the necessary synaptic structures are present [10]. Similar anatomical structures were correlated with physiological mechanisms underlying feedback synapse in turtles [1] and in Tiger salamanders [8, 14].

The involvement of feedback in relative contrast enhancement and color coding has been reported in turtles [1], Tiger salamanders [8], and fish [15]. In the primate OPL similar data is lacking but it is generally believed that feedback plays similar role. The role of feedback in invariant contrast adaptation has been more difficult to verfiy. In Tiger salamanders it has been shown that feeback from HC can instantaneously shift the cone operating point to be in register with ambient light intensity of the surround; this preserves relative contrast invariance [13]. We present here an anatomically and physiologically plausible model of the primate OPL that allows for computational verfication of the role of feedback.

A Primate OPL Model

Our primate OPL model consists of two layers representing cones and the HC1's. Functional view of the model is shown in Fig. 22.1.

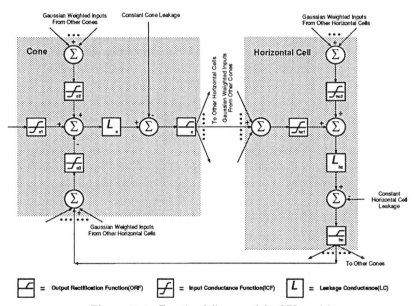

Figure 22.1 Functional diagram of the OPL model.

The sign of the various signals are actually opposite that of their respective physiological correlates (e.g. hyperpolarization is represented by signal increment in the model). The mathematical formalism of the model is presented in Eqns. 22.1 through 22.4.

$$O_c(t+1) = ORF(ICF(G_{l-c1}, G_{l-c2}, S_{l-c}) + ICF(G_{c-c1}, G_{c-c2}, S_{c-c}) - ICF(G_{hc-c1}, G_{hc-c2}, S_{hc-c}) - L_c O_c(t) - K_c) \quad (22.1)$$

Where: $S_{l-c} \equiv$ incident light intensity; $S_{c-c} \equiv$ Gaussian weighted sum of cone-cone inputs; $S_{hc-c} \equiv$ Gaussian weighted sum of HC1-cone inputs; $G_{l-c1}, G_{l-c2} \equiv$ incident light gain parameters; $G_{c-c1}, G_{c-c2} \equiv$ cone-cone gain parameters; $G_{hc-c1}, G_{hc-c2} \equiv$ HC1-cone gain parameters; $L_c \equiv$ cone voltage dependent leakage modulation parameter; $K_c \equiv$ constant cone leakage.

$$O_{hc}(t+1) = ORF(ICF(G_{c-hc1}, G_{c-hc2}, S_{c-hc}) + ICF(G_{hc-hc1}, G_{hc-hc2}, S_{hc-hc}) - L_{hc} O_{hc}(t) - K_{hc}) \quad (22.2)$$

Where: $S_{c-hc} \equiv$ Gaussian weighted sum of cone-HC1 inputs; $S_{hc-hc} \equiv$ Gaussian weighted sum of HC1-HC1 inputs; $G_{c-hc1}, G_{c-hc2} \equiv$ cone-HC1 gain parameters; $G_{hc-hc1}, G_{hc-hc2} \equiv$ HC1-HC1 gain parameters; $L_{hc} \equiv$ HC1 voltage dependent leakage modulation parameter; $K_{hc} \equiv$ constant HC1 leakage.

$$ICF(G_1, G_2, I) \quad = \quad G_2 \tanh(G_1 I) \tag{22.3}$$

Where: $G_1 \equiv$ lumped pre-nonlinearity gain; $G_2 \equiv$ lumped post-nonlinearity gain.

$$ORF(I) \quad = \quad \begin{cases} I & \text{if } I \geq 0 \text{ and } I \leq 1 \\ 1 & \text{if } I > 1 \\ 0 & \text{if } I < 0 \end{cases} \tag{22.4}$$

Cone input connections are modelled using three transfer functions: hyperpolarization due to incident light, coupling to neighboring cones, and depolarizing input due to feedback from neighboring HC1's. The cone output is the sum of these three **Input Conductance Functions** (ICF's) that differ only in parameters (see Eqn. 22.3). HC1's receive only hyperpolarizing inputs from neighboring cones and other HC1's. Each of the inputs to HC1 is also characterized by unique ICF. Each ICF operates on a Gaussian weighted sum of inputs over a two dimensional RF (see Table 22.1 for the actual RF sizes).

Parameter	Value	Parameter	Value
cone-cone RF diameter	7 cones	G_{hc-c2}	1.000
HC1-HC1 RF diameter	27 HC1's	G_{c-hc1}	0.132
cone-HC1 RF diameter	3 cones	G_{c-hc2}	1.000
HC1-cone RF diameter	3 cones	G_{hc-hc1}	0.530
G_{l-c1}	0.662	G_{hc-hc2}	1.000
G_{l-c2}	1.750	L_c	0.000
G_{c-c1}	0.221	L_{hc}	0.000
G_{c-c2}	2.000	K_c	0.300
G_{hc-c1}	0.883	K_{hc}	0.000

Table 22.1   Model Parameters

At the cone output the sum of the 3 ICF's and 2 leakages (fixed and voltage dependent) is rectified by the **Output Rectification Function** (ORF) (Eqn. 22.4). The output of a HC1 is determined in a similar manner as the cone output.

The ICF is a lumped model of the actual synaptic conductance characteristics that might vary from synapse to synapse. This lumped model is adequate to identify the *robust* physiological characteristics of the OPL. A more serious limitation is the fact that all signal delays in the model are equal and constant (i.e. all capcitances have the same value). This limitation was introduced intentionaly to simplify the model and facilitate the analysis of the steady-state behavior.

Simulation Results

All simulations were conducted using *RetSim*, a retinal network simulator developed within the UCLA-SFINX simulation environment [9]. Separate experiments were conducted to demonstrate two aspects of the contrast processing properties of the primate OPL model, namely invariant contrast adaptation and relative contrast enhancement. A model with a 128x128 cones and 42x42 HC1's was used in all simulations. Identical setup was used for all experiments (Fig. 22.2) except for the stimulus configuration. In all cases the center spot diameter and the annulus inner diameter was equal to 7 cone diameters and the annulus outer diameter was equal to 27 cone diameters.

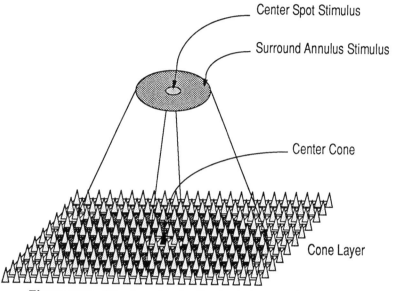

Center Spot Stimulus

Surround Annulus Stimulus

Center Cone

Cone Layer

Figure 22.2   Stimulus setup for the simulation experiments.

Figure 22.3 shows the center cone response to a stimulus with a 1.5 LU center-surround reflectance difference maintained under various illumination settings. The response curve can be divided into 4 sub-regions, region I (-3 to -2 LU), region II (-2 to 0 LU), region III (0 to 3 LU),and region IV (3 to 7 LU). Region III is of interest since invariant contrast adaptation occurs here. The invariance is seen as the near constant response over the region. However, region III does not really show the entire extent of the invariant contrast adaptation response since the region bounds are responses to a 1.5 LU difference stimulus. At a given operating point the linear cone response domain is about 4 LU (see Figure 22.4) thus, assuming operating point is 1.5 LU above cut-off response, the correct extent of the invariant contrast adaptation region is -1.5 (=0-1.5) to 5.5 (=3+2.5) LU; the model is capable of invariant contrast adaptation over 7 LU (=5.5-(-1.5)). Region I characterize the sub-threshold reponse of cones. Region II is the transition region from sub-threshold to super-threshold response regions. Region IV basically illustrates the gradual failure of feedback control.

To show the high contrast sensitivity of the model, a series of experiments were done where the surround is fixed at a given illuminance level while the

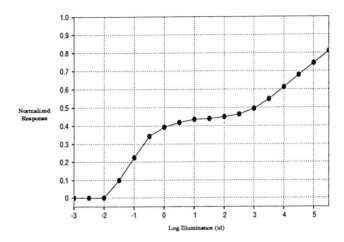

Figure 22.3   Center cone response to stimulus with a constant 1.5 LU center-surround intensity difference under various uniform illumination settings. The abscissa indicates the center spot illuminance.

center light intensity is varied from -3 to 7 LU in 0.5 LU increments. The center cone response for 7 different surround settings were *recorded* and are shown in Figure 22.4. The feedback from HC1 shifts the cone response curve to achieve invariant contrast adaptation while preserving high contrast sensitivity.

Discussion

The cone response curves shown in Figure 22.4 correspond surprisingly well to actual primate extracellular recordings [16]. Traditionally, three basic mechanisms have been hypothesized to explain cone light adaptation, namely response compression, pigment bleaching, and cellular adaptation involving network interactions between cones and HC's. Our model essentially supports the cellular adaptation hypothesis.

Results shown in Figure 22.3 correlate well with psychophysically determined foveal threshold-versus-intensity (TVI) curves [4] up to 4 td. Clearly, cone response over the invariant contrast adaptation region is not perfectly constant thus it does not agree Weber-Fechner law completely. This fact is actually consistent with the observed phenomenon of *more light means better sight*. It is possible that invariant contrast adaptation underlies the observed Weber-Fechner-like behavior but further study is necessary.

Invariant contrast adaptation fails at higher luminance levels although perhaps not as dramatic as shown in Figure 22.3. Cone pigment bleaching is known to be a signficant sensitiviy reduction mechanism at high illuminance levels [6]

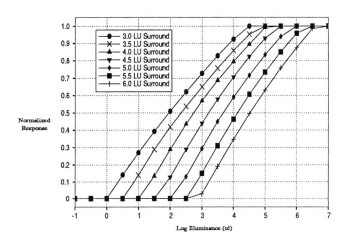

Figure 22.4 Center cone response to stimulus center-surround luminance difference under various fixed surround settings.

which ameliorates the effect of invariant contrast adaptation failure. Data from [16] shows that cone pigment bleaching becomes significant at illuminance levels above 4 td which is roughly the onset point of invariant contrast adaptation failure. The net effect of bleaching is more than 1 LU of the operating curve shift at high illuminance levels [16]. Indeed, even wider invariant contrast adaptation region can be achieved if pigment bleaching mechanism is utilized.

In summary, the simulation results appear to be quantitatively consistent with available physiological and psychophysical data on primate foveal photopic vision; luminance contrast processing is accomplished no later than the first synaptic stage (i.e. in OPL). We claim, the computational model of the OPL forms a bridge among the anatomical, physiological and psychophysical data.

REFERENCES

[1] D.A. Baylor, M.G.F. Fuortes, and P.M. O'Bryan. Receptive Fields of Single Cones in the Retina of the Turtle. *J. Physiol*, 214:265–294, 1971.

[2] B.B. Boycott, J.M. Hopkins, and H.G. Sperling. Cone Connections of the Horizontal Cells of the Rhesus Monkey's Retina. *Proc. R. Soc. Lond. B*, 229:345–379, 1987.

[3] R.F. Dacheux and E. Raviola. Physiology of HI Horizontal Cells in Primate Retina. *Proc. R. Soc. Lond. B*, 239:213–230, 1990.

[4] W.S. Geisler. Evidence for the Equivalent-Backgrounds Hypothesis in Cones. *Vision Research*, 19:799–805, 1979.

[5] S. Grossberg and D. Todorovic. Neural Dynamics of 1-d and 2-d Brightness Perception: A Unified Model of Classical and Recent Phenomena. *Perception and Psychophysics*, 43:241–277, 1988.

[6] D.C Hood and M.A. Finkelstein. Sensitivity to Light. In K.R. Boff, L. Kaufman, and J.P. Thomas, editors, *Handbook of Perception and Human Performance*, volume 1, pages 5/1–5/66. Wiley, New York, 1986.

[7] E.H. Land and J.J. McCann. Lightness and Retinex Theory. *Journal of the Optical Society of America*, 61(1):1–11, January 1971.

[8] A. Lasansky. Synaptic Action Mediating Cone Responses to Annular Illumination in the Retina of the Larval Tiger Salamander. *J. Physiol.*, 310:205–214, 1981.

[9] E. Mesrobian and J. Skrzypek. A Software Environment for Studying Computational Neural Systems. *IEEE Transactions on Software Engineering*, pages 575–589, July 1992.

[10] E. Raviola and B. Gilula. Intramembrane Organization of Specialized Contacts in the Outer Plexiform Layer of the Retina. *J. Cell Biol.*, 65:192–222, 1975.

[11] J. Röhrenbeck, H. Wässle, and B. Boycott. Horizontal Cells in the Monkey Retina: Immunocytochemical Staining with Antibodies against Calcium Binding Proteins. *Eu. J. Neurosci*, 1:407–420, 1989.

[12] Stanley J. Schein. Anatomy of Macaque Fovea and Spatial Densities of Neurons in Foveal Representation. *J. Comp. Neurol.*, 269:479–505, 1988.

[13] J. Skrzypek. Feedback Synapse to Cone and Light Adaptation. In D Touretzky, editor, *Advances in Neural Information Processing Systems, NIPS 3*, San Mateo, CA, 1990. Morgan Kaufman.

[14] J. Skrzypek and G. Wu. Neither DoG nor LoG Fits the Receptive Field of a Vertebrate Cone. In F. Eeckman, editor, *Analysis and Modeling of Neural Systems 2*, page [In Press]. Kluwer Academic, 1992.

[15] W.K. Stell. Functional Polarization of Horizontal Cell Dendrites in Goldfish Retina. *Invest. Ophthal. Visual Sci.*, 15:895–907, 1976.

[16] J.M. Valeton and D. van Norren. Light Adaptation of Primate Cones: An Analysis Based on Extracellular Data. *Vision Research*, 23:1539–1547, 1983.

[17] H. Wässle, B. Boycott, and J. Röhrenbeck. Horizontal Cells in the Monkey Retina: Cone Connnections and Dendritic Network. *Eu. J. Neurosci*, 1:421–435, 1989.

[18] G. Westheimer. Spatial Interaction in Human Cone Vision. *J. Physiol.*, 190:139–154, 1967.

23

A COMPUTATIONAL MODEL OF THE *LIMULUS* EYE ON THE CONNECTION MACHINE

Daryl R. Kipke[1]
Erik D. Herzog
Robert B. Barlow, Jr.

Syracuse University, Syracuse, New York, 13244-5290

ABSTRACT

We describe a computational model of the *Limulus* lateral eye to investigate the relationship between the neural coding of the retina and the visual performance of the animal. The network model, implemented on the massively parallel Connection Machine, computes the firing rates of optic-nerve fibers for simulated visual scenes. Intensity-response functions computed for individual nerve fibers that project to the brain agree with experimental measurements. Responses computed for the entire array of retinal neurons show that moving targets enhance the modulation of optic-nerve activity.

1 INTRODUCTION

What is the neural code the eye sends to the brain? How is it generated by the optics, anatomy, and physiology of the eye? How is it related to visual performance? These are fundamental questions of early visual processing. Answers may come from animals with relatively simple visual systems such as the horseshoe crab, *Limulus polyphemus*. Although *Limulus* does not see very well--the eye has low spatial acuity and slow temporal responses--vision plays a significant role in its natural behavior; males use vision to locate females during the mating season [5].

The lateral eye of *Limulus* is a relatively simple structure that lends itself to physiological manipulation and computational modeling. The eye is composed of

---

[1] Address correspondence to Daryl Kipke, Department of Chemical, Bio, and Materials Engineering, Arizona State University, Tempe, Arizona 85287-6006, or e-mail to kipke@asuvax.asu.edu

approximately 1,000 ommatidia whose orientation and optical properties determine the resolution and light-collecting properties of the eye. The optic axes of adjacent ommatidia diverge at an average of 5.5° and an eye can see approximately 60% of the hemisphere on one side of the body with little binocular overlap [13].

Each ommatidium is composed of 10 to 15 photoreceptors and a single spike-generating output cell, the eccentric cell. The photoreceptor membrane has both light-sensitive and voltage-sensitive conductances, and the cells respond with an increase in conductance leading to membrane depolarization [16]. The overall conductance change is the sum of many discrete conductance events each associated with the absorption of a single photon [10]. The summed conductance change is sensed by the dendrite of the second-order eccentric cell where it forms the generator potential for optic-nerve activity. The eccentric cell response is inhibited by its own spike activity through a process of self inhibition and by the activities of neighboring eccentric cells via lateral inhibition [18]. The lateral inhibitory signals are transmitted by eccentric-cell processes in the neural network of the eye [11]. Spikes are generated through active conductances in the axon. The properties of the eye are modulated by a circadian clock located in the brain [2,4].

Here we present a cell-based model for computing neural responses generated by the *Limulus* eye. We limit our analysis to the eye in its daytime state. This project is part of a research program that combines computational, physiological, and behavioral experiments to investigate the neural coding of visual information. An earlier version of the model has been presented [7].

2 MODEL DESCRIPTION

We implemented the model on a massively parallel supercomputer, the Connection Machine model CM-2 (Thinking Machines Inc., Cambridge MA) operated by the Northeast Parallel Architectures Center at Syracuse University. Of the 32,768 processors available on the CM-2, our simulations use either 8,192 or 16,384 processors. We constructed the model in C* , the data parallel dialect of the C programming language. Simulations are run with a time step of 10 ms. The fully connected model with all processes enabled requires about five minutes to run a three second simulation.

We represent the eye with a two-dimensional parallel variable[1] having 32 rows and 64 columns (2048 total elements) with each element corresponding to a retinal unit,

---

[1] Parallel variables are multi-dimensional variables that are mapped onto the parallel processors of the CM-2. Their elements are manipulated simultaneously with a single statement.

i.e., an ommatidium. The model is oversized (2048 units vs. 1000 ommatidia) as a tradeoff between the size constraints of parallel variables (powers of 2) and a desire to maintain a 2:1 width to height ratio. We assign each model unit an acceptance angle of 6 degrees and an unique optic axis given by azimuth and elevation angles relative to the cardinal axes (latero-medial and antero-posterior) of the eye. Optic axes of neighboring units diverge by 4 degrees.

We represent the visual field of the eye with a second parallel variable (128 rows and 256 columns) within which we construct simulated scenes. The model eye spans $\pm 128°$ horizontally about the latero-medial axis and $\pm 64°$ vertically about the antero-posterior axis. Each point in the scene can be independently set to a simulated light intensity that varies over a 6 log-unit range.

At each 10 ms time step, we assign to each model unit the simulated light intensity, $I_{eff}(t)$, integrated over the unit's visual field and then use this signal to update its optic-nerve activity, $r(t)$ (Figure 1). Four functional blocks corresponding to the four major physiological processes in the eye--phototransduction, spike generation, self-inhibition, and lateral inhibition--determine the input-output properties of every unit.

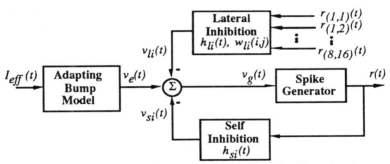

Figure 1. Signal flow for a single model unit. The signals are described in the text.

Phototransduction is an excitatory process and is modeled as a single-compartment membrane model that corresponds to the photoreceptors and the eccentric cell of an ommatidium. The membrane equation is given by

$$-C_m \frac{dv_e(t)}{dt} = g_L\big(v_e(t) - E_{Na}\big) + g_{Na}\big(v_e(t) - E_{Na}\big) + g_K\big(v_e(t) - E_K\big) +$$
$$g_{leak}\big(v_e(t) - E_{rest}\big) - I_{inj}$$

where $v_e(t)$ is the membrane potential, C_m is the membrane capacitance, g_{leak} is the leakage conductance, E_{rest} is the membrane resting potential, and E_{Na} and E_K are the equilibrium potentials for sodium and potassium, respectively. We represent the

light-sensitive sodium conductance, g_L, with the adapting bump model of
photoreceptor response [10,19]. Light determines the rate, height, and duration of
discrete conductance events, i.e., quantum bumps, which sum to establish the
amplitude and time course of g_L (Figure 2). Voltage-sensitive conductances, g_{Na} and
g_K, determine the membrane's active properties. The unit can be driven by light or
a current source, I_{inj}.

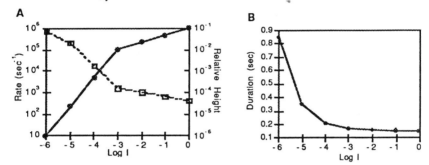

Figure 2. Adapting bump model of phototransduction. **A.** Bump rate
(filled circles, solid line) and relative bump height (open squares, dotted
line) as a function of light intensity. Bump height is the product of
relative height and the resting membrane conductance. **B.** Bump
duration as a function of light intensity. The bump time course is an
exponential with a time constant equal to the bump duration.

The optic-nerve firing rate for a model unit, $r(t)$, is calculated from its generator
potential, $v_g(t)$, with a piece-wise linear transform[2] based on measurements of the
steady-state spike generation process [3]. The generator potential is the difference
between the excitatory potential, $v_e(t)$, and inhibitory potentials resulting from self
and lateral inhibition,

$$v_g(t) = v_e(t) - v_{si}(t) - v_{li}(t)$$

where $v_{si}(t)$ is the self inhibitory input and $v_{li}(t)$ is the lateral inhibitory input. The
self inhibitory input is obtained by convolving $r(t)$ with a first-order low-pass filter
having a time constant of 400 ms. The lateral inhibitory input for unit (i,j) is
obtained by convolving the output of the Hartline-Ratliff equation that describes the
steady-state lateral inhibitory input [18] with a second low-pass filter,

---

[2] For depolarization less than 3 mV, slope is 1.0; depolarization between 3 mV and 6 mV,
slope is 0.8; depolarization between 6 mV and 20 mV, slope is 1.7; depolarization greater
than 20 mV, slope is 0.25; spontaneous firing rate is 5 spikes/sec.

$$v_{li}^{(i,j)}(t) = h_{li}(t) * \left(\sum_{m=-4}^{3} \sum_{n=-8}^{7} w_{li}(m,n) r_{i-m,j-n}(t) \right)$$

where $h_{li}(t)$ has a time constant of 200 ms and $w_{li}^{(i,j)}$ is a weight matrix that describes the inhibitory field. The inhibitory field around each unit is elliptical with the major axis along the anterior-posterior axis of the eye. Inhibition is strongest 3-5 units from the reference unit and it decreases smoothly with increasing distance [1]. The computed steady-state lateral inhibitory input to unit *(i,j)* is summed over the optic-nerve activity of 128 of its nearest neighbors (8 rows and 16 columns). The dynamics of self inhibition and lateral interaction are estimated from measurements of the frequency responses of these processes in the excised eye [18]. The lateral inhibitory network is implemented on the Connection Machine with a three-dimensional parallel variable that allocates a virtual processor for every connection between ommatidia.

3 RESULTS

We have carried out simulations with the model in the "daytime" state, i.e., physiological parameters measured during the day or with the optic nerve cut to abolish the brain's input to the eye. We also have carried out simulations with and without self and lateral inhibition, and with current injection into the eccentric cell to mimic light activation.

Figure 3 (solid curve) shows the intensity-response function for steady-state discharge computed for a single unit without lateral inhibitory inputs. Spike rate, measured one second after the onset of the light flash varies from 5 to 40 spikes/sec as light intensity varies from $\log I = -6$ to 0. A plateau in the region from $\log I = -5$ to -3 characterizes both the computed function and that recorded from a single optic-nerve fiber *in situ* (dashed curve). The optic-nerve function has the same shape and range as that for the photoreceptor potential [3].

Figure 4 compares the responses of the model eye (32x64 retinal units) for a moving and stationary simulated scene of a cylindrical target located underwater on a sandy bottom at a distance of 50 cm which is close enough to be seen by *Limulus* [17]. The gray level of each element in the array represents the spike rate of the corresponding model unit. When the target is stationary, the computed response rates across the array are nearly uniform regardless of whether the ommatidia views the water, sand, or target (Figure 4A). The exceptions are a band of units with elevated spike rates viewing the region immediately outside of the target and a narrow band with decreased spike rates viewing the region immediately inside the target. The small changes in response rate near the target borders represent the enhancement of contrast by lateral inhibition [18].

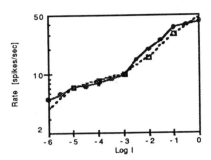

Figure 3: Computed (solid curve) and measured (dashed curve) intensity-response functions for optic-nerve activity.

When the target is moving the computed response rates are lower for receptors inside the leading (right) edge of the target and higher after the trailing edge (Figure 4B). Slower movement of the target produces a smaller modulation of response rates.

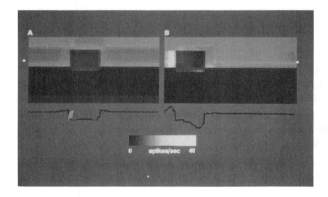

Figure 4: A. The computed response of the eye model to a stationary target 1.5 sec after the simulation onset. The left and top edges of the 2-D array represent the anterior and dorsal margins of the field of view. Firing rate is represented with a gray scale. The trace (bottom) shows the activity of units along the horizontal line marked by the asterisk. The 35x50 cm target is 50 cm from the eye. Light intensity levels: water at log I=-2.0; target at log I=-2.3; sand at log I=-2.6. **B.** The computed response to the same target moving at 1.0 m/s from left to right (target wraps at the scene edges) 2 sec after the simulation onset.

4 DISCUSSION

We developed a quasi cell-based model of the lateral eye to explore the coding of visual information in *Limulus*. At present, we model the eye's daytime state with parameters estimated from measurements of both *in situ* and excised eyes. Our approach is to model the eye as realistically as possible. A previous model used a linear transfer-function to describe the input-output characteristics of retinal processing [9]. While the general linearity of the eye's functional properties motivates this top-down approach, the relative simplicity of the eye's structure motivates our alternative bottom-up approach.

We model the individual neurons of the eye and the connections between them with the goal of learning how neurons interact to compute the output code of the retina. The model also serves an important role as an experimental tool--a computational "preparation"--which can include new laboratory data and test the level of our knowledge of retinal processes. The model in its present form describes with transfer-functions several physiological processes of the eye. We will refine the model by replacing transfer functions with more realistic components and representing the eye's dynamics [8,12], noise [6], and efferent modulation [14] with greater precision.

How can the model be used to relate the retinal code to visual performance? We hypothesize that the spatial and temporal coding properties of the retina tune the eye to specific stimuli in the animal's natural environment. The optic-nerve then transmits the encoded information to the central visual pathways of the brain which may further process the eye's input before signaling motor pathways to generate overt behavior. We will simulate underwater scenes for which visual performance has been measured [17] and analyze the optic-nerve responses computed across the entire retinal array for information relating directly to known visual performance. We hope this approach will reveal coding properties of the retina underlying visually-guided behavior.

Acknowledgment

We thank Professor Ernest Sibert of Syracuse University for his help implementing the neural model on the Connection Machine. This work is supported by the Mabel E. Lewis Postdoctoral Fellowship for Vision Research, the CASE Center at Syracuse University, and NSF grant BNS 9012069 and NIH grant EY-00667.

REFERENCES

[1]   Barlow, R.B., Jr. Inhibitory fields in the *Limulus* lateral eye. *J. Gen. Physiol.* 54:383-396, 1969.

[2]   Barlow, R.B., Jr. Circadian rhythms in the *Limulus* visual system. *J. Neurosci.* 3:856-870, 1983.

[3]   Barlow, R.B., Jr. and E. Kaplan. Properties of visual cells in the lateral eye of *Limulus in situ*: Intracellular recordings. *J. Gen. Physiol.* 69:203-220, 1977.

[4]   Barlow, R.B., Jr., S.J. Bolanowski Jr., and M.L. Brachman. Efferent optic nerve fibers mediate circadian rhythms in the *Limulus* eye. *Science* 197:86-89, 1977.

[5]   Barlow, R.B., Jr., L.C. Ireland, and L. Kass. Vision has a role in *Limulus* mating behavior. Nature 296:65-66, 1982.

[6]   Barlow, R.B., Jr., E. Kaplan, G.H. Renninger, and T. Saito. Circadian rhythms in *Limulus* photoreceptors. I. Intracellular studies. *J. Gen. Physiol.* 89:353-378, 1987.

[7]   Barlow, R.B., Jr., R. Prakash, and E. Solessio. The neural network of the *Limulus* retina: From computer to behavior. *American Zoologist* in press, 1992.

[8]   Batra, R. and R.B. Barlow, Jr. Efferent control of temporal response properties of the *Limulus* lateral eye. *J. Gen. Physiol.* 95:229-244, 1990.

[9]   Brodie, S.E., B.W. Knight, and F. Ratliff. The spatiotemporal transfer function of the *Limulus* lateral eye. *J. Gen. Physiol.* 72:167-202, 1978.

[10]   Dodge, F.A., B.W. Knight, and J. Toyoda. Voltage noise in *Limulus* visual cells. *Science* 160: 88-90, 1968.

[11]   Fahrenbach, W.H. Anatomical circuitry of lateral inhibition in the eye of the horseshoe crab, *Limulus polyphemus*. *Proc. Royal Soc. of London, Section B.* 225:219-249, 1985.

[12]   Grzywacz, N.M., P. Hillman, and B.W. Knight. The amplitudes of unit events in *Limulus* photoreceptors are modulated from an input that resembles the overall response. *Biol. Cybern.* 66:437-441, 1992.

[13]   Herzog, E.D., and R.B. Barlow, Jr. The *Limulus*-eye view of the world. *Vision Research* in press, 1992.

[14]   Kaplan, E. and R.B. Barlow, Jr. Properties of visual cells in the lateral eye of *Limulus in situ:* Extracellular recordings. *J. Gen. Physiol.* 66:303-326, 1975.

[15]   Kaplan, E., R.B. Barlow, Jr., G. Renninger, and K. Purpura. Circadian rhythms in *Limulus* photoreceptors; II. Quantum bumps. *J. Gen. Physiol.* 96:665-685, 1990.

[16]   Millecchia, R. and A. Mauro. The ventral photoreceptor cells of *Limulus*. II. The basic photoresponse. *J. Gen. Physiol.* 54:310-330, 1969.

[17]   Powers, M.K., R.B. Barlow, Jr., and L. Kass. Visual performance of horseshoe crabs day and night. *Visual Neuroscience* 7:179-189, 1991.

[18]   Ratliff, F. *Studies on Excitation and Inhibition in the Retina*. The Rockefeller University Press, New York, 1974.

[19]   Wong, F. and B.W. Knight. Adapting-bump model for eccentric cells of Limulus. J. Gen. Physiol. 76:539-557, 1980.

DOES SYNAPTIC FACILITATION MEDIATE MOTION FACILITATION IN THE RETINA?

Norberto M. Grzywacz[1]
Franklin R. Amthor[2]
Lyle J. Borg-Graham[3]

[1]*Smith-Kettlewell Institute, 2232 Webster Street, San Francisco, CA 94115*
[2]*Dept. Psychol. and NRC, Univ. Alabama, Birmingham, AL 35294*
[3]*Massachusetts Institute of Technology, E25-201, Cambridge, MA 02139*

ABSTRACT

Responses of retinal directionally selective ganglion cells to preferred-direction motions are larger than expected by the sum of responses to stationary stimuli. Experiments on directional selectivity argue against this motion facilitation being based on disinhibition, threshold, voltage-dependent conductance decrease, or depolarizing shunting inhibition. We postulate that motion facilitation is mediated by the synaptic facilitation of an amacrine contact to ganglion cells. Compartmental model simulations of a stained amacrine cell show that this model is feasible. In particular, it accounts for responses to apparent motions with and without antagonists to $GABA_A$ receptors.

24.1 MOTION FACILITATION

Directionally selective (DS) cells respond strongly to motion in a particular (preferred) direction and weakly to motion in the opposite (null) direction[1,2]. This directionality may arise from facilitation of preferred-direction motion signals and/or inhibition of null-direction signals[1,2]. Facilitation is apparent when the response to a two-slit preferred-direction apparent motion is larger than the sum of the responses to the slits presented individually.

24.2 INADEQUACIES OF PREVIOUS MODELS

Our recent electrophysiological data[2] indicate that four models that have been proposed for preferred-direction facilitation (Figure 1) do not account for it. The first model (A; Disinhibition) suggests that facilitation is really just a reduction of inhibition seen during null-direction motion. A second model (B; Threshold-Supralineartity) postulates that activation of early inputs during preferred-direction motion raises the responses closer to or above some threshold onto which latter inputs are added. A third model (C; Voltage Dependent Conductance Decrease) proposes that depolarization early in the motion increases the membrane resistance via a voltage-dependent channel, thus increasing the voltage elicited by synaptic currents produced late in the motion[3]. Finally, a fourth model (D; Depolarizing Shunting Inhibition) postulates that the inhibitory mechanism causes facilitation, since it might work through a slighly depolarizing shunting-inhibition synapse[4].

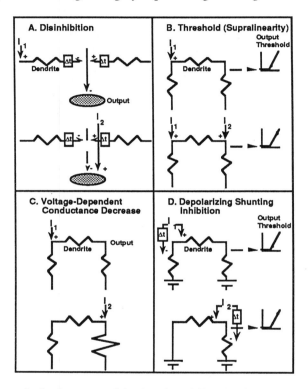

Figure 1: Inadequate models of preferred-direction facilitation.

24.3   SYNAPTIC-FACILITATION MODEL

We propose a model in which preferred-direction facilitation occurs when a sequence of activated inputs propagates towards the distal tip of an amacrine-cell dendrite where its excitatory output synapse is located[5]   The model assumes that the early portions of the motion stimuli produce, through depolarization, an agent (possibly calcium) that can facilitate the next synaptic release of the output synapse[7]. This release is gated by depolarization elicited during the late portions of the motion[8]. (Other mechanisms are possible for both the enhancing agent and gating, leading Grzywacz and Amthor[2] to suggest the class of gated-enhancer models for motion facilitation.)

Another element incorporated in the model is shunting inhibition of the presynaptic terminal, which can account for null-direction inhibition.

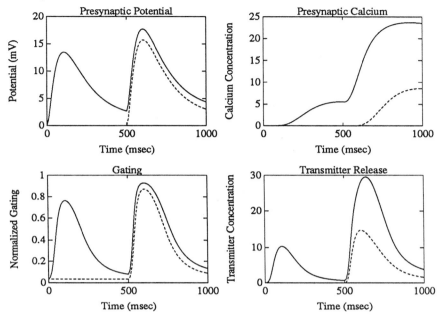

Figure 2: Temporal behavior of a gated-enhancer model based on synaptic facilitation.

The curves in Figure 2 are the results of simulations and calculations with the model. (For this paper, we performed compartmental-model calculations with a starburst amacrine cell recovered by Tauchi and Masland[6] using Lucifer Yellow. The simulator was Surf-Hippo, a new neuron simulator, which is particularly well suited for retinal studies[3].) The stimulus was an apparent motion with the first slit delivered at 0 msec and the second at 500 msec (solid lines), or a slit presented alone in the position of the second slit of the motion at 500 msec (dashed line). The top-left panel shows the presynaptic potential, which does not display facilitation. The top-right panel shows the buildup of calcium in the presynaptic terminal; it is this accumulation after the first slit that causes facilitation in the model. The gating of transmitter release (bottom-left panel) follows the presynaptic potential closely. This gating causes the transmitter release, which is greatly facilitated by motion (bottom-right panel).

24.4 RESULTS: FACILITATORY TIME COURSE

The time course of facilitation was studied by varying the delay between the slits in the apparent motion (Figure 3). The measure of facilitation used was the integral of the response 500 msec after the second slit, with the response to the first slit after the delay between the slits subtracted. The inter-slit distance was 100 μm. Facilitation rises in about 500 msec and falls in about 2 sec in the model (solid line), as in the rabbit-retina as seen electrophysiologically. This time course is much slower than the time course of the response to the single slit (dashed line in Figure 3). In the model, the slow rise and fall times of facilitation are due to the slow accumulation and removal of the enhancer agent (possibly calcium), whereas the fast time course of the single-slit response is due to the gating mechanism (possibly depolarization).

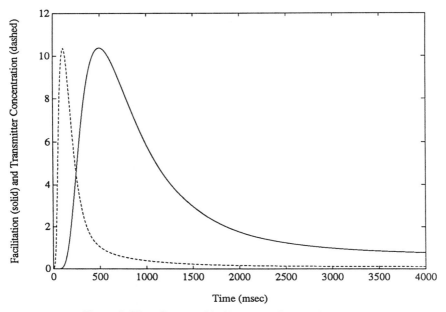

Figure 3: Time Course of facilitation and transmitter release.

24.5   RESULTS: GABA$_A$ ANTAGONISTS

To study facilitation in the absence of inhibition, we removed the inhibition with GABA$_A$ antagonists (in particular, picrotoxin) and recorded the responses to preferred and null-direction apparent motions in the turtle retina[9]. Both the physiology and model (see simulation results in Figure 4) showed that when inhibition is absent, facilitation can appear in null-direction responses. This facilitation can be larger than that occurring in the original preferred direction. In the model, null-direction facilitation can be stronger than preferred-direction facilitation, since during null-direction apparent motion the first slit is closer to the releasing terminal of the amacrine cell than the first slit during preferred-direction motion. Hence, because of the cable properties of the dendrite, the depolarization, and thus calcium accumulation due to the first slit of the null-direction motion can be larger than that of the preferred direction.

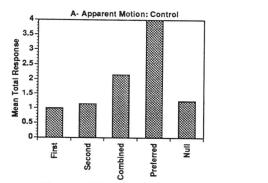
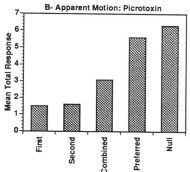

Figure 4: Effects of GABA_A antagonists on motion facilitation and inhibition.

24.6 RESULTS: OTHER SIMULATIONS

Other simulations demonstrated that the model also accounts for the spatial and contrast properties of motion facilitation.

Acknowledgement

This research was supported by a grant to F.R.A. and N.M.G. from ONR (N00014-91-J-1280), by grants to N.M.G. from NEI (EY08921-01A1), NSF (BNS- 8809528), and a NIH core grant to Smith-Kettlewell (EY06883), and by grants to F.R.A. from NEI (EY05070) and Sloan Foundation. L.B.G. was supported by a postdoctoral fellowship from MIT's McDonnell-Pew Center for Cognitive Science.

REFERENCES

1 Barlow, H. B. and Levick, W. R. J. Physiol. Lond. 178: 477-504, 1965.

2 Grzywacz, N. M. and Amthor, F. R. Facilitation in directionally selective ganglion cells of the rabbit retina. J. Neurophysiol. In Press. 1992.

3 Borg-Graham, L. J. and Grzywacz, N. M. In: Single Neuron Computation, edited by T. McKenna, J. Davis, and S. F. Zornetzer. Orlando, Florida, USA: Academic Press, 1992, p. 347-375.

4 Torre, V. and Poggio, T. Proc. R. Soc. Lond. B, 202: 409-416, 1978.

5 Famiglietti, E. V. J. Comp. Neurol. 309: 40-70, 1991.

6 Tauchi, M. and Masland, R. H. Proc. R. Soc. Lond. B. 223: 101-119, 1984.

7 Katz, B. and Miledi, R. J. Physiol. Lond. 195: 481-492, 1968.

8 Hochner, B., Parnas, H., and Parnas, I. Nature, Lond. 342: 433-435, 1989.

9 Smith, R. D., Grzywacz, N. M., and Borg-Graham, L. J. Effects of GABA_A antagonists on preferred motion direction of turtle ganglion cells Submitted for publication. 1992.

25

FAST-MOST INFORMATION EXTRACTION BY RETINAL Y-X CELLS

Zhaoping Li

Rockefeller University
1230 York Avenue, New York, NY 10021
zl@venezia.rockefeller.edu

ABSTRACT

In mammalian retina, the Y (or M) ganglion cells respond more transiently, have larger receptive fields, and are less spectral selective than the X (or P) cells. A "fast-most" model is introduced to show that the differences between the cells can be explained by assigning different functional goals to them. The goal of the Y cells is to extract as *fast* as possible the *minimum* amount of information necessary for quick responses. In contrast, the X cells are to extract as *much* information as possible.

1   MOTIVATIONS FOR DIFFERENT VISUAL GOALS

Earlier works ([8], [1]) proposed that the retinal ganglion cells in cats or monkeys are designed to transmit the maximum information possible. They explained the properties of the retinal X (or P) cells, but do not explain another type of ganglion cells — the Y cells in cats or M cells in monkeys ([3], [9]). (The distinctions between cats and monkeys are left outside the paper). Compared to X cells, Y cells have lower spatial resolution, more transient responses, less spectral sensitivity, and a larger conduction velocity. Unlike X (P) cells, Y (M) cells send axons to Superior Culliculus involved in eye movements.

It is generally believed that while the X channel analyzes the *fine* structures in the visual image, the Y channel does the *initial* analysis of the *gross* structure

[6]. In the present study, such different goals are formulated explicitly — while the X cells extract as *much* visual information as possible; the Y cells are modeled to extract as *fast* as possible, only the *minimum* amount of information needed for quick responses. It will be shown that the X and Y receptive fields are optimal to achieve their respective goals. Briefly, the sustained response by X cells integrates inputs, and thus enhances the signal-to- noise but sacrifices the speed. Therefore, the survival need to quickly locate relevant visual objects for visual feedback to, e.g., motor tasks, can only be served by the Y cells, whose transient response leads to fast reaction but "inaccurate" information. Denoting the X path as the "most" path (referring to extracting as much information as possible), then the Y path is the "fast" path. Roughly, the Y path detects "where", and then the X path recognizes "what", the objects are.

This "fast-most" model suggests that information preservation is not necessarily the goal for every pathway in the sensory systems — even at the early processing stages. *Fast* extraction of *minimum necessary* information is an important goal as well. It is consistent with the shorter latency of the Y cell signals to arrive in the cortex ([5], [2]). At higher visual levels, X-Y paths remain segregated ([10]), feeding the fast information back to the "most" path at lower visual levels can enhance the extraction of "more" information by attention. More details of the work can be found in [7].

2   INFORMATION EXTRACTION VS. IMPULSE RESPONSE

Let the visual input and noise at time t be $S(t)$ and $N(t)$ respectively. The ganglion cell with impulse response $A(t)$ and intrinsic noise $N_\delta(t)$ has outputs $O(t) = \sum_{t'} A(t-t')(S(t')+N(t')) + N_\delta(t)$. (Only linear response properties of the cell are considered for simplicity). Let $I(a;b)$ be the amount of information carried by a about b, we ask how much of the new information brought in by $S(0)$ at time 0 is extracted by outputs $O^t \equiv \{O(t), O(t-1), O(t-2), ..., O(-\infty)\}$ over all time up to $t \geq 0$. The answer is $I_{out}(t) \equiv I(O^t; S^{t'=0}) - I(O^t; S^{t'=-1})$, where S^t is defined analogously to O^t.

By causality, $I_{out}(t < 0) = 0$. It increases with t and saturates as $t \to \infty$. A large $I_{out}(t)$ for small t or a short saturation time for I_{out} are signitures for a fast extraction; while a larger $I_{out}(\infty)$ means more total information extraction. $I_{out}(t)$ depends on $A(t)$. For (a trivial) example, if the output receives inputs with a delay δt, i.e., $A(t \leq \delta t) = 0$, then $I_{out}(t \leq \delta t) = 0$ by causality.

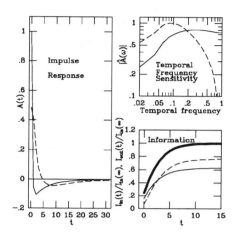

Figure 1 Results of "fast" and "most" cells in thin solid and dashed curves, respectively. Both impulse responses $A(t)$ have short time excitation and long time inhibition, but of different durations. The temporal frequency sensitivity $\hat{A}(\omega)$ is the Fourier transform of the impulse responses $A(t)$. The information plot shows $I_{out}(t)$ for "fast" and "most" cells (thin solid and dashed curves), and in thick solid curve is the input information $I_{in}(t)$, analogous to $I_{out}(t)$.

For their goals, X and Y cells should have impulse responses $A(t)$ to enlarge, respectively, $I_{out}(\infty)$ and $I_{out}(small\ t)$, or, with some output cost constraint C, to minimize, respectively, E^{most} and E^{fast}:

$$E^{most} \equiv C - \lambda^{most} I_{out}(\infty)$$

$$E^{fast} \equiv C - \lambda^{fast} F \equiv C - \lambda^{fast} \sum_t f(t)(I_{out}(t) - I_{out}(t-1))$$

where F is a rough measure of the fast information characterized by $f(t) = e^{-t/T}$ or other monotounously decreasing functions, T approximates the animal reaction time. Constants λ^{fast} and λ^{most} weight the information extraction goals against the output costs.

3   FROM "FAST-MOST" TO Y-X

The impulse responses are obtained by performing a gradient descent of E^{fast} (or E^{most}) on the space of $A^{fast}(t)$ (or $A^{most}(t)$) modelled with finite duration d with $0 \leq t \leq d-1$. (The input signals are approximated as gaussian with correlations $< S(t')S(t'-t) > \equiv R(t) = S^2 e^{-t/\tau}$ and noises are assumed white.) Figure 1 shows that the fast cell has shorter durations in both the excitory and the inhibitory parts of its impulse response, i.e., more transient, and is thus more sensitive to higher temporal frequencies. At about the same output cost, the fast cell is faster ($I_{out}^{fast}(0) > I_{out}^{most}(0)$) while the "most" cell eventually extracts more information ($I_{out}^{fast}(\infty) < I_{out}^{most}(\infty)$). Thus, the transient and sustained responses are required by the "fast" and "most" goals, respectively. This is even more so if one considers X-Y cooperation, when the most path concentrates only on the information missed by the fast one ([7]).

Figure 2 The top and bottom rows are the reconstructions of a retinal input image by the X and Y paths, respectively, in a model retina at time 0 ms, 10 ms, 40 ms, and 400 ms (from left to right, 10 ms per time step) after the image's onset, assuming no synaptic delays. The input layer is presented with spatial-temporal white noises until at time 0 when a natural image is added and stays on, retinal outputs from the X and Y paths are obtained using sustained-small and transient-large receptive fields, respectively, plus output white noises. The signal-to-noise at each input pixel at each time is around 1.5. At t, a homunculus (the brain) at the output reconstructs the inputs of time 0 using only the outputs before and up to time t from the X or Y path. It knew beforehand nothing of the inputs (including the image presentation time and the fact that the image is static) except the receptive field structures and the statistical properties in inputs and noises. A better reconstruction means more information extracted from the inputs. Note the goseling at its mother's lower right.

The impulse responses can be obtained for stimuli (or gratings) of all spatial frequencies, taking into account that signals of higher f have smaller signal powers (decreases as $1/f^2$ [4]) and shorter correlation times. In addition, λ^{fast} decreases with f to reflect the "minimum information" requirement for the "fast" cells. From the results, the center-surround receptive fields can be derived, demonstrating the lower spatial resolution of the "fast" cells. Color-blindness of the "fast" cells can be understood similarly [7].

4   CONCLUSIONS AND DISCUSSIONS

The agreements between the model consequences and the physiological observations supports the "fast-most" goals for Y-X cells — the Y cells extract the *minimum* information needed for responses as *fast* as possible, while the X cells extract as *much* information as possible. Figure 2 demonstrates such goal differences — while the Y path immediately detects objects, it never sees clearly; the X path sees well eventually, although slowly.

The "fast-most" model implies that the Y path is more likely to be concerned with visual attention task, and is likely the source of feedbacks to the X path in the lower visual levels. It stresses that retinal X and Y paths have different goals, and it is an alternative view to that of Van Essen and Anderson [10] who proposed that the retinal X and Y paths are complementary specializations to the same goal of transmitting the most information to the brain.

REFERENCES

[1] Atick J.J., Redlich A. N., Neural Comp. 1990 v. 2 p. 308-320. and v. 4 p. 196-210 (1992). also Atick J.J., Li Z, Redlich A.N. v. 4 p. 559-572 (1992)

[2] Cleland B.G. , Levick W.R, and Sanderson K.J. (1973) J. Physiol. 1973 v. 228, p. 649-680.

[3] Derrington A.M, Lennie P. J. Physiol. 1982 v. 333 p. 343-366, and v. 357 p. 219-240 (1984).

[4] Field D.J. J. Opt. Soc. Am. 1987 v. 4 p. 2379-2394

[5] Ikeda H., Wright M.J. J. Physiol. 1972 v. 227 p. 769-800.

[6] Kandel E.R., Schwartz J. H. "Principles of Neural Science" 1985 Second Ed. Elsevier

[7] Li Z. International Journal of Neural. Systems. 1992 v. 3(3)

[8] Linsker R, Computer 1988 21 v. 3 p. 105-117. and "Advances in Neural Information processing Systems I", ed D. Touretzky p. 186-194.

[9] Shapley R, and Perry V. H. Trends in Neurosci. 1986 v. 9 p. 229-235.

[10] Van Essen D. C, Anderson C. H. In "Introduction to Neural and Electronic Networks" 1990 eds S.F. Zornetzer, J.L. Davis, C. Lau. Academic Press, Orlando, Florida.

26

AN ADAPTIVE NEURAL MODEL OF EARLY VISUAL PROCESSING

Harry G. Barrow
Alistair J. Bray

School of Cognitive and Computing Sciences,
University of Sussex, Brighton BN1 9QH, UK

ABSTRACT

We describe an adaptive computer model of the mammalian visual system, from retina to primary visual cortex. With real world images as input, the basic model robustly develops oriented receptive field patterns, Gabor-function-like fields, and smoothly-varying orientation preference. Extensions of the model show: simultaneous development of oriented receptive fields, retinotopic mapping, and ocular dominance columns; development of "color blobs"; and development of "complex" cells.

26.1 MODELING EARLY VISION

We are developing computer models of processing in the mammalian visual system, including retina, lateral geniculate nucleus (LGN) and primary visual cortex (V1), to investigate the nature and function of neural architecture and circuits. A central hypothesis in this research is that activity-induced adaptation plays an important role in the development of the system, and that in this way generally-applicable cortical circuitry becomes specialised for visual processing. Our models combine the use of detailed neural populations, as in von der Malsburg's early model of visual cortex [1] with the processing of real image data in several stages, as in Barrow's model of receptive field development [2].

In modeling complex neural systems, computational resource limitations inevitably force a trade-off between the number of cells and the level of detail in the simulation. We must reach a compromise which captures key aspects of the behaviour of neurons but permits experiments with very large populations. The compromise we have made has been amply justified by the emergent similarities between the model and the biological system and by the robustness of the results to variation of model details.

26.1.1   The Basic Model

The structure of the basic model is illustrated in Figure 26.1.

Input: To present inputs to our model we first choose randomly from a small set of stored grey-level images of natural scenes. We then choose randomly a fragment (typically 15×15 pixels) from within the image as input to the retina.

Retina: Responses of retinal X-cells are modeled by first convolving the input image with a difference-of-Gaussians kernel. The responses for on-center cells are computed by adding a constant background level and then clipping at zero to remove negative values. Responses for off-center cells are computed by subtracting the kernel output from a background level and clipping at zero.

LGN: The on-center and off-center channels are kept separate in the LGN. For each channel we again convolve with a difference-of-Gaussians kernel and clip at zero to remove negative values.

Cortex: We model processing in layer IV of the cortex with populations of excitatory and inhibitory cells, with population sizes in the ratio of 4:1 (typically using arrays of 50×50 excitatory cells). All cells receive inputs from (15×15) patches of the on- and off- layers of the LGN and from neighboring cortical cells of both types. Inhibition ranges further than excitation, and both have a Gaussian modulation representing connection probability.

Cortical cells are modeled with a standard membrane potential equation: $C dv/dt = (v_+ - v)g_+ + (v_- - v)g_- + (v_r - v)g_r$, where conductances g_+ and g_- depend linearly upon summed excitatory and inhibitory input respectively. Cells fire when membrane potential reaches a threshold value, v_θ, and reset the potential to v_0. The firing rate, $f = 1/(\tau_0 + \tau \log((v_\infty - v_0)/(v_\infty - v_\theta)))$, where $v_\infty = (v_+ g_+ + v_- g_- + v_r g_r)/(g_+ + g_- + g_r)$, $\tau = C/(g_+ + g_- + g_r)$, and τ_0 is an absolute refractory time. This is a non-linear output function with a hard threshold but a continuous response surface.

Input connections from the LGN obey a variant of the Hebbian adaptation rule for synaptic weights, w_{ij}: $dw_{ij}/dt = \alpha y_j(p_i x_i - w_{ij})$, where x_i is the input signal value, y_j the output value and p_i is a constant representing the connection probability. Lateral connections are fixed in strength.

We repeatedly present an image fragment, allow activity to stabilize, and adapt the synaptic weights. Figure 26.2 shows the weight patterns for a patch of cortex after some tens of thousands of presentations. The weight patterns strongly resemble Gabor functions, and their preferred orientations vary smoothly across the cortical array, but with occasional "pinwheel" discontinuities.

26.1.2   Ocular Dominance

In related research, Goodhill found that a similar model provided with inputs from two (uncorrelated) retinal arrays simultaneously developed ocular dominance stripes and retinotopic mapping as well as receptive field patterns [3].

26.1.3   Color Blobs

In a variant of our model, we used natural color images as input and three classes of X-cells in the retina. For the kernel of the first class, both Gaussians were sensitive to broad-band luminance; for the second, the small central Gaussian was red-sensitive and the larger one green-sensitive; for the third, the smaller Gaussian was green-sensitive and the larger one red-sensitive. (In these initial experiments, we ignored the *blue-yellow* channel, which involves only about 6% of ganglion cells.) Outputs to the LGN were on- and off-channels for the *luminance, red-green* and *green-red* classes. In the LGN itself, the 6 channels were processed individually with another difference-of-Gaussians convolution, as in the basic model. The cortical stage received inputs from all 6 channels but was otherwise unchanged.

After adaptation, we found that the model had developed "color blobs", as shown in Figure 26.3. A small patch of the excitatory cortical array is depicted with three weight patterns for each cell, corresponding to the three color classes. As can be seen, we find islands of cells which are unoriented but color-sensitive in a sea of cells which are oriented but color-insensitive [4]. We also find that many blob cells are double-opponent.

26.1.4   Complex Cells

In a natural visual environment with moving images, simple cells with the same orientation but differing retinal locations will be activated in sequence. A mechanism quite like that of classical conditioning would give rise to typical complex cell behaviour: A complex cell which initially responds to a line in one position could learn to respond also to the line in its preceding positions.

To model the adaptation of complex cells we use a variant of a rule investigated by Sutton and Barto in their study of classical conditioning [5]:
$$dw_{ij}/dt = \beta(y_j - \bar{y}_j)(p_i \bar{x}_i - w_{ij}) \quad \text{and} \quad \tau_y d\bar{y}/dt = y - \bar{y}, \quad \tau_x d\bar{x}/dt = x - \bar{x}$$
where $\bar{y}_j$ is a short-term time average, $\bar{x}_i$ is a longer-term time average, an "eligibility trace". Adaptation only occurs if $(y_j - \bar{y}_j) > 0$. Each complex cell necessarily receives inputs from a large number of simple cells. In these experiments, we used an array of 240 × 240 simple cells to provide input to a small group of 8 or 16 competing complex cells (each having 72,000 weights).

We repeatedly swept randomly-oriented long lines across the retina, adapting both the simple and the complex cells. Figure 26.4 shows the resulting synaptic weights for one of the complex cells. The grid-like pattern shows that the cell responds to widely distributed simple cells with one particular orientation. It responds to either long lines with specific orientation but general position, or to a texture with a given element orientation [6].

26.2 CONCLUSIONS

The models we have described incorporate some key characteristics of mammalian visual processing, especially the non-linearities introduced by on- and off- retinal channels, excitatory and inhibitory cells, combination of excitation and inhibition via membrane potentials, and repetitive firing. They develop many of the features of mammalian vision, including Gabor-function-like receptive fields, smooth variation of orientation across the cortex, ocular dominance columns, retinotopic mapping, and color blobs. Remarkably, the same basic mechanism appears to underlie all of these phenomena.

Acknowledgement

We gratefully acknowledge that this work has been supported by a grant from the UK Science and Engineering Research Council and the Ministry of Defence.

REFERENCES

[1] C. von der Malsburg. Self-organisation of orientation sensitive cells in the striata cortex, *Kybernetik*, 14:85–100, 1973.

[2] H. G. Barrow. Learning receptive fields. In *IEEE First International Conference on Neural Networks*. IEEE, 1987.

[3] G. J. Goodhill. *Correlations, Competition and Optimality: Modelling the Development of Topography and Ocular Dominance*. PhD thesis, School of Cognitive and Computing Sciences, University of Sussex, 1992.

[4] H. G. Barrow and A. J. Bray. Activity-induced "color blob" formation. In I. Aleksander and J. Taylor, editors, *Artificial Neural Networks, 2*. North-Holland, 1992.

[5] R. S. Sutton and A. G. Barto. Time-derivative models of Pavlovian reinforcement. In J. W. Moore and M. Gabriel, editors, *Learning and Computational Neuroscience*. MIT Press, 1989.

[6] H. G. Barrow and A. J. Bray. A model of adaptive development of complex cortical cells. In I. Aleksander and J. Taylor, editors, *Artificial Neural Networks, 2*. North-Holland, 1992.

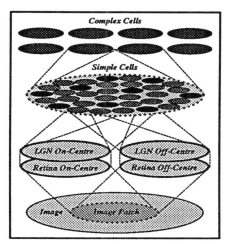

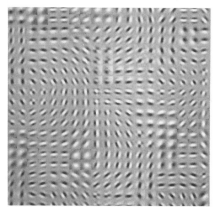

Figure 26.1 The basic model.
The on- and off-center channels from the retina are separately processed in the LGN. The cortex has excitatory and inhibitory cell populations, each of which receives inputs from the LGN and from neighboring cortical cells.

Figure 26.2 Simple cell weights.
A patch of model cortex showing oriented synaptic weight patterns resembling Gabor functions. Their preferred orientation varies smoothly across the cortex. (Weights for the off- channel have been subtracted from those of the on-channel.)

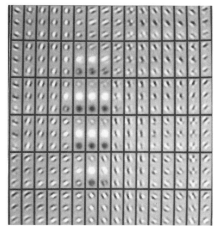

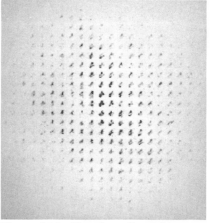

Figure 26.3 A "Color Blob".
A patch of cortex showing weight patterns for 3 color channels: *(from top)* broadband, red-green, green-red. (Weights for each off- color channel have been subtracted from those of the corresponding on-channel, as in Figure 2.)

Figure 264 Complex cell weights.
Weights for one cell showing connections to 240 × 240 simple cells. The regular pattern indicates connections to simple cells with similar orientations over a wide area. The complex cell responds to a long line in any location or to a textured field.

A LOCAL MODEL FOR TRANSPARENT MOTIONS BASED ON SPATIO-TEMPORAL FILTERS

James A. Smith and Norberto M. Grzywacz

Center For Biological Information Processing,
E25-201 MIT, Cambridge, MA 02139, and
Smith-Kettlewell Institute, 2232 Webster St. San Francisco, CA 94115.

ABSTRACT

Psychophysical studies suggest that humans perceive moving multiple sine patterns as transparent when the underlying component gratings are of sufficiently different speed, spatial frequency, orientation, and contrast. We propose a gradient inhibition model that explains these phenomena using only local intensity or contrast information. Computational results show a contrast dependent transition for splitting/coherence of sine plaid images. This model suggests that some transparency information can be locally computed and subsequently fed into global vision modules.

The detection of multiple motions from an image sequence of transparent objects is computationally difficult. One has to decide when to combine motion information and when to process the information separately. Consider images such as sine plaids, which can be defined as intensity fields I:

$$I(t, \vec{x}) = I_1 \sin \left(\vec{\omega}_{\vec{x}1} \cdot \vec{x} + \omega_{t1}t \right) + I_2 \sin \left(\vec{\omega}_{\vec{x}2} \cdot \vec{x} + \omega_{t2}t \right)$$
$$= I_1 \sin \left(\vec{\omega}_{\vec{x}1} \cdot (\vec{x} - \vec{v}_1 t) \right) + I_2 \sin \left(\vec{\omega}_{\vec{x}2} \cdot (\vec{x} - \vec{v}_2 t) \right),$$

where t and $\vec{x}$ stand for temporal and spatial locations, and $\vec{\omega}_x, \omega_t$, and $\vec{v}$ stand for the sine waves' spatial frequencies, temporal frequency, and velocity, respectively. These images can validly be interpreted as due to either two distinct overlapping objects or a single coherent object (Figure 1). Psychophysical studies have shown that sine plaids tend to cohere when their components have sufficiently similar orientations, speeds, spatial and temporal frequencies, and contrasts [E. Adelson and J. Movshon, *Nature, Lond.* **300** 1982, p.523; L. Welch, *Ph.D. thesis, Berkeley* 1990, p.80]. In this paper we ask the question of whether such spatially homogeneous phenomena can be explained by a spatially local mechanism. By local, we mean an area on the order of the size of a receptive field in primary visual cortex. We propose a gradient inhibition (GI) model based on spatio-temporal (motion energy) filters which satisfies this requirement.

Our model is based on previous work by N. Grzywacz and A. Yuille [*Proc. Roy. Soc. Lond.* **B 239** 1990, p.129]. Physiological studies report that cells

Figure 1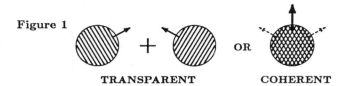

in primary visual cortex are tuned to specific spatial and temporal frequencies. Their responses can be approximated by the convolution of the image with spatio-temporally tuned filters. Grzywacz and Yuille modeled this tuning with a set of units

$$A(\Omega_t, \Omega_r, \theta_\Omega) = \|I * G(\Omega_t, \Omega_r, \theta_\Omega)\|^2 \,, \tag{1}$$

where I is the image, $*$ stands for convolution, and G is a Gabor filter

$$G\left(\Omega_t, \Omega_r, \theta_\Omega; \vec{x}, t; \sigma_r, \sigma_t\right) = \exp\left[\frac{-|\vec{x}|^2}{2\sigma_r^2} + \frac{-t^2}{2\sigma_t^2} + i\left(\vec{\Omega}_x \cdot \vec{x} + \Omega_t t\right)\right]. \tag{2}$$

The parameters $\Omega_t, \Omega_r, \theta_\Omega$ indicate the filter's spatial frequency, temporal frequency, and orientation tuning, with $\vec{\Omega}_x = (\Omega_x, \Omega_y) = (\Omega_r \cos\theta_\Omega, \Omega_r \sin\theta_\Omega)$. Motion information is consequently represented as variations in the 3D space of responses $A(\Omega_t, \Omega_r, \theta_\Omega)$. Grzywacz and Yuille subsequently proved that if the image is that of a single translating object $I(t, x, y) = I(\vec{x} - \vec{v}t)$, then the distribution of $A(\Omega_t, \Omega_r, \theta_\Omega)$ will be unimodal and have local maxima for those filters whose parameters satisfy the planar equation

$$\vec{\Omega}_x \cdot \vec{v} + \Omega_t = 0. \tag{3}$$

They have thus restated the problem of finding the correct velocity as the computation of a best fit plane.

We can extend this model by first showing that for an image I_T of two transparent moving objects $I_T = I_1(\vec{x} - \vec{v}_1 t) + I_2(\vec{x} - \vec{v}_2 t)$, the total response A_T is

$$A_T(\Omega_t, \Omega_r, \theta_\Omega) = \|I_T * G(\Omega_t, \Omega_r, \theta_\Omega)\|^2 \approx A_1(\Omega_t, \Omega_r, \theta_\Omega) + A_2(\Omega_t, \Omega_r, \theta_\Omega) \tag{4}$$

when the underlying images and motions $\vec{v}_1$ and $\vec{v}_2$ are sufficiently different. Under such conditions, the response space will be bimodally distributed with peaks for Ω's that satisfy either

$$\vec{\Omega}_x \cdot \vec{v}_1 + \Omega_t = 0 \qquad \text{or} \qquad \vec{\Omega}_x \cdot \vec{v}_2 + \Omega_t = 0.$$

We note that the magnitude of each local maximum, and hence the magnitude of the falloff (gradient) of the response surrounding each maximum, is proportional to the intensity or contrast of each gradient. In addition, we make the following assumption: areas of very large gradients in response space partition the space into regions that get responses from different objects. This assumption is the basis of our GI model.

In Figure 2 below, we describe the model with schematic drawings for two moving plaid images. The component gratings of one plaid (left) have identical contrasts, while the other (right) has components whose contrasts are different. A version of the GI model can be approximately outlined in 5 steps:

1. Compute the response of the spatio-temporal units $A(\Omega_t, \Omega_r, \theta_\Omega)$. The set of responses $A(\Omega_t, \Omega_r, \theta_\Omega)$ for such convolutions with sinusoidal plaid images is depicted in Fig. 2A. This picture represents the responses for a 1-d slice in the three dimensional $A(\Omega_t, \Omega_r, \theta_\Omega)$-space in the θ_Ω direction that passes through the two local maxima. Note that the two peaks in the left panel (equal contrast) have similar heights, while those in the right panel (different contrast) have different heights.

2. Normalize the convolution responses (Fig. 2B). Compute

$$N(\Omega_t, \Omega_r, \theta_\Omega) = \sum_{\vec{\Omega}' \epsilon \{ \text{ local neighborhood of } \vec{\Omega} \}} A(\Omega_t', \Omega_r', \theta_\Omega'),$$

and then let $B(\vec{\Omega}) = A(\vec{\Omega})/N$ ($\vec{\Omega}$ is shorthand for $(\Omega_t, \Omega_r, \theta_\Omega)$). This normalization step makes the model invariant to intensity or contrast scaling. Note in Fig. 2B that the height of the larger peak in the right panel (corresponding to the normalized response from the high contrast grating) is higher than either of the peaks on the left.

3. Identify locations in $B(\Omega_t, \Omega_r, \theta_\Omega)$-space where the gradients are above a predetermined threshold Θ (Fig. 2c). We choose Θ so that the maximum gradients for equal contrast plaids are below threshold, while those for plaids with components of significantly different contrast are above threshold. We use these large gradient regions to partition the normalized response space into sub-regions.

4. Perform a weighted summation of the responses in the low gradient regions. Within each such region, compute

$$D(\vec{\Omega}) = \sum_{\vec{\Omega}' \in \{ \text{ local low gradient region } \}} B(\vec{\Omega}')w(\vec{\Omega}', \vec{\Omega}).$$

We construct the weighting factor $w(\vec{\Omega}', \vec{\Omega})$ so that textured images will have maximal response in the intersection of constraints (IOC) velocity. Fig. 2D depicts these summations in the two component directions (top) and the IOC direction (bottom). The IOC summation for the symmetric plaid is larger than that of the components since its summation region

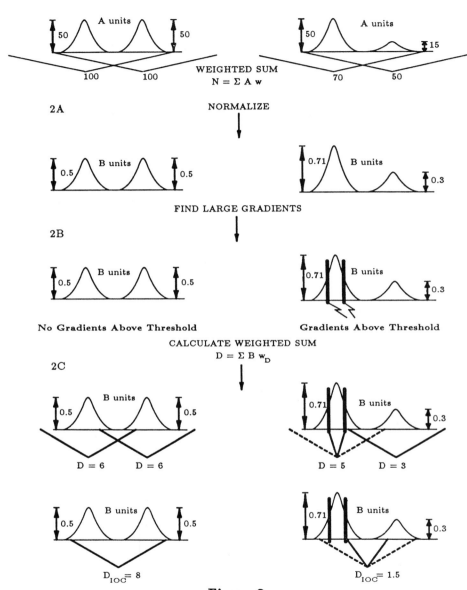

Figure 2

includes responses from both peaks. In the example of different contrasts, though, the IOC summation is much smaller; the summation region does not cross the large gradient regions (solid vertical lines). The response in the IOC direction is consequently much lower than those of the component directions.

5. Perform local winner-take-all computations on the candidate velocity $D(\vec{\Omega})$ units computed in Step 4. From this set of values $\{(\Omega_t, \Omega_r, \theta_\Omega)_i\}$ we get our predicted velocities $\{(v, \theta)_i\} = \left\{(\frac{\Omega_{t_i}}{\Omega_{r_i}}, \theta_{\Omega_i})\right\}$.

We implemented a version of the GI model and tested it on a variety of sinusoidal plaid patterns. Figure 3 shows the results of these experiments for three different sets of plaids. In all of the experiments we presented plaids for which each component grating moved with an angle $\pm\theta$ with respect to the intersection of constraints direction. In each experiment we examined the models predicted motions as we varied the angle θ and the relative contrasts of the two components. The spatial frequencies of the gratings were held fixed within each set, but were varied between the three sets. In each figure, the circles and crosses represent patterns for which the model predicted transparent and coherent motion, respectively. These data qualitatively agree with the literature in that transparency is more likely to be perceived for plaids when the component gratings have differing contrasts and orientations. For high contrast ratios and small angles θ, the model predicts that patterns will be perceived as single motions biased in the direction of the higher contrast grating (triangles in the figure).

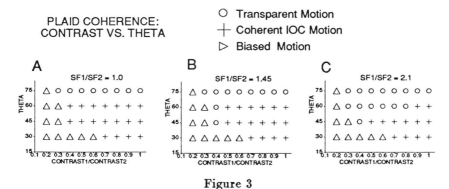

Figure 3

The GI model suggests that motion transparency information can be gleaned from local measurements. It is a first step that can ultimately be fed into optical flow and other global vision modules.

N.M.G. was supported by a grant from AFOSR (92-005564) and a core grant to Smith-Kettlewell from NIH (EY06883). J.A.S. was supported by a grant from the Johnson & Johnson Foundation Research Fund.

A NEURAL MODEL OF ILLUSORY CONTOUR PERCEPTION

Brian Ringer and Josef Skrzypek

Machine Perception Laboratory, Computer Science Department, University of California, Los Angeles, California 90024

ABSTRACT

Illusory contours result from occlusion by a surface whose border is not defined by a continuous discontinuity in any image attribute. Available explanations assume that contrast information on both sides of a gap is sufficient to complete an illusory contour. Using a computer simulation we evaluated this hypothesis by making explicit physiological and psychophysical constraints. Our results indicate that a solution to illusory contour perception may lie with a recurrent network which integrates information about visual surfaces with available contour information

Introduction

Illusory contours are boundaries of perceptually occluding surfaces that can be seen across a physically homogeneous region. Although the phenomenon has been studied extensively, the neural structures which give rise to illusory contours remain a matter of controversy. Physiological evidence from single cell recordings in rhesus monkeys suggest that some aspects of illusory contours are detected preattentively at relatively early stages of visual processing [8]. These results led to a hypothesized model in which illusory contours are completed through integration of responses from hypercomplex cells signaling line terminations and corners [4]. However, simulation results [7] showed that this model, like others relying strictly on local contrast information [5, 2], produced consistent completion of illusory contours between coincidently aligned line terminators, which contradicts human psychophysical results.

Our model of illusory contour perception uses surface occlusion and depth to disambiguate potential boundary completions. Occluding surfaces are reconstructed (filled-in) from image elments which best define the presence of a foreground object (corners, points of curvature and contours). Our model can account for 1) Oriented contour perception, 2) Perceptual depth effects of illusory surfaces, and 3) Increased brightness of illusory surfaces.

Computational Model

Our network consists of multiple layers of neuron-like nodes (cells), who's output is a positive real number correlated with average firing frequency. Early layers of the network are computational models of simple, complex and hypercomplex cells which filter out oriented "features" (edges, line terminations, etc.) to be aggregated at later stages. The desired cell property and sensitivity is controlled by the strength and topology of the connections within the cell's receptive field.

Our model is based on a hypothetical neural circuit located in area V2 [4], involving two sets of striate inputs, one sensitive to luminance edges, and the other sensitive to illusory contours. Both inputs converge on binocular cells which we call "General Contour Neurons" (GCNs). Line terminations are explicitly grouped to yield relative depth information. Both illusory curvatures as well as straight segments are detected to better define illusory surfaces. Recurrent excitation from layers of "surface neurons" is used to modulate the GCN response to line terminations (see Fig. 1). Each part of the model (feedforward, recurrent and GCN) is described below.

Our model of edge detection (simple cells) is a difference of offset Gaussian (DOOG) function [9]. Line terminations [1] are detected using a difference of responses from two simple cells of the same orientation tuning and different spatial extents, with a slight offset between the two receptive field centers to generate asymmetrical end-stopping inhibition. At each spatial position, feature responses are normalized by those of similar features at all other orientations (ie. edges compete with other edge responses across all orientations).

Illusory contours are completed by General Contour Neurons only where line terminations are present. These terminations are grouped to yield information about the relative position of the occluding illusory surface. Illusory surfaces usually overlay a discontinuous background structure; knowing the direction of

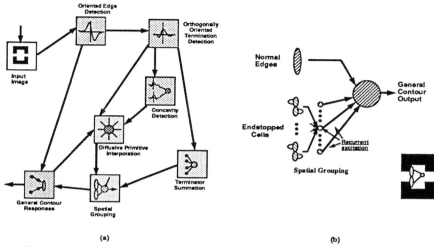

Figure 1 (a) The network architecture. Inputs to the General Contour Neurons are filtered by early layers of the network. Terminator outputs feed into a spatial grouping layer. General Contour responses are a combination of luminance edges and these spatial groupings. (b) The General Contour Neuron. Recurrent excitation from surface neurons modulates the response of the cell to line terminator inputs.

the line terminations (from hypercomplex cells with asymmetrical end inhibition) provides foreground/background data. To produce a response, a GCN requires evidence about line terminators on each side of the receptive field center, covering both sides of a perceptual gap. The neighborhood size used in our model over which to integrate line terminator information was between 1.0 and 2.5 visual degrees, in correspondence with psychophysical measurements of the maximum size gap which can be perceptually crossed by an illusory contour [5]. In addition to GCNs tuned to straight illusory contours as described above, our model includes a mechanism for the detection of illusory concavities (corners).

The second component controlling the activities of the GCNs is recurrent excitation derived from estimates about visual surfaces present in a scene, as signaled by a layer of surface neurons. These surface neurons are driven by contour features which best define foreground surfaces (corners, line terminations and contours). Each surface neuron feeds activity to its eight neighbors according to a simple diffusion equation [6]. Contour features decrease the interaction between neighboring cells, providing a graded shunting of the diffusion process [2].

Surface neuron activity levels determine the recurrent excitation fed to the GCNs. Facilitation is provided when this response is high at the border of a proposed (through line termination evidence) illusory contour. The response of the illusory contour component of the General Contour Neuron is a non-linear function of the strength of the termination responses lying within its receptive field; surface neuron activities shift this response profile curve. Where the surface gradient is strong, it is possible for only limited terminator responses to fire the cell; where it is weak or absent, very strong terminator responses are required to produce a response. Further, the surface activity gradient must be compatible with the depth gradient of the proposed illusory contour completion. The final layer of General Contour Neurons signals all contours, real and illusory detected at each orientation and spatial position (see Fig. 1).

Simulation Results

Simulations presented in this paper were performed using the UCLA-SFINX network simulator [3], which allows the construction and simulation of large scale fixed or variable connection networks, with graphics based methods for presenting input image patterns to the network and displaying results. Input patterns were 128x128 pixel grey level (256 grey levels) images, representing an area of 16 square visual degrees.

Our model is able to detect all luminance based and illusory contours in a variety of well-known illusory contour patterns. Additionally, the network produced results which compare well with human psychophysics in cases involving potentially ambiguous completions. Simulation outputs for a variety of illusory contour stimuli are shown in Fig. 2. The network provides good completions of contours bounding illusory surfaces, as well detecting the boundary between an abutting grating.

REFERENCES

[1] A. Dobbins, S. W. Zucker, and M. S. Cynader. *Nature*, 329(6138):438–441, October 1987.

[2] S. Grossberg and E. Mingolla. *Psychological Review*, 92(2):173–211, April 1985.

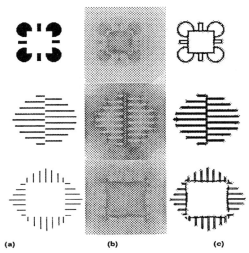

Figure 2 Simulation results. (a) Shows a series of illusory contour input images. (b) The final steady state surface neuron activity levels. (c) The responses of all layers of General Contour Neurons at steady state. Higher response levels are indicated by darker areas.

[3] E. Mesrobian and J. Skrzypek. *IEEE Transactions on Software Engineering*, 18(7):575–589, July 1992.

[4] E. Peterhans, R. von der Heydt, and G. Baumgartner. In *Visual Neuroscience*, pages 343–351, Cambridge, 1986. Cambridge University Press.

[5] R. Shapley and J. Gordon. In *The Perception of Illusory Contours*, pages 109–115. Springer-Verlag, 1987.

[6] J. Shrager, T. Hogg, and B. A. Huberman. *Science*, 236:1092–1094, 1987.

[7] J. Skrzypek and B. Ringer. In O. M. Omidvar, editor, *Progress in Neural Networks*, volume IV. Ablex Publishing Corporation, 1992. in press.

[8] R. von der Heydt, E. Peterhans, and G. Baumgartner. *Science*, 224(4645):1260–1262, June 1984.

[9] R. A. Young. Technical Report GMR-5323, General Motors Research Laboratories, 1986.

ADAPTIVE RECEPTIVE FIELDS FOR TEXTURAL SEGMENTATION

Edmond Mesrobian and Josef Skrzypek

Machine Perception Laboratory, Computer Science Department,
University of California, Los Angeles, California 90024

ABSTRACT

Textural segmentation plays an important role in the figure-ground discrimination process. Evidence from neuroscience and psychophysics suggests that the segregation of texture patterns composed of oriented line segments is strongly influenced by the orientation contrast between the patterns [13]. We describe a neural network architecture for textural segmentation that can adaptively delimit the boundaries of uniformly textured regions. Simulation results are presented to demonstrate the segmentation capablities of the architecture.

1 INTRODUCTION

The human visual system can easily perform textural segmentation, discriminating objects from a background using local information such as figural "features" and global information such as luminance. Evidence from neurophysiology and psychophysics suggests that the segregation of texture patterns is strongly influenced by the orientation contrast between the patterns [13, 9].

Many computational models of textural segmentation are based on localized spatial frequency processing mechanisms modeled after visual cortical cells [7, 6]. Our model is a 3D neural architecture, faithful to functional neuroanatomy of the vertebrate vision, that can adaptively delimit the boundaries of an uniformly textured region. Cells in the early layers of the model, with odd-symmetric (OS) and even-symmetric (ES) receptive fields (RF) are sensitive to contrast changes at certain preset orientations, widths, and sizes. The outputs of these cells are normalized and integrated to produce a local measure of orientation energy. The local orientation energy estimates can then be grouped into uniformly textured regions [1] using Gestalt [14] rules, implemented as neurons with adaptive RF's (ARF's). The basic notion is that the cell's RF can be dynamically modulated according to "rules" that capture intrinsic properties of a given texture. Using local and global "feature" variance estimates, ARF's encode uniformly textured regions (surfaces) by interpolating orientation energy estimates across the image; the interpolation is modeled as a diffusive process. Orientation contrast boundaries are detected at gradients across the interpolated regions to produce the final segmentation. The results suggest that variance-like measures computed over RF sizes on the order of 2-3 times the fundamental frequency of the feature can provide valuable contextual information for modulating RFs of cells engaged in encoding uniformly textured regions. This is consistent with the psychophysical results which suggest that the aggregation of textural features is limited by a critical distance between textural elements [10].

2 COMPUTATIONAL MODEL

The computational model consists of three major components (Fig. 1): local "feature" analysis network, a region encoding network, and a global boundary discrimination network. Interactions between these networks results in segmentation of the textured image. The first network extracts a primitive set of contrast-normalized textural elements using ON and OFF simple cells, sensitive to spots, edges and slits of light. Cells sensitive to bright and dark spots are modeled as RFs with concentric, center-surround antagonistic zones [5]. RF's encoding steps or slits of light are modeled after OS and ES simple cells [4], using difference of offset Gaussians (DOOG) [15]. OS (ES) RFs are constructed by the linear combination of two (three) offset Gaussians. Cell responses are half-wave rectified and passed through a non-linearity. All cells produce graded responses which are contrast normalized using interactions modeled after retinal mechanisms of light adaptation and sensitivity control [11]; a spatial frequency selective cell is normalized by the collective responses of cells tuned to different orientations and neighboring spatial frequencies (an octave apart). A complex cell, signaling the concept of orientation without strict reference to position or feature contrast polarity [4], provides a measure of orientation energy, and computes an average of contrast normalized simple cell responses (i.e., the same orientation, but different phase polarities). Layers of region en-

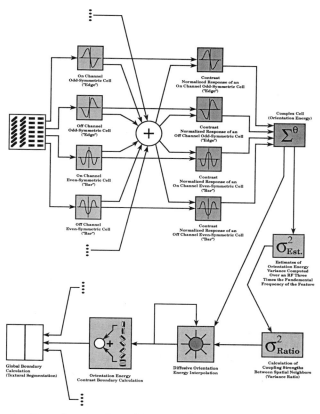

Figure 1 Neural network architecture for textural segmentation.

coding (RE) cells construct perceptual surfaces of uniform orientation energy using Gestalt rules of similarity and proximity. The surfaces are "aggregated" by "diffusively" interpolating local orientation energy estimates (complex cell output) across the uniform region. The RF's of RE cells are adaptively modulated by inter-neurons which calculate variance ratios across image locations, thereby providing a gating mechanism to control the diffusive process. Our use of only agonistic interactions between RE cells is motivated by the desire to enhance continuity of uniformly textured regions.

The final stage of our architecture detects the local orientation contrast (OC) boundaries around an uniformly textured region and then determines the global boundaries between differently textured regions. Significant orientation contrast boundaries are detected by cells with odd-symmetric RFs tuned to different orientations and phases. A local OC cell pools together local estimates of orientation contrast given by RF's tuned to a range of orientations and contrast steps. The local estimates of OC boundaries are then processed by a layer of cells which signal the "global" boundaries. A global boundary (GB) cell simply computes a point-wise average response across a the OC cell responses.

3 SIMULATION RESULTS

We have simulated the neural architecture for textural segmentation on an IBM RS 6000 workstation using SFINX, a neural network simulator developed at UCLA [8]. We used 128×128 grey scale images of natural [2] and artificial textures. All simulation results presented in this paper are displayed as grey scale images with *black* representing a zero response and *white* representing a maximal response. While the network operates over a range of scales, we present simulation results using "features" processed only at one scale.

Humans are capable of preattentively segregating the psychophysical texture in Fig. 2 into two parts: a target texture pattern (line elements at 45°) and a background pattern (line elements at 135°). For the sake of brevity, we have omitted the responses from cells at various stages of the architecture identified with unique icons in Fig. 1 (but see [12] for details). Consistent with neurophysiological data from simple and complex cells sensitive to texture patterns [3], our model explains the preattentive textural segregation localized to areas V1, V2, and perhaps V3 or V5 using ARF's to capture context-dependent response characteristics [3]. ARF-like dynamics can occur at many processing stages including contrast normalization and orientation contrast detection. Consequently, our model postulates the existence of region encoding cells that encode the uniformity of a region (surface) based on inputs from complex cells.

REFERENCES

[1] J. Beck, K. Prazdny, and A. Rosenfeld. In J. Beck and B. Hope, editors, *Human and Machine Vision*, pages 1–38. Academic Press, New York, 1983.

[2] P. Brodatz. *Textures, A Photographic Album for Artists and Designers.* Dover Publications Inc., New York, 1966.

[3] C. D. Gilbert and T. N. Wiesel. *Nature*, 356:150–52, March 1989.

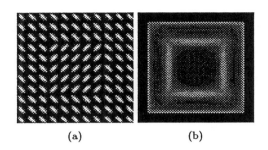

(a)   (b)

Figure 2 Textural segmentation example: (a) Texture image; (b) Segmentation result.

[4] D. H. Hubel and T. N. Wiesel. *Journal of Physiology (London)*, 160:106–154, 1962.

[5] S. W. Kuffler. *J. Neurophysiol.*, 16:37–68, 1953.

[6] M. Landy and J. Bergen. *Vision Research*, 31(4):679–691, 1991.

[7] J. Malik and P. Perona. *Journal Optical Society America A*, 7(5):923–932, 1990.

[8] E. Mesrobian and J. Skrzypek. *IEEE Transactions on Software Engineering*, 18(7):575–589, July 1992.

[9] H. C. Nothdurft. *Vision Research*, 31(6):1073–1078, 1991.

[10] D. Sagi and B. Julesz. *Spatial Vision*, 2:39–49, 1987.

[11] J. Skrzypek. In D. Touretzky, editor, *Advances in Neural Information Processing Systems, NIPS-3*, pages 391–396, San Mateo, CA, 1991. Morgan Kaufman.

[12] J. Skrzypek and E. Mesrobian. *Spatial Vision*, 1992. submitted.

[13] D. C. Van Essen, E. A. DeYoe, J. F. Olavarria, J. J. Knierim, J. M. Fox, D. Sagi, and B. Julesz. In D. M. K. Lam and C. Gilbert, eds., *Neural Mechanisms of Visual Perception*, Chapter 6, pages 137–154. Portfolio Publishing, Woodlands, Texas, 1989.

[14] M. Wertheimer. *Psychologische Forschung*, 4:301–350, 1923. Translation in A Source Book of Gestalt Psychology, W. D. Ellis, ed., New York: Harcourt, Brace, 1938.

[15] R. A. Young. Technical Report GMR-5323, General Motors Research Laboratories, 1986.

CORTICAL MECHANISMS FOR SURFACE SEGMENTATION

Paul Sajda and Leif H. Finkel

*Department of Bioengineering, University of Pennsylvania,
Philadelphia, Pennsylvania 19104-6392*

30.1 INTRODUCTION

Physiology has shown that the neural machinery of "early vision" is well suited for extracting edges and determining orientation of contours in the visual field. However, when looking at objects in a scene our perception is not dominated by edges and contours but rather by *surfaces*. Previous models have attributed surface segmentation to filling-in processes, typically based on diffusion. Though diffusion related mechanisms may be important for perceptual filling-in [4], it is unclear how such mechanisms would discriminate multiple, overlapping surfaces, as might result from occlusion or transparency. For the case of occlusion, surfaces exist on either side of a boundary and the problem is not to fill-in the surfaces but to determine which surface "owns" the boundary [1][3]. This problem of boundary "ownership" can also be considered a special case of the binding problem, with a surface being "bound" to a contour.

We propose a model of intermediate-level visual processes responsible for surface segmentation. The basic function of the model is to bind contours to surfaces by determining the *direction of figure* for points along a contour. We present computer simulations showing how these cortical processes, as part of a larger model of depth-from-occlusion [1], discriminate multiple and/or occluding surfaces. Finally, we test the model with ambiguous and "illusory" surfaces, and show that the model behaves in a manner consistent with human perception.

30.2 DIRECTION OF FIGURE

The "direction of figure" (DOF) is defined at each point of a contour as the

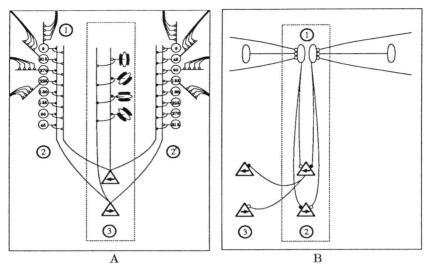

Figure 30.1   Two mechanisms for computing the direction of figure, **A**
DOF-stellate and **B** DOF-bipolar.

direction towards that surface which "belongs" to the contour. The DOF of
a contour segment is represented as the distributed activity of units within a
"column" (we have used columns consisting of units tuned to eight different
directions). The neural circuitry within a DOF column can be divided into
two mechanisms; DOF-stellate and DOF-bipolar. Since long-range horizontal
connections [2] play a critical role, the names of the two mechanisms have been
chosen to reflect the pattern of these connections. Figure 1 is a simplified
schematic of the two circuits (not all units are shown). The dashed rectangles
represent units within a single DOF column, assumed here to be operating on
a vertical segment of contour. All units have an associated phase (Φ); points
belonging to the same contour have the same phase [1]. In discussing how the
two mechanisms work, we assume all input to the column originates from units
responding to other segments of contour in the visual scene.

DOF-stellate

For figures having object contours which are not excessively convoluted, the
DOF-stellate mechanism determines direction of figure by comparing the number of intersections made by two sets of rays (represented in the model by
integration of input over long-range horizontal connections). Each set of rays
emanates in all directions, forming a "stellate" pattern, and the two sets have
origins on either side of the contour. The side of the contour having the set of

rays with the greater number of intersections represents the direction of figure. This mechanism involves the following processes (refer to figure 1A);

1. Units at (1) receive input (originating from a particular direction), via long-range horizontal connections, and are excited by suprathreshold input with phase Φ.

2. A population of type (1) units, receiving inputs spanning N different directions, creates a "stellate" pattern of connectivity. This population functions as a context-dependent temporal filter, where population activity depends upon the number of directions from which input, having a phase Φ, is received.

3. A pair of DOF units at (3), representing directions orthogonal to the contour, are "enabled" via orientation selective units. Units at (3) compute the difference between population activity at (2) and (2)'. The unit having a net positive input represents the local DOF.

DOF-bipolar

The DOF-bipolar mechanism assumes that parallel segments of contour, located in a direction orthogonal to their orientation, will tend to have DOFs toward one another. In addition, the closer the parallel contours are to one another the greater their influence. The mechanism involves the following processes(refer to figure 1B);

1. A pair of orientation selective units at (1) receives input from units having the same orientation preference via long-range horizontal connections running orthogonal to the units' orientation tuning. The connection strength between (1) and its inputs is a function of distance. These units are only excited by suprathreshold input with phase Φ.

2. The "enabled" pair of DOF units computes the difference between the activity of units at (1). The unit having a net positive input represents the local DOF at (2).

3. Units at (3) receive excitatory input from DOF units at (2) having the opposite DOF tuning. The unit at (3) which is most excited represents the local DOF at (3).

30.3   SIMULATIONS

The model was built and simulated using the NEXUS neural simulation environment, developed at the University of Pennsylvania. Figure 2A is an image taken from a CCD camera and input into the model. Figure 2B is an enlarged view of DOF activity near the cup/pen boundary. The DOF is shown as an oriented arrowhead, where the orientation represents the DOF tuning of the unit most activated in the column. The vertical contour has a DOF which is toward the surface of the cup, indicating that surface is bound to the contour. This surface binding is then utilized by other processes in our model of depth-from-occlusion to determine relative depth, which is shown in figure 2C.

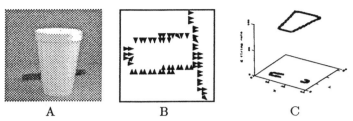

Figure 30.2 **A** Input image, **B** direction of figure near cup/pen boundary, **C** relative depth computed using surface segmentation.

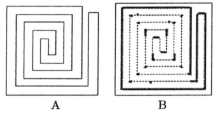

Figure 30.3 **A** Figure which human subjects are unable to discriminate the surface for brief stimulus presentations. **B** Simulations showing the model also has difficulty discriminating the surface of the spiral.

Figure 3A is an example of a stimulus in which human subjects report difficulty segmenting the surface for brief stimulus presentations. Figure 3B shows simulation results. The dashed segments indicate points at which 1) no DOF unit in a column is above threshold or 2) the DOF activity in a column oscillates between two units. Though the model can discriminate the surface near the perimeter of the figure, it has difficulty determining the surface of the spiral for the interior points. This is in contrast to surface segmentation models based on diffusion. Such models check for global consistency by utilizing boundaries to constrain the filling-in process. However, the human visual system does not appear to check for global consistency and therefore these models do not accurately reflect the cortical processes responsible for surface segmentation.

Finally, our model also responds to illusory surfaces (figure 4A). Figure 4B shows DOF activity near the L-shaped section of a pac-man inducer. After the first cycle the L-shaped contour is bound to the surface of the pac-man. However, after the second cycle, an illusory contour has been generated, resulting in a change in ownership of the contour–the contour is now owned by the surface of the illusory square. Our model predicts that the binding of contours to surfaces, in particular a change in the binding, plays a crucial role in the perception of subjective contours.

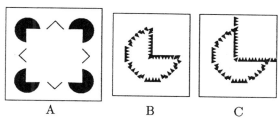

A   B   C

Figure 30.4 A Stimulus generating an illusory square. **B** (cycle 1) L-shaped segment bound to surface of the pac-man. **C** (cycle 2) L-shaped segment bound to surface of the illusory square.

30.4 SUMMARY

We have presented a model of cortical processes for surface segmentation. The model utilizes long-range horizontal connections and a columnar structure to bind contours to surfaces by determining the local direction of figure. The system can 1) segment multiple and/or occluding surfaces 2) account for cases in which surfaces cannot unambiguously be discriminated and 3) discriminate illusory surfaces. Our model has only addressed the case of surface segmentation in static scenes. Dynamic scenes may also utilize motion information for surface computation, and therefore our model could be extended by the addition of a DOF-motion mechanism. Thus, in general, the computation of direction of figure can be considered a local integration, occurring within a column, of surface information arising from different visual modalities.

REFERENCES

[1] Finkel, L.H., and Sajda, P., "Object Discrimination based on Depth-from-Occlusion," *Neural Computation*, Vol. 4, 1992.

[2] Gilbert, C., "Horizontal integration and cortical dynamics," Neuron, Vol 9, pp. 1-13, 1992.

[3] Nakayama, K., and Shimojo, S., "Toward a neural understanding of visual surface representation," Cold Spring Harbor Symposia on Quantitative Biology, Vol. LV, pp. 911-924, 1990.

[4] Ramachandran, V., and Gregory, L., "Perceptual filling in of artificially induced scotomas in human vision," Science, pp. 699-702, 1992.

DIRECTION SELECTIVITY USING MASSIVE INTRACORTICAL CONNECTIONS

Humbert Suarez[1], Christof Koch[1],

and Rodney Douglas[1,2]

[1] *Computation and Neural Systems Program 216-76,
California Institute of Technology, Pasadena, CA 91125, USA.* [2] *MRC
Anatomical Neuropharmacology Unit, University of Oxford, Oxford, UK.*

ABSTRACT

Almost all models of orientation and direction selectivity in visual cortex are based on feed-forward connection schemes. However, the majority of the synaptic input on any cortical cell is provided by other cortical cells. Based on the canonical microcircuit of Douglas and Martin (1991), we model the behavior of a network of pyramidal and stellate cells receiving geniculate input using massive excitatory and inhibitory synaptic interconnections. In particular, we show how this circuit endows simple cells with direction selectivity over a large range of velocities.

Models of direction selectivity in primary visual cortex generally overlook two important constraints. **1.** 80% of excitatory synapses on cortical pyramidal cells originate from other pyramidal cells. **2.** Intracellular *in vivo* recordings in cat simple cells by Douglas, et al. (1988) failed to detect significant changes in somatic input conductance during stimulation in the null direction.

One very attractive solution incorporating these two constraints was proposed by Douglas and Martin (1991) in the form of their *canonical microcircuit*: for motion of a visual stimulus in the preferred direction, weak geniculate excitation excites cortical pyramidal cells to respond moderately. This relatively small amount of cortical excitation is amplified via excitatory cortico-cortical connections. For motion in the null direction, the weak geniculate excitation is vetoed by weak inhibition (mediated via an interneuron) and the cortical loop is never activated.

The model consists of a retino-geniculate and a cortical stage. The former

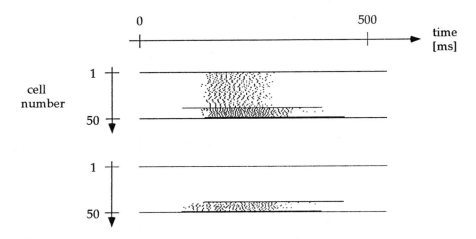

Figure 1   Response of the pyramidal (cells 1-40) and inhibitory neurons (cells 41-50) to a bar moving at 2 deg/sec in the preferred (above) and null (below) directions of motion. Each dot in the figure represents one spike in a neuron whose number can be found on the y axis, at a time indicated on the x axis, all for the same stimulus trial.

includes a center-surround type of spatial receptive field and a bass-pass type of temporal filtering. The output of the LGN—action potentials—feeds into the cortical module consisting of 40 pyramidal (excitatory) and 10 inhibitory neurons. Each neuron is modeled using 3-4 compartments whose parameters reflect, in a simplified way, the biophysics and morphology of visual cortical neurons. To obtain direction selectivity, the receptive field of the geniculate input to the inhibitory cortical neurons is slightly displaced in space from the geniculate input to the pyramidal neurons. We here only simulate 1-D patterns in the ON system. The connections between pyramidal and interneurons reflect the connectivity of the canonical microcircuit, and the time courses of the post-synaptic potentials (p.s.p.'s) are consistent with physiologically recorded p.s.p.'s. The e.p.s.p.'s in our model arise exclusively from non-NMDA synapses, and the i.p.s.p.'s are of two types, $GABA_A$ and $GABA_B$.

Results

The response of all 50 model cortical neurons during presentation of a 50 % contrast bar moving at 2 deg/sec in both the preferred and null directions is

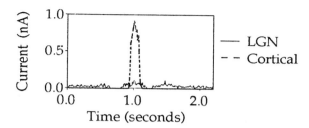

Figure 2 Total synaptic current in a simulated pyramidal cell due to the geniculate input and due to the excitatory cortico-cortical feedback during stimulation by a 50 % contrast bar moving at 2 deg/sec in the preferred direction. The effect of the cortical loop is to amplify the LGN input by an order of magnitude

summarized in the raster-plots of Fig. 1. All of the pyramids show strong direction selectivity, while the smooth neurons show a weak form of direction preference. The amplification by the cortical loop is visualized in Fig. 2: the small amount of excitatory geniculate current is amplified by one order of magnitude via the excitatory cortico-cortical feedback. For motion in the null direction, only this small amount of excitatory current needs to be shunted, necessitating only a relative modest (about 50%) increase in the somatic input conductance (Fig. 3b). From the intracellular *in vivo* recordings of Douglas *et al.* (1988; Fig. 3a), we know that the somatic input conductance during null direction stimulation increases by less than 20%. However, the total excitatory current during motion in the preferred direction is so large (Fig. 2) that the weak inhibition sufficient to veto the weak geniculate excitation fails to effectively reduce the cell's discharge (Fig. 1). Finally, in the absence of excitatory intracortical connections, the geniculate input current must be much higher, to drive the cell's discharge to 100 Hz and higher. The large excitatory current (in excess of 1 nA) requires a large inhibitory conductance input to sink this current, here a peak somatic input conductance change close to 400 % (Fig. 3c).

In the model, cortical processsing transforms the velocity response of the geniculate input by sharply decreasing the response at high velocities and shifting the peak response down to very low velocities (Fig. 4). This occurs through the combined action of the excitatory and inhibitory intracortical connections. The majority of cells in cat area 17 show such a velocity-low pass behavior (Orban, 1984). The slow $GABA_B$ inhibition is responsible for direction selectivity at small velocities. Direction selectivity in the model is remarkably strong and invariant with velocity, persisting over two orders of magnitude with a direction selectivity index close to 100 % over the full range.

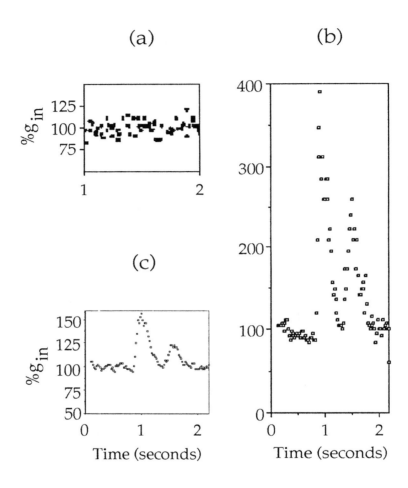

Figure 3 Somatic input conductance during stimulation by a bar moving in the null direction, as a function of time, for a direction-selective pyramidal neuron. (a) Data from Douglas *et al.*, 1988). (b) Model without intracortical excitatory connections (c) The full-blown model

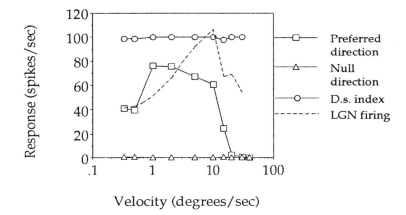

Figure 4 Peak firing rate versus velocity for the pyramidal neurons in the model, during stimulation by a 50 % contrast bar. Also shown is the direction index ($(P - NP)/P$) and the peak geniculate firing rate as a function of the stimulus velocity.

Conclusion

Our model describes well the observed velocity-tuning curve of layer 17 simple cells using a simple anatomical connectivity principle. Direction selectivity is computed by a Barlow and Levick (1965) like mechanism, followed by cortical amplification (in the preferred direction). However, our current model is quite complex, with dozens of free parameters defying easy interpretation and plausible predictions. We are now attempting to understand models that incorporate such massive cortical feedback from an analytical point of view.

Acknowledgements. This research is supported by NIH, ONR, the McDonnell Foundation and by a NSF Presidential Young Investigator Award to CK.

References

Barlow, H.B., and Levick, W.R. (1965)*J. Physiol.* **178**, 477-504.

Douglas, R.J., and Martin, K.A.C. (1991)*J. Physiol.* **440**, 735-769.

Orban, G.A. (1984) Neuronal operations in the visual cortex. Springer, Berlin.

Douglas, R.J., Martin, K.A.C., and Whitteridge, D. (1988) *Nature* **332**, 642-644.

32

CORRELATIONAL-BASED DEVELOPMENT
OF DISPARITY SENSITIVITY

Gregory S. Berns, Peter Dayan,
and Terrence J. Sejnowski

Computational Neurobiology Laboratory
Salk Instititute and Howard Hughes Medical Institute
P.O. Box 85800, San Diego, CA 92186-5800

ABSTRACT

A correlational-based model of development of disparity sensitivity is proposed. The weights between two one-dimensional input layers and a single cortical layer were modified by a linear Hebb rule using fixed correlation matrices both within and between eyes and fixed cortical connections. With local correlations, the delayed presentation of a slight amount of between-eye correlation led to the development of both binocular cortical cells with the left and right receptive fields aligned, i.e. zero disparity and monocularly dominated cells, which tended to have non-zero disparity preferences.

32.1 INTRODUCTION

The development of response properties of neurons in the mammalian visual cortex depends on visual experience during a critical period (1). This plasticity is dependent on the statistics of the pattern of neural activity on both the geniculocortical inputs and the cortical neurons themselves (2, 3, 4, 5).

The mechanism of depth perception has been extensively studied, and the disparity selectivity of cortical neurons probably plays a central role. Disparity is the relative difference in position in the two retinas on which an image is cast, and disparity sensitivity refers to the ability of cortical cells to detect relative image displacements between the two eyes. The random-dot stereogram showed that disparity is sufficient, although not necessary, to perceive depth (6). Many disparity selective neurons are maximally stimulated when presented with either convergent, divergent, or zero disparity (7, 8). In this study we address the problem of how the disparity sensitivity of cortical neurons develops.

We assume that development of disparity sensitivity is activity dependent and driven by correlations both within and between the retinas. Several different models have been proposed exploring how visual cortex structures such as ocular dominance columns (9, 10, 11, 12) and orientation selectivity (13, 14) might develop by virtue of the correlations in retinal activity. In these models, orientation and ocular dominance are not explicitly specified but emerge through competition and cooperation for cortical synapse sites, with the more correlated cells reinforcing their synapses and the uncorrelated cells weakening theirs. We sought to identify the conditions that would allow the development of a population of cortical cells sensitive to a range of disparities within this simple framework. In general, we find that this type of model allows the development of cortical cells with a range of disparity selectivity and that by allowing development to occur in two phases, corresponding to prenatal and postnatal periods, the experimentally observed relationship between ocular dominance and disparity emerges (15, 16, 17).

32.2 METHODS

Initial visual development was modeled with two one-dimensional input layers, representing the retinas of the left and right eyes, fully connected with synaptic weights to a one-dimensional cortical layer of the same size (Fig. 1). This is the same as in Miller et al. (10, 18) except that only a few columns of cortex were simulated, the arbor function of the retinal cells was flat, and the model was one-dimensional. Fixed lateral connections were used to represent the influence of one cortical cell on another.

A linear Hebb rule was used to model the changes in synaptic strength between the retinas and the cortex using correlations C^{LL} and C^{RR} within the eyes, and C^{LR} and C^{RL} between the eyes. A lower bound of zero was imposed on the weights. Weights were also normalized using a combination of subtractive and multiplicative procedures, thus keeping the total input to a cortical cell constant. A transition from predominately subtractive to multiplicative normalization was necessary to stabilize mixtures of both monocular and binocular cells.

The form of the correlation matrices as well as the fixed cortical interaction matrix were Gaussian. C^{LL} and C^{RR} were equal to each other and constructed such that each retinal point was locally correlated with its neighbors with a Gaussian distribution. C^{LR} and C^{RL} were also Gaussian and equal to each other but with four times the variance and 0.2 times the amplitude. The cortical interaction matrix, which represented fixed intracortical connections, was generated by a difference of short and long range Gaussians, giving a "Mexican hat" influence function. The width, or standard deviation, of each Gaussian function relative to the layer size was: 0.05 for the same-eye correlation function; 0.10 for the between-eye function; and 0.05 for the positive component of the cortical interaction function and 0.15 for the negative component (but with 1/9 amplitude).

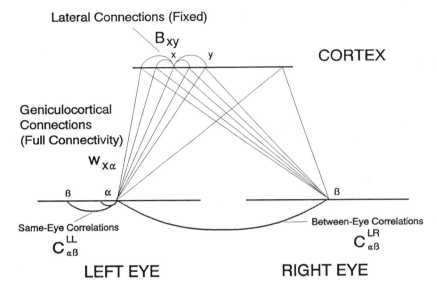

Figure 32.1 Schematic diagram of the model of visual cortex development. The model consists of a one-dimensional layer for each eye fully connected to a one-dimensional cortex of the same size, also containing fixed lateral connections. A linear Hebb rule was used to modify weights using correlations both within and between eyes instead of specific input patterns.

Binocularity is obviously necessary to develop disparity sensitive cells with a single cortical layer. Miller (10, 18) analyzed the development of ocular dominance, and Dayan and Goodhill (19) showed how binocularity could arise from between-eye correlations. However, there are two trade-offs. If the between-eye correlations are too small, all cells become monocularly dominated; conversely, if the between-eye correlations are too strong, then the cells all become responsive to both eyes, but with zero disparity. Although it is possible to balance both the same-eye and between-eye correlations and adjust the normalization procedure so that both monocularity and binocularity are equally favored, this is unstable. If between-eye correlations are introduced only after development has commenced, some cells will have progressed too far towards monocularity to be perturbed. The cortical interaction function does not affect this outcome and simply ensures that cortical cells are generally consistent with their neighbors. We therefore searched for conditions that would produce patches of relative ocular dominance, involving cells of nonzero but smoothly varying disparity, with boundaries of cells responsive to both eyes, but with zero disparity.

The amplitude of the correlation matrices was varied to model three different developmental paradigms. In the first model, the between-eye correlations were set to zero ($\mathbf{C^{LR}=C^{RL}=0}$), corresponding to an animal whose visual cortex development is completely prenatal. In the second model, correlations were added between the eyes, which might occur in an animal whose visual cortex development is completely postnatal. The third model had two phases: one phase with only same-eye correlations, and the second with both same-eye and between-eye correlations, a circumstance in which visual cortex development has both prenatal and postnatal components. The amplitude of the between-eye correlation relative to the same-eye was, respectively in these three paradigms, 0.0, 0.2, and 0.2. Retinal and cortical layers had 60 cells each, and periodic boundary conditions were adopted to avoid edge effects. For all results reported here, initial weights were randomly assigned and ranged from 0.49 to 0.51, but the same results were obtained with a range of initial weights of 0.4 to 0.6. Computer simulations were run on a Sun SparcStation 2, a complete model taking 800 iterations, or three hours of run time, to develop to a stable pattern.

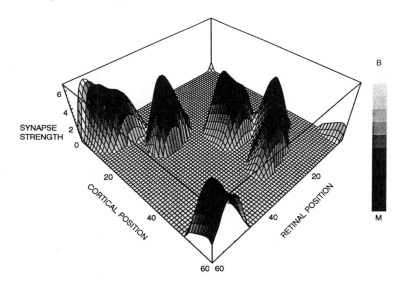

Figure 32.2 Final development of geniculocortical connections in a model with same-eye correlations only. The height represents the total input from a given retinal location in both eyes to a given cortical cell. Monocularity is given by shading (black=monocular).

Ocular dominance was calculated for each cortical cell as the difference between the total synaptic input from the right eye and the total synaptic input from the left eye. This was normalized to the total synaptic strength, which was constant. Ocular dominance for each cortical cell was defined by *(R-L)/(R+L)* where *R* was the total input from the right eye and *L* was the total input from the left eye. Thus ocular dominance ranged from -1.0 (completely dominated by the left eye) to 1.0 (completely dominated by the right eye). Disparity sensitivity for a given cortical cell was calculated by first determining the left and right receptive fields, including both the direct and intracortical inputs. The receptive fields were then summed at each retinal position and the peak response in the field noted. This was then repeated with successive shifts of the receptive fields, the shift with the maximum peak response representing the best disparity. It should be noted that this method cannot define disparity for completely monocular cells since shifting the receptive fields has no effect when one is uniformly zero.

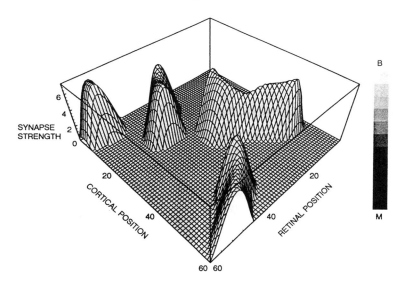

Figure 32.3 Final geniculocortical connections in a model with both same-eye and between-eye correlations throughout development. The presence of between-eye correlations early in development led to an exclusively binocular cortex.

32.3 RESULTS

In geniculocortical development with same-eye correlations only, the first feature to appear was the localization of the receptive fields, which appears as peaks in Fig. 2. The scale of the peaks in the retinal direction was determined by the width of the correlation function and was seen to correspond to the Gaussian width of the same-eye correlations. As discussed by Miller (18), the cortical scale of the peaks was determined by the width of the cortical interaction function, which in these simulations was the same as the width of the same-eye correlations. The peaks were organized along diagonal bands, which reflected the tendency of the model to form topographics maps because of the cortical interaction function. In a mature network, nearly all the receptive fields were found to be monocular, illustrated by the dark peaks. Most of the cortical cells were completely dominated by one of the eyes, and the periodicity of ocular dominance across the cortex corresponded to the width of the cortical interaction function. The presence of between-eye correlations throughout development led to a cortex full of binocular cells (Fig. 3). The ocular dominance was effectively 0.0 across the cortex, and most cells had zero disparity.

The two-phase development led to a mixture of both monocular and binocular cortical cells. Approximately half of the cortical cells were monocularly dominated, but there were zones of binocularity at the transition between left and right eye dominance (Fig. 4). The pattern of ocular dominance and disparity was similar to the other two paradigms but with a relatively even distribution of monocular and binocular cells. The scatter plot in Fig. 5 shows that the binocular cells tended to have zero disparity while the more monocular cells had nonzero disparity.

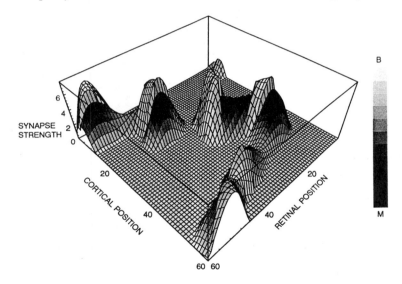

Figure 32.4 Final geniculocortical connections in a model with two-phase development. Initial same-eye correlations were followed by the addition of between-eye correlations, leading to the development of both monocular and binocular cells.

32.4 DISCUSSION

The model of visual cortex development described here should be interpreted as a highly simplified representation of activity-dependent development. First, we made a number of architectural simplifications to reduce the complexity of a two-dimensional cortex to a one-dimensional problem; second, we have used a linear version of Hebb's rule of synapse modification. However, this model does address the experimental evidence of species-dependent differences of visual development. Monkeys have a high proportion of monocular cells in area 17, particularly layer IV, whereas cats are known to have predominantly binocular cells in area 17. Furthermore, the binocular cells in area 17 of the cat tend to be of the tuned excitatory type, i.e. cells with best disparity of zero (15). The species difference can be explained most simply by timing differences (Michael Stryker, personal communication). Primate visual cortex development begins prenatally and, according to our model, should be driven by locally correlated activities within each eye. However, between-eye correlations are not present, and so lead to monocular cortical cells. The cat, however, has a greater proportion of its visual development postnatally when binocular correlations are present because both eyes are stimulated similarly. Thus the cat is expected to have a greater proportion of binocular cells, and furthermore, these cells should have zero disparity if the cats' vergence apparatus is intact.

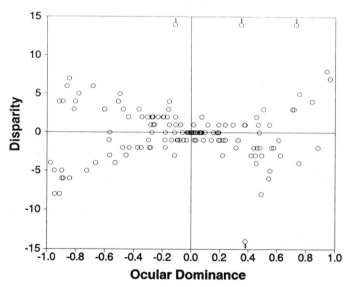

Figure 32.5 Scatter plot of disparity vs. ocular dominance for three simulations using different initial conditions. There is a tendency for cells with balanced ocular dominance to have best disparities near zero.

These predictions emerge from our model, as does the finding in the cat (15, 16, 17) that the binocular cells tend to have best disparity of zero, while the more monocular cells have less of a preference for zero disparity. In the linear Hebb model described here, we could only generate such a relationship between ocular dominance and disparity using a two-phase paradigm. Although it was possible to achieve a mixture of monocular and binocular cells with single-phase paradigms, such as those using same-eye anti-correlations (18), it was not possible to achieve a systematic relationship between ocular dominance and disparity similar to that obtained with the two-phase model. Our model is based on a number of simplifying assumptions; however, we believe that the main conclusions will hold for more general correlation-based models of development.

We thank Dr. K. Miller for generous advice regarding correlation-based models of development and Drs. K. Miller and M. Stryker for helpful comments during the preparation of this manuscript. This work was supported by the Howard Hughes Medical Institute and the Science Education Research Council.

32.5 REFERENCES

1. Wiesel, T. N., Hubel, D. H. (1963) *J. Neurophysiol.* **26**, 978-993.
2. Rakic, P. (1977) *Phil. Trans. R. Soc. (Lond.) B* **278**, 245-260.
3. Stryker, M. P., Strickland, S. L. (1984) *Invest. Ophthalmol. Vis. Sci. (suppl.)* **25**, 278.
4. Stryker, M. P., Harris, W. (1986) *J. Neurosci.* **6**, 2117-2133.
5. Shatz, C. J. (1990) *Neuron* **5**, 745-756.
6. Julesz, B. (1960) *Bell Syst. Tech. J.* **39**, 1125-1162.
7. Barlow, H. B., Blakemore, C., Pettigrew, J. D. (1967) *J. Physiol. (Lond.)* **193**, 327-342.
8. Poggio, G. F., Fischer, B. (1977) *J. Neurophysiol.* **40**, 1392-1405.
9. Willshaw, D. J., Von der Marlsberg, C. (1976) *Proc. R. Soc. (Lond.) B* **194**, 431-445.
10. Miller, K. D., Keller, J. B., Stryker, M. P. (1989) *Science* **245**, 605-615.
11. Montague, P. R., Gally, J. A., Edelman, G. M. (1991) *Cerebral Cortex* **1**, 199-220.
12. Swindale, N. V. (1980) *Proc. R. Soc. (Lond.) B* **208**, 243-264.
13. Linsker, R. (1986) *Proc. Natl. Acad. Sci. USA* **83**, 8390-8394.
14. Miller, K. D. (1992) *NeuroReport* **3**, 73-76.
15. Ferster, D. (1981) *J. Physiol.* **311**, 623-655.
16. Gardner, J. C., Raiten, E. J. (1986) *Exp. Brain. Res.* **64**, 505-514.
17. LeVay, S., Voigt, T. (1988) *Vis. Neurosci.* **1**, 395-414.
18. Miller, K. D. (1990) in *Neuroscience and Connectionist Theory*, eds. Gluck, M. A., Rumelhart, D. E. (Lawrence Erlbaum Assoc., Hillsdale, NJ), pp. 267-353.
19. Dayan, P., Goodhill, G. J. (1992) in *Advances in Neural Information Processing Systems 4*, eds. Moody, J. E., Hansen, S. J., Lippman, R. P. (Morgan Kaufmann, San Mateo), pp. 000-000.

Dynamical Control of Visual Attention Through Feedback Pathways: A Network Model

Janani Janakiraman[*+]
K.P. Unnikrishnan[+*]
[*]AI Laboratory, University of Michigan, Ann Arbor, MI 48109
[+]Computer Science Dept., GM Research Labs., Warren, MI 48090

1 Introduction

Scientists investigating selective attention have to confront two issues.

- The static aspects: How the brain selects a few objects from the retinal image for recognition and further processing.

- The dynamic aspects: How attention is rapidly shifted from one of these objects to the next.

Psychologists and neuroscientists have attempted to specify the levels in the sensory pathway at which these operations take place and the mechanisms for efficiently carrying out these processes.

In the mammalian visual system, sensory information is relayed through LGN to the primary cortex. Different levels of feature extraction take place in these areas and the information gets passed on to the association cortical areas. Learned associations between patterns are stored in the association cortical areas. Feedback is a ubiquitous feature in this sensory pathway and many of these areas are reciprocally connected [Van Essen, 1985].

We present a model to investigate the role of these feedback pathways in visual attention. In this paper we look at the feedback from association cortical areas to V1 and from V1 to LGN. Previous studies of similar models have addressed the static aspects of attention [Harth, Unnikrishnan, and Pandya, 1987; Harth, Pandya, and Unnikrishnan, 1990; Unnikrishnan, 1987] and have

[0]This research was supported in part by a grant from GM Research Laboratories

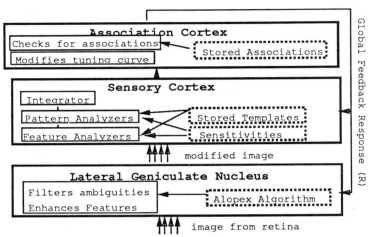

Figure 1: Block Diagram

shown that the feedback pathways are capable of suppressing noise in the input, completing incomplete patterns and selecting one pattern from many. Here we specifically address the dynamic aspects. Using simulations of the model, we show that temporal associations stored (learned) at the association cortical areas can affect the dynamics of sensory processing at lower centers. We also hypothesize on possible mechanisms that will allow rapid shifts of attention. The neural algorithm (Alopex) used in our simulations is easily carried out by known neural circuitry [Nine and Unnikrishanan, 1992; Sekar and Unnikrishnan, 1992].

2   The Model

Figure 1 shows a schematic of the system. We consider a small patch of the visual field and represent it as a 16 x 16 array. This forms the input to LGN. The output of LGN is analyzed by a bank of 16 feature analyzers modeled after the oriented line detectors found in V1 [Hubel and Wiese l, 1979]. Outputs of these feature analyzers are used by pattern analyzers whose templates are defined as conjunctions of features. Associations between different patterns are defined in the association area. Figure 2 for example shows an association matrix. The outputs of pattern analyzers are combined to form a simple scalar response. The tuning characteristics of pattern and feature analyzers are determined by a set of parameters. The single scalar response is used to generate a filter at LGN and modify the tuning properties of the pattern analyzers in the sensory cortex. Details of this model are given elsewhere [Janakiraman and Unnikrishnan, 1992].

Patterns	P1	P2	P3	P4
P1	-1	1	0	0
P2	0	-1	1	0
P3	0	0	-1	1
P4	1	0	0	-1

Figure 2: Association Matrix

3 Results

We have investigated the convergence of the system under a number of conditions. Here we report the results under a few of these conditions. The first set of these were to investigate if dynamically changing the tuning curves helps in the convergence of the system. In all these experiments, the system was given completely random inputs and one of the pattern analyzers was favored by increasing its sensitivity. Figure 3a shows the output of LGN after 2700 iterations, when the system was run with a fixed set of tuning curves for the pattern analyzers. Figure 3b shows the LGN output after 2700 iterations, when the tuning curves of the pattern analyzers sharpened during the run. We can see that dynamically changing the tuning curves helps the system to converge rapidly.

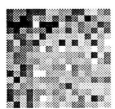   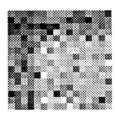

Figure 3a   **Figure 3b**

Our next set of experiments were done to see if the system could shift its focus from one pattern to another and cycle through the set of patterns defined in the association matrix. Our system was incapable of doing this when the tuning curves were static. Figure 4 shows the behavior of the system with dynamically changing the tuning curves. We can see that the system is capable of converging on to each one of the patterns and then shifting to the next pattern as defined in the association matrix.

4 Discussion

Previous studies of similar models [Harth, Unnikrishnan, and Pandya, 1987; Harth, Pandya, and Unnikrishnan, 1990; Unnikrishnan, 1987] have shown that

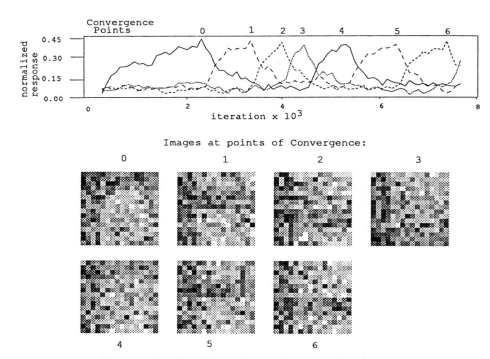

Figure 4: Cycling through associated patterns

systems of the sort described here can indeed converge on the patterns defined and invert the normal process of sensory information flow. The current set of experiments were designed to ask the following questions.

- If temporal associations between patterns are defined in the association cortex, is the system capable of converging to each one of the patterns and then rapidly shift to the next one?

- In such a system what are the most important parameters?

Results shown in figure 4 clearly indicates that the system is capable of dynamically shifting its attention. From this we would like to propose that the role of some of the feedback pathways may be to dynamically alter sensory processing at lower levels, based on previously learned associations. The most interesting result from our simulations is that the crucial parameter in such a system is the tuning curve of pattern recognizers. If the system starts with broadly tuned detectors which become progressively narrower (this process being repeated as the system converges to successive patterns), rapid convergence to individual patterns are possible. In this respect it is most interesting that recent neurophysiological experiments in V4 and IT have indicated that tuning curves of single neurons in this area change during matching-to-sample tasks

[Desimone *et. al.*,1990]. This has lead Desimone e t al. (1990) to speculate that this phenomenon may play a role in static aspects of visual attention. Based on our simulation, we would like to propose that changing the tuning properties of feature and pattern analyzers may be a dynamic phenomenon carried out through feedback pathways and may play a crucial role in the more dynamic aspects of visual attention. This would suggest that if the same kind of experiments reported by Desimone et al. are done on an animal which has to rapidly shift its attention from one pattern to another, one would see the dynamic change of tuning curves during each of its shifts.

References

[1]   Desimone R., Wessinger M., Thomas L., Schneider W.: Attentional Control of Visual Perception: Cortical and Subcortical Mechanisms, *Cold Spring Harbor Symposia on Quantitative Biology* **55**, pp 963 - 971, 1990.

[2]   Harth E., Pandya A.S., Unnikrishnan K.P.: Optimization of cortical responses by feedback modification and synthesis of sensory afferents - A model of perception and rem sleep, *Concepts in Neuroscience* Vol I, No. 1 , pp 53 - 68, 1990.

[3]   Harth, E., Unnikrishnan, K.P. and Pandya, A.S., 1987: The inversion of sensory processing by feedback pathways: a model of visual cognitive functions, *Science* 237, pp 187-189.

[4]   Hubel D.H., Wiesel TN: Brain Mechanisms of vision,*Sci. Am.* 24, pp 150 - 162, 1979.

[5]   Janakiraman, J., Unnikrishnan, K.P.: A Feedback Model of Visual Attention, *IJCNN* Vol III, pp 541-546, June 1992.

[6]   Nine, H.S. and Unnikrishnan, K.P., The role of subplate feedback in the development of ocular dominance columns (this volume), 1992.

[7]   Sekar, N.S., and Unnikrishnan, K.P.: The Alopex algorithm is biologically plausible, *Abstr. Learning and Memory meeting, Cold Spring Harbor Laboratory*, 1992..

[8]   Unnikrishnan K.P.: A Hierarchical Model of Visual Perception, *Ph.D. Thesis, Syracuse University*, 1987.

[9]   Van Essen, D.C.: Functional organization of primate visual cortex ,*Cerebral Cortex*, Vol 3 , A. Peters and E. G. Jones, Eds. (Plenum Press, New York, 1985).

34

THE GATING LATTICE: A NEURAL SUBSTRATE FOR DYNAMIC GATING

Eric O. Postma
H. Jaap van den Herik
Patrick T. W. Hudson

Computer Science Department, University of Limburg,
PO Box 616, 6200 MD Maastricht, The Netherlands

ABSTRACT

We propose the Gating Lattice as a model of the neural substrate for attentional selection. Dynamic gating requires the selection of patterns at an initial stage and their routing over one or more subsequent stages. It is assumed to underlie attentional processing. A network of stacked Gating Lattices realizes dynamic gating and copes with the problem of translational invariance in visual object recognition.

1 BACKGROUND

The retinal image captures the visual scene in a high-dimensional representation. This becomes manifest in the neural organization of primary visual cortex (V1), i.e., the main recipient area of retinal signals. For each retinal position, V1 contains cells tuned to parameters, such as orientation, spatial frequency, and color. Selection from the many signals available at V1 has to be made appropriately. This can be realized by gating the signal flow through pyramidal projection cells. Such selective gating has been hypothesized to be effectuated by interneurons (e.g., [1]). In particular, the GABAergic axoaxonic or *chandelier* cell, which contacts the initial segment of the axon of pyramidal neurons in cerebral cortex, is believed to control pyramidal output to other cortical areas ([2]). Our gate model is inspired by this chandelier-pyramidal cell interaction.

2 GATING LATTICE

The *gate model* shown in Figure 1a, contains the basic units of gating. A binary "interneuron" (solid circle) controls signal flow through a "pyramidal neuron" (triangle) in an all-or-none fashion. The control signal h excites the interneuron, whose inhibitory output G gates signal flow through the pyramidal cell. An active interneuron closes the gate, i.e., $out = 0$, whereas an inactive interneuron opens it, i.e., $out = in$.

A *Gating Lattice* (GL) is formed by coupling gates on different sublattices through short-range inhibitory connections ([3]). The GL selects one of several two-dimensional input patterns and gates it towards a next stage without distorting the topographical structure. The number of input patterns selectable for gating is determined by the number of sublattices in the GL. Here we concentrate on a GL with two sublattices, A and B. For this GL the control signal h_i^X of gate i on sublattice $X \in \{A, B\}$ is defined as $h_i^X = \sum_{j \neq i}^4 G_j^Y + H^X$, with $Y \in \{A, B\}$, $Y \neq X$, H^X is a global control signal for sublattice X, and the summation is over the four nearest neighbors. Figure 1b shows a top view of the GL in three characteristic states: no winning sublattice, sublattice A winning, sublattice B winning. The gating of an input pattern is represented in the GL as the opening of all gates on the *winning sublattice*. When gates are updated with Glauber's ([4]) function, the GL is equivalent to an antiferromagnetic Ising Lattice (e.g., [5]) so that its global behavior can be described with statistical mechanics. For particular values of the global parameters, the GL relaxes to an intermediate state (e.g., the left state in Figure 1b) if $|H^A - H^B| < \theta$, or to one of the winning-sublattice states if $|H^A - H^B| \geq \theta$. The value of the threshold θ depends (amongst others) on a global *temperature* parameter.

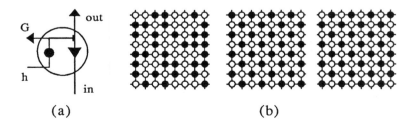

(a) (b)

Figure 1 (a) The gate model containing an interneuron (solid circle) and a pyramidal neuron (triangle). (b) A patch of a Gating Lattice in three characteristic states. The two states on the right are winning-sublattice states. Shaded circles represent open gates.

3 STACKED GATING LATTICES

A single GL can only select one out of two subpatterns from an input pattern. Gating any subpattern out of a large input pattern, as required for translation-invariant object recognition, is realized by stacking a number of GLs into a multilayer structure. The shifter-circuits interconnection scheme proposed by [6] provides an efficient way of interconnecting stacked GLs. This scheme allows for the selection of one out of 2^L subpatterns when L GLs are stacked into a *rewiring network* (Figure 2). Moreover, it preserves the topographical structure of the gated patterns. Combining the rewiring network with a *control network* (cf. [7]) results in local control of subpattern selection. The competition among each pair of gates in the subtrees of the control network is resolved by their corresponding patch of GL in the rewiring network. Sparse bidirectional connectivity between both networks (not shown in Figure 2) ensures that a concatenated sequence of open gates in the control network (concatenated arrows) is linked to an equivalent sequence in the rewiring network. To achieve our aim of modeling visual selective attention, we have combined the compound neural network described above with an ART classifier ([8]). The new model realizes translation-invariant object perception by gating contiguous subpatterns towards a recognition module. Multiple matching modules (the rectangular boxes at the input in Figure 2) compare subpatterns with the expectation pattern (generated by the ART classifier) in parallel. Each matching module delivers a single *match value* to the control network, the largest of which reaches the top. Simultaneously, the corresponding part of the input (shaded) is rewired to the input of the ART classifier.

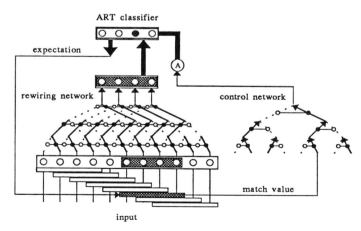

Figure 2 The model for translation-invariant object recognition. The connected circles are Gating Lattices (side view).

4 SIMULATIONS

We have performed Monte-Carlo simulations of our model with GLs of two sublattices. This restricted model copes with translations in one direction only. (Translations in two directions can be accomplished in a straightforward way by adding a GL to each layer or using a single GL with three or more sublattices, cf. [3].) The input of the simulations has been a number of input images of 11 x 4 pixels, each containing a single 4 x 4 target pattern. Thus, the best-matching subpattern obviously was in one of 8 positions.

5 RESULTS

Figure 3 illustrates the typical responses to an input image. Each rectangular box represents a "snapshot" of the input area, showing the distribution of *sample points*. A sample point (black square) is a point of the input area which is connected by a sequence of open gates (see Figure 2) to one of the 4 x 4 inputs of the ART classifier. The first snapshot shows the initial distribution of sample points; all GLs are randomly initialized and, consequently, the sample points are widely distributed over the input area. The two subsequent snapshots show the movement of the sample points towards the target as well as their clustering around the target. The target's position is indicated by the horizontal bar. After the third snapshot, we have moved the target to the left. The next three snapshots show that the sample points now rapidly move to the new target position. We have observed that, in general, the time the sample points need to reach the new target is dependent on the number of patterns in the input image and their similarity to the target.

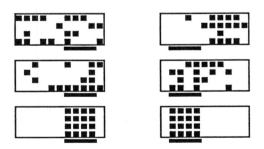

Figure 3 Snapshots of the sample-point distribution for the model.

6 CONCLUSIONS AND FURTHER WORK

Our model of visual attention has been shown to be capable of translation-invariant object recognition. In our simulations we saw that the model adaptively allocates its sample points to a target pattern in the input. The target selection proceeded by matching stored knowledge against inputs and is therefore entirely top-down ([9]; see also [10] for a similar approach). Further work concentrates on combining this top-down control with ("pre-attentive") bottom-up control. Additionally, the model will be extended to realize scale-invariant object recognition. Moreover, a spatial memory will be added enabling the model to store and execute scanning sequences.

REFERENCES

[1] Marin-Padilla, M., "The pyramidal cell and its local-circuit interneurons: A hypothetical unit of the mammalian cerebral cortex," Journal of Cognitive Neuroscience, Vol. 2, 1990, pp. 180–194.

[2] Somogyi, P. and Martin, K.A.C., "Cortical circuitry underlying inhibitory processes in cat area 17," In D. Rose and V.G. Dobson (Eds.), Models of the Visual Cortex, pp. 514–523. New York, NY: John Wiley and Sons, 1983.

[3] Postma, E.O., van den Herik, H.J., and Hudson, P.T.W., "Dynamic selection through gating lattices," In Proceedings of the International Joint Conference on Neural Networks, Baltimore, 1992.

[4] Glauber, R.J., "Time-dependent statistics of the Ising model," Journal of Mathematical Physics, Vol. 4, 1963, pp. 294–307.

[5] Plischke, M. and Bergersen, B., "Equilibrium statistical physics," Englewood-Cliffs, NJ: Prentice-Hall, 1989.

[6] Anderson, C.H. and Van Essen, D.C., "Shifter circuits: A computational strategy for dynamic aspects of visual processing," Proceedings of the National Academy of Sciences USA, Vol. 84, 1987, pp. 6297–6301.

[7] Koch, C. and Ullman, S., "Shifts in selective visual attention: towards the underlying neural circuitry, " Human Neurobiology, Vol. 4, 1985, pp. 219–277.

[8] Carpenter, G. A. and Grossberg, S., "A massively parallel architecture for a self-organizing neural pattern recognition machine, " Computer vision, graphics, and image processing, Vol. 37, 1987, pp. 54–115.

[9] Biederman, I. and Cooper, E. E., "Evidence for complete translational and reflectional invariance in visual object priming," Perception, Vol. 20, 1992, pp. 385–593.

[10] Olshausen, B., Anderson, C.H., and Van Essen, D.C., "A neural model of visual attention and invariant pattern recognition," Computation and Neural Systems Memo 18, California Institute of Technology, Pasadena, California, 1992.

SECTION 4

AUDITORY SYSTEMS

The auditory section presents a wide variety of modeling from single hair cell models to brain stem maps in the vertebrate auditory system. Sound localization and frequency tuning of receptors are the topics presented in this section.

HAIR CELL MODELLING TO EXPLORE THE PHYSIOLOGICAL BASIS OF TUNING IN THE LOWER VERTEBRATE EAR

David Egert

UC Berkeley/UC San Francisco Bioengineering Program
230 Bechtel Center, University of California, Berkeley, Berkeley, CA 94720

ABSTRACT

I developed a model of an isolated acoustic sensory hair cell, including membrane dynamics, to explore the role of hair cell resonance in the tuning processes of lower vertebrates and to ascertain if the frequency response of a single hair cell could adequately explain the tuning observed in primary afferents from intact end-organs of the bullfrog. Driving the modelled hair cell with small-signal sinusoidal force, I observed the transfer relationship between bundle stimulation and membrane potential. The observed transfer relation was of low dynamic order and was not consistent with the transfer relations typically seen in bullfrog auditory afferents. Hence I suspect the tuning involves interactions between several hair cells, perhaps facilitated by mechanical linkages.

35.1 INTRODUCTION

Lewis, R.S. (1984) proposed a molecular model encompassing channel kinetics and diffusion processes to explain the resonant behavior seen in isolated bullfrog saccular hair cells in response to current clamp steps. This model's frequency response exhibits a relatively sharp peak, a shallow roll-off (-20 dB/decade), and a total phase lag of about 90 degrees. This differs significantly from the typical tuning curves we observe *in vivo* from bullfrog saccular axons which regularly exhibit wide pass-bands with a steep roll-off (>80 dB/decade) (see figure 1) and linear phase for over a cycle. To explore saccular tuning and hair cell resonance in the intact, functioning sacculus, I have developed a model that extends Lewis's resonance model, adding features that provide a more complete representation of the hair cell, and have characterized how each of these features affects the hair cell's frequency response.

35.2 Results

Adding the Transduction Channel:
The initial step in constructing a more complete hair cell model was the addition of the transduction channel, which transforms mechanical movement of the stereociliary bundle into a membrane conductance change. This was originally done by Hudspeth and Lewis (1988b).   Although   the transduction channel conductance is a non-linear, sigmoidal function of bundle displacement,   for   small   signal displacements it is approximately linear. Adding the transduction channel and driving the hair cell with sinusoidal displacement did not significantly alter the hair cell's frequency response.

Adding Bundle Dynamics:
The next step was to add bundle dynamics to the model using stiffness and damping values obtained by Howard and Ashmore (1986) and Howard and Hudspeth   (1988).   The   physical

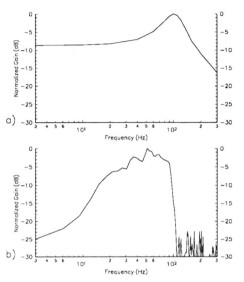

Figure 1. a) Frequency response using the R.S. Lewis saccular hair cell resonance model.   b) Typical Revcor derived frequency response of an *in vivo* bullfrog saccular axon (courtesy of Xiaolong Yu).

dynamics of the stereo-ciliary bundle added another pole to the hair cell transfer relation which increased the roll-off by another -20 dB/decade, where the input consisted of driving the model with small-signal sinusoidal force applied at the bundle tip and the output was taken as the membrane potential. Although this roll-off is slightly closer to what we observe in actual recordings of saccular fibers, it is still far too shallow.

Effect of Membrane Potential on Hair Cell Resonance:
Although the hair cell resonance is often thought of as analogous to a simple LRC (inductance-resistance-capacitance) electrical circuit, it differs in some fundamental ways. In their physiological experiments on isolated saccular hair cells, Hudspeth and Lewis (1988a) saw a change in both the resonant frequency and the damping of the hair cells' response as different stimuli shifted the hair cells' membrane potential. I studied the effect of membrane potential on frequency response by constructing hair cell models with varying resting potentials and obtaining their small-signal transfer functions. As the cell became more depolarized, the damping decreased while the resonant frequency increased. These results were consistent with the physiological

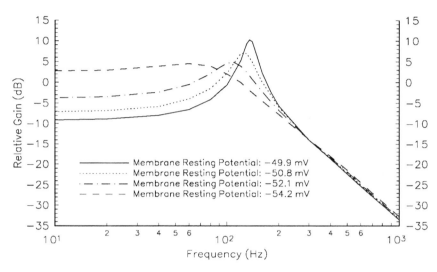

Figure 2. The effect of resting potential on the frequency response of the hair cell model (transfer relation - input: force on bundle; output: membrane potential).

data and demonstrate an important non-linearity in the hair cell response if its membrane potential varies by values on the order of a millivolt (see figure 2). However, this linear system analysis also makes clear that despite the change in resonant frequency and damping, the basic form, roll-off and phase behavior of the frequency response remains unchanged.

Adding Reverse Transduction (Voltage to Force):
Electro-mechanical transduction has been observed in hair cells from the frog and the turtle, as well as in mammals and seems likely to play a role in tuning. In the bullfrog, the ciliary bundle is seen to move back and forth in response to fluctuations of the hair cell's membrane voltage (Assad et al., 1989). I added reverse transduction to the model implemented as a force on the hair bundle proportional to the membrane voltage. The addition of reverse transduction altered the damping and shifted the peak of the frequency response of a solitary hair cell, with the stronger reverse transduction causing more damping and a higher resonant frequency (see figure 3). Reversing the sense of the reverse transduction had the opposite effect, undamping the system and moving the resonant frequency to a lower value. However, as with altering the membrane potential, adding reverse transduction did not change the basic nature of the tuning curve.

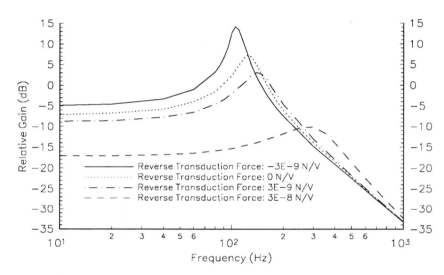

Figure 3. The effect of reverse transduction on the frequency response of the hair cell model (transfer relation: force on bundle to membrane potential). The electromechanical transduction force is defined as pushing away from the kinocilium as the cell is depolarized (Denk and Webb, 1992) and is given in Newtons per Volt of membrane potential change.

35.3 Discussion

After extending the hair cell model to include all the major features presently recognized that I felt could affect tuning, the revised hair cell model still did not exhibit a frequency response in accord with the tuning we observe in primary afferents from the intact bullfrog sacculus. This suggests that an isolated hair cell, despite the tuning due to its electro-chemical resonance, is insufficient to account for saccular tuning. On its own, the solitary hair cell exhibited tuning characteristic of a low dynamic order system. Furthermore, the high dynamic order tuning we see *in vivo* cannot be derived by a simple summation of the outputs of these low order hair cells (Lewis, E.R., 1988). It is more likely that the high order tuning results from a cascade type connection, with the output of one hair cell driving the input of another. This scenario is feasible given the existence of electro-mechanical transduction and the interconnections of the hair bundles through the overlying otoconial membrane. This hypothesis, proposed by Lewis, E.R. (1988), and the role of the other mechanical dynamics of the sacculus, are the focus of my current work. (Research supported by NIH grant DC00112.)

References

Assad J.A., Hacohen, N., Corey D.P. (1989) Voltage dependence of adaptation and active bundle movement in bullfrog saccular hair cells. Proc. Natl. Acad. Sci. USA Vol.86:2918-2922.

Denk W. and Webb W.W. (1992) Forward and reverse transduction at the limit of sensitivity studied by correlating electrical and mechanical fluctuations in frog saccular hair cells. Hearing Research 60:89-102.

Howard J. and Ashmore J.F. (1986) Stiffness of sensory hair bundles in the sacculus of the frog. Hearing Research 23:93-104.

Howard J. and Hudspeth A.J. (1988) Compliance of the hair bundle associated with gating of mechanoelectrical transduction channels in the bullfrog's saccular hair cell. Neuron, Vol. 1: 189-199.

Hudspeth A.J. and Lewis R.S. (1988a) Kinetic analysis of voltage- and ion-dependent conductances in saccular hair cells of the bull-frog, *Rana catesbeiana*. J. Physio., 400:237-274.

Hudspeth A.J. and Lewis R.S. (1988b) A model for electrical resonance and frequency tuning in saccular hair cell of the bull-frog, *Rana catesbeiana*. J. Physio. 400:275-297.

Lewis E.R. (1988) Tuning in the bullfrog. Biophys. J. 53:441-447.

Lewis R.S. (1984) The ionic basis of frequency selectivity in hair cells of the bullfrog's sacculus. Ph.D. Dissertation, California Institute of Technology, Pasadena, CA, U.S.A.

MECHANISMS OF HORIZONTAL SOUND LOCALIZATION AT LOW AND HIGH FREQUENCY IN AVIAN NUC. LAMINARIS

W. Edward Sullivan

Princeton University, Princeton, N.J. 08544

36.1 INTRODUCTION

Since Cajal [1], neuroscientists have been impressed with the variety of neuronal form and even without experimental support, the idea that structure and function must be related has not been seriously challenged over this time. The problem is that knowledge of a cell's anatomy, its pattern of synaptic input and its distribution of voltage-sensitive ion channels, does not enable us to infer with certainty what the cell computes, i.e., what its function is. However, if we know or can reasonably hypothesize what a cell does, we can begin to investigate the reasons for its morphology. That is, as Marr [2] realized, structure-function questions only become tractable once computation and algorithm are understood.

For many animals, interaural disparity in sound wave timing cues horizontal location. An algorithm to compute binaural time difference proposed by Jeffress [3] has received strong support [4-9]. In this model, time information from each ear travels along delay lines through an array of neurons acting as "coincidence detectors", firing maximally for coherent inputs. Binaural time tuning is determined by array position: a time code has been transformed to a place code. Low frequency (up to 100's of Hz) coincidence detectors are bipolar neurons with a few, long highly branched dendrites, each side receiving input from one ear. Their axons are initially unmyelinated. In chickens, higher frequency cells have shorter, less branched but more numerous dendrites [10]. In barn owls, where 5 - 9 kHz frequencies are processed, these cells are adendritic [7], and their thicker initial axon has a myelin sheath abutting the soma [11]. These anatomical differences must be related to the constraints on coincidence detection at low and high frequency.

36.2 FUNCTION OF BIPOLAR DENDRITES

Although the adendritic barn owl cells seem unusual, this may reflect a need to avoid high frequency information loss through low-pass filtering. Why then do low frequency cells have dendrites, i.e., what can be done with dendrites that is not possible in an iso-potential cell? Fig. 1 shows a potential role based on two physiological facts: 1) that dendrites provide electrical isolation and 2) that post-synaptic voltage saturates at some equilibrium potential By moving voltage closer to their equilibrium, excitatory synapses reduce the effectiveness of other nearby inputs. Consequently, two inputs that are far apart, but equidistant from the soma - by being on opposite dendrites - produce greater somatic depolarization than two inputs to the same site (fig. 1). This produces what I call a "dendritic advantage" and allows bilateral coincidences to be selected, a function that cannot be done in an iso-potential cell. The bilateral excitation advantage can be more than 10 mV and may therefore enhance binaural time comparison given variable synaptic input.

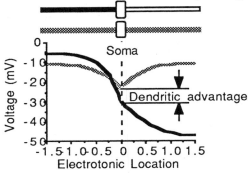

Figure 1: Shows steady state voltage vs. position (re. the dendrite length constant) in a simple dendrite model excited bilaterally (grey curve) at 1/2 the strength of unilateral excitation (black curve). Dendritic advantage is the voltage difference between these two conditions at the soma.

Dendritic advantage increases with length, excitatory conductance and synaptic isolation from the soma, and so is maximal in cells with few, long, relatively thin and highly branched dendrites. Increased length, peripheral branching and distal excitation enhance bilateral voltage contrast both by increasing voltage saturation on the excited dendrite and by providing greater current shunting to the opposite side. This makes bilateral excitation relatively more effective. These optimal models look like low frequency cells of the chick, but do not resemble higher frequency cells. This can be resolved by looking at the model's response to periodic

Figure 2: Normalized maximum dendritic advantage shows how optimal length changes with stimulus frequency. Excitatory conductance were sine functions of the relative wavelengths shown in the inset

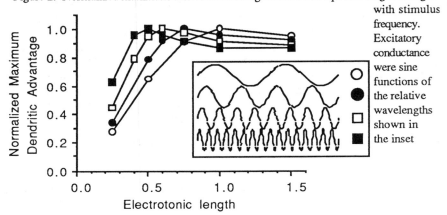

excitation where both depolarization and repolarization must be considered. At higher frequency, there is less time between excitatory inputs for repolarization to occur. Residual depolarization reduces bilateral voltage contrast and is more severe in cells with longer dendrites that can be more easily depolarized distally. Thus, if Rm does not decrease to offset this problem, optimal dendrite length decreases as best frequency increases(fig. 2), similar to what is seen in the chicken. If resistance decreases with best frequency, a length gradient is also found, but electrotonic length is now conserved while physical length decreases. These two assumptions make different predictions about other parameters such as optimal dendrite number [12] suggesting that membrane time constant is at least partially conserved.

36.3 ACTIVE MECHANISMS FOR uSEC RESOLUTION:

Humans and other species can distinguish μsec changes in interaural time disparity at low frequencies. What role might single neuron physiology play in this acuity? To examine this, a model consisting of a soma, axon initial segment and myelinated axon was used (fig.3). Temporal selectivity was assessed by finding the maximum delay between two subthreshold inputs capable of causing a spike for a range of subthreshold strengths. Smaller maximum delays less sensitive to changes in synaptic strength, indicate greater temporal selectivity. The results suggest that μsec time resolution can be obtained by a combination of anatomical and biophysical factors. Models showing greatest temporal selectivity had high densities of voltage sensitive potassium channels in a long, thin unmyelinated axon hillock. Their strong delayed rectification enhances the Hodgkin-Huxley (H-H) model's selectivity against the slow depolarization experienced in the elongated proximal axon. Spike generation therefore requires rapid excitation. These models had high threshold where both sub- and suprathreshold excitation produce saturating voltages. For this reason, peak or integrated depolarization did not affect spike probability. Rather, in the most selective models, initial depolarization rate is the sole determinant of spike output. The long initial segment serves as a serial temporal filter in that each distal-ward segment receives a successively slowed voltage change. The first node of the myelinated axon, which has a normal, sodium

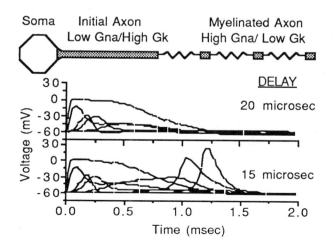

Figure 3: Results of simulations with model designed to study mechanisms of microsecond time resolution. Two sub-threshold inputs to the soma were used (0.6 x single threshold). Curves are voltages at proximal to distal compartments in the model (Left to Right).

channel dense membrane can then act as a threshold detector of the spatial-temporally filtered synaptic input. Although parameters for sub-μsec selectivity can theoretically be found, it is not likely that real neurons do this because if a detector is too sharply tuned, it will be rarely activated due to time jitter in its synaptic input. Nevertheless, these results indicate that biophysical mechanisms may be important in temporal "hyperacuity" phenomena.

In the above mechanism, temporal sharpening requires delaying and low-pass filtering, so that the model cannot fire rapidly and is easily inactivated by repetitive inputs. This property is not detrimental at low frequency, but in the barn owl, such filtering seems to be avoided. The thick myelinated initial axon segment of these adendritic cells provides close electrical coupling between synaptic input and spike output. Simulations suggest that in this case, coincidence detection involves the generation and detection of a high frequency synaptic signal proportional to the phase difference between inputs from the two sides. With the greater resolution of high frequency signals, the owl has less need to improve temporal selectivity. What it requires instead are mechanisms to improve high frequency responsiveness.

36.4   HIGH FREQUENCY MECHANISMS:

These simulations were done in three steps. First, spike trains having the phase-locking and spike interval properties of magnocellular cochlear nucleus fibers were generated pseudo-randomly. Synaptic input was simulated by assuming that each spike in the afferent population produces a unitary conductance change in the laminaris cell and that all synapses operate in parallel (since the barn owl cell can be treated as iso-potential). The ability of spiking mechanisms to respond in a phase-selective fashion was explored with an H-H model to which parameter variations were applied. The results reveal several pre- and post-synaptic properties that appear to be required for phase comparison above 5 kHz.

Synchronized phase locking of all afferents, which occurs at the best binaural phase difference, produces a synaptic conductance modulation at the phase-locking frequency, which disappears when the two halves of the cell's input are locked to opposite phases. In addition to this phase dependent modulation, a broadband fluctuation of conductance occurs due to the stochastic nature of the input spike trains. The phase-dependent modulation depends on synaptic input number, spike rate, phase locking strength and on the rise time of the excitatory conductance change (which must be less than 200 μsec above 5 kHz). In contrast, noise power is de-termined by a complex relation between EPSC duration and spike rate variability (measured by the interspike interval distribution's coefficient of variation). For EPSC's much shorter than the mean spike interval, interval variation has little effect on synaptic noise. For long EPSCs, noise increases with interval variability. Another synaptic property that carries no timing information is the average or steady component of the summed input. This is related to cumulative input rate and EPSC duration. Short EPSCs therefore maximize the information bearing signal relative to steady excitation and noise. Large numbers of synaptic inputs are also important since the phase dependent signal increases linearly with number while noise typically increases more slowly [13].

The problem for the barn owl laminaris cell is therefore to detect rapid modulations in its synaptic input. Simulations with a modified H-H model suggest that this can be done by increases in channel density, rate kinetics and in the proportion of voltage sensitive potassium channels. Increases in channel density lower the cell's active membrane time constant while potassium channels provide a high-pass filtering mechanism. At a minimum, the laminaris cell should not elim-inate the information bearing signal by electrical filtering but if this signal is weak, it may need to be extracted from background noise. Potassium channels do this by selecting against slow or low frequency depolarizations, analogous to their proposed function in low frequency coincidence detection described above. In add-ition to channel density, increases in gating speed were evaluated. At low density, faster gating had little effect because the speed at which the membrane can react to rapidly changing synaptic currents is limited by its time constant. Increased conductance thereby allowed faster kinetics to be used, so that good high frequency responsiveness could be obtained with smaller changes in each. Once the constraint on membrane time response was removed, increases in gating speed were found to be more effective in producing further improvements in high frequency response.

High frequency phase comparison thus requires a synaptic signal that is detectable by the post-synaptic cell. The former is enhanced by using short synaptic durations with rapid rise times and the latter by fast membranes and channels. However, since we are dealing with a serial process, the actual system may involve a compromise with less radical changes to both synaptic input and spike output mechanisms.

The problems of binaural phase comparison are different at low and high freq-uencies and the large structural differences may reflect an equal diversity of bio-physical mechanism. While the structure-function question is complicated by noting that even cells with similar functions can have different structures due to quantitative differences in their task, by focusing on well defined functions, clear examples of anatomical / physiological relationships may be found. Furthermore, some of these mechanisms may not be unique to binaural phase comparison. The dendrite model for example, provides a general means by which sensitivity to specific input combinations can be produced, and this can work equally well within as it does between dendrites. In addition, coincidence detection itself may have many applications, such as directional selectivity in the visual system (which may involve "anti-coincidence" as well). More abstractly, if two stochastic inputs are subjected to coincidence detection, output rate is proportional to the product of the input rates [14]. Greater temporal selectivity enables greater dynamic range for such multiplicative neural operations. Finally, the demonstration that potassium channels provide selectivity for voltage change by reducing responsiveness to steady input suggests their role in these and other temporal operations.

1) Cajal, S.R. y (1909) "Histologie du Systeme Nerveux".
2) Marr (1981) "Vision", W.H. Freeman, N.Y.
3) Jeffress, L.A. (1948) J. Comp. Physiol. Psychol. 41: 35-39.
4) Goldberg, J.M. and Brown, P.B. (1969) J. Neurophysiol. 32: 613-636.
5) Young, S.R. and Rubel, E.W. (1983) J. Neurosci. 3: 1373-1378.
6) Sullivan, W.E. and Konishi, M. (1986) Proc. Natl. Acad. Sci. 83: 8400-8404.
7) Carr, C.E. and Konishi, M. (1990) J. Neurosci. 10: 3227-3246.
8) Overholt, E.M., Rubel, E.W. and Hyson, R.L. (1992) J. Neurosci. 12: 1698-1708.
9) Yin, T.C.T. and Chan, J.C.K. (1990) J. Neurophysiol. 64: 465-488
10) Smith, D.J. and Rubel, E.W. (1979) J. Comp. Neurol. 220: 199-205.
11) Carr, C.E. and R.E. Boudreau (1992) J. Comp. Neurol. (submitted).
12) Sullivan (1992) J. Neurophysiol. (submitted).
13) Sullivan (1992) J. Neurosci. (in prepaaration).
14) Srinivasan, M.V. and Bernard, G.D. (1976) Biol. Cybern. 21: 227-236.

A NEURONAL MODELING SYSTEM FOR GENERATING BRAINSTEM MAPS OF AUDITORY SPACE

Bruce R. Parnas
Edwin R. Lewis

University of California, Berkeley, CA 94720

ABSTRACT

We have developed a model for the mammalian medial superior olive (MSO) based on the concept of neuronal cross-correlation. The input to the model is derived from a representation of the cochlea and auditory nerve. We show that the model is capable of discriminating the difference in the time of arrival of signals at the two ears not only for single tones, but for a single delayed tone within a tone complex as well.

37.1 INTRODUCTION

The nuclei in the auditory brainstem undoubtedly generate maps which partition auditory space along dimensions corresponding to parameters of interest in the perception of sound. One such nucleus, the medial superior olive (MSO) maps the spatial location of a sound source (as measured by the difference in time of arrival of the signal at the two ears) into a firing pattern across a population of neurons. This pattern of activity is tonotopically arranged within the nucleus so that the interaural time difference (ITD) as a function of frequency is encoded. One model proposed for the function of the MSO in decoding the ITD is that of cross correlation performed by coincidence detector neurons (Jeffress, 1948). This scheme is quite popular and has also been found in more recent models for the MSO (Colburn, et al., 1990).

We have developed a model for the MSO based on the coincidence detection concept. The input for the model derives from a modeling framework for the auditory periphery (Parnas, et al., 1991) with the signals relayed by spike initiator models (Parnas and Lewis, 1992). The input is received directly from the model VIIIth nerve under the assumption that the intervening subnucleus (the anteroventral cochlear nucleus (AVCN) acts as a simple relay station). This is obviously an oversimplification, but the nature of the synapses in that structure make this a reasonable first-pass assumption.

37.2 MSO MODEL

The MSO model we have developed is intended to examine the preservation of timing information as the signal is passed through a bank of lowpass filters with sharp high-frequency edges (a model for the cochlea) and converted into spike trains by the spike initiator models. The current implementation is aneuronal employing, instead, an algorithmic cross-correlation. Future implementations of this module will see this algorithm replaced by models for MSO neurons. The present simplified version allows a quick exploration of the influence of the auditory front end on signal timing. The system model is shown in figure 1. There are 100 cochlear filter channels and each is connected to 10 auditory nerve fibers. The noise processes in each of the 1000 spike initiators are independent. The responses from all 10 fibers are summed at each time interval as a measure of the

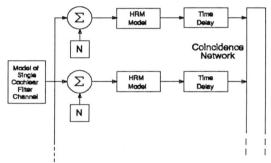

Figure 1. Block diagram for the auditory front-end and MSO system. Each of 100 filter channels has 10 auditory nerves. Each nerve has independent noise. The delayed outputs feed a coincidence network.

number of units actively firing at any timestep. These are collected in array with filter frequency versus timestep. The correlation time is t_h, the approximate acoustic delay the head. A good value for t_h is 500 s for an average-sized human head. For each ear, at a given timestep, values of this summed activity for the previous t_h are stored in a vector. The resulting cross-correlation is thus a function of τ, the lag variable, and of t, the time variable:

$$MSO(\tau,t) = \int_{t-t_h}^{t-\tau} ear1(t')\, ear2(t'-\tau)\, dt'$$

where

$$\tau \in [-t_h, t_h] \quad , \quad ear1(t), ear2(t) = 0 \; t \notin [-t_h, 0]$$

and ear1(•) and ear2(•) are the time signals arriving at the two ears. Thus, for any time, t, a new MSO(τ) function is created at each frequency in the tonotopic map. This results in a dynamic mapping of ITD onto the MSO structure.

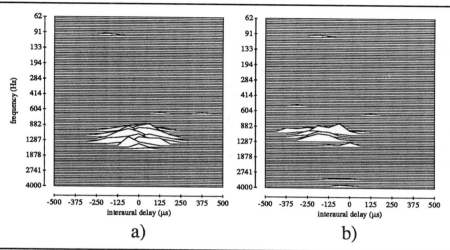

a)   b)

Figure 2. MSO images for a single 1 kHz tone taken at 15.5 ms during a 37.5 ms simulation. a) same signal to both ears. b) signal delayed 200 s to contralateral ear.

37.3 RESULTS

This simple MSO model was tested using tonal stimuli in order to evaluate its signal localization properties. Figure 2 shows the response of the MSO model to a single 1 kHz tone. Panel a shows the presentation of the same signal to both ears and in panel b the signal to the contralateral ear is delayed by 200 s. The noise to each of the ears is different, even in the situation where the stimulus is the same. In a) the activity is localized to the region around 1 kHz. The spread is due to the fact that the cochlear filters are essentially lowpass, rather than bandpass. Since the signal is the same to both ears one would expect a delay of 0 s to be favored and, indeed, the peak of the activity is centered at 0 s. In b) the frequency region of the signal is as in a), but the peak of activity has shifted to about -200 s, indicating that the signal

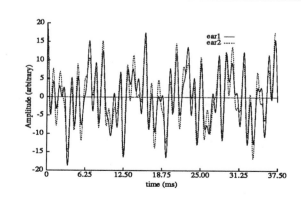

Figure 3. Waveforms composed of 135, 437, 753 and 1000 Hz signals. For ear1, all signals are in cosine phase. For ear2, the 753 Hz tone is delayed by 200 s.

arrives first at the ipsilateral ear or, conversely, is delayed to the contralateral ear by 200 s. It should be noted that, due to the form of the cochlear filters and the presence of intrinsic noise, one should not expect a clear single peak in the MSO image, but rather a locus of activity around a particular delay value. It seems to be an approximate symmetry in the image that indicates no delay and a deviation from this symmetry that indicates a non-central spatial position.

We now turn to a situation more like one that the auditory system must face: a complex stimulus waveform. We have created an input which consists of 4 tones at nonharmonically-related frequencies (135, 437, 753 and 1000 Hz). In one case none of the signals are shifted; in the other, the 753 Hz tone is delayed by 200 s to the contralateral ear. The two waveforms are shown in figure 3. The signals are superimposed to show that it is very difficult to separate them by eye and impossible to determine that one component has simply been time shifted. Figure 4 shows the response to the stimulus paradigm of figure 3, taken at two different time slices. In

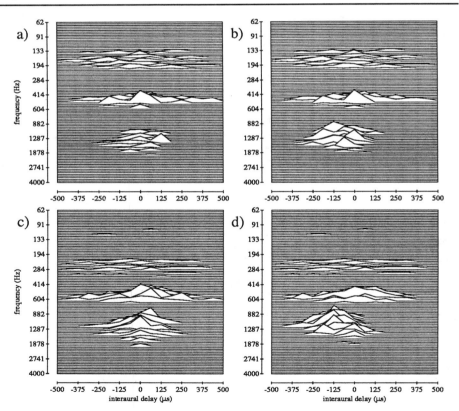

Figure 4. MSO responses to the stimulus paradigm of figure 3. Left panels: same signal to both ears. Right: 753 Hz tone delayed 200 s to contralateral ear. Top: time = 30.50 ms. Bottom: time = 34.50 ms.

the left panels, a and c, the signal is the same (ear1) to both ears. In the right panels, b and d, the contralateral ear receives the altered signal, ear2. At each of the timesteps, different portions of the signal energy spectrum are shown. In the left panels, the energy in all bands is centered about a 0 s delay. In the right panels we see energy in filters above 1 kHz. This is due to the low-frequency tails of the higher cutoff-frequency filters. In these bands, the (unshifted) 1 kHz energy dominates, and these signals are still centered at 0 s. As we approach the filters near the (shifted) 753 Hz signal, we see that the energy peak shifts to the left, toward the delay value of 200 s of this component. The shift is gradual, not sharp, due to the fact that the nature of the cochlear filters allows signal energy from a band of frequencies into each channel. Thus there is shifted and unshifted energy in the same channels. Below the 753 Hz band, the signals are centered about 0 s. While the two time slices show similar information, it is important to note that the image created by the MSO is likely a continuously moving landscape of energy without clearly defined peaks, but simply general symmetries and asymmetries.

37.4 CONCLUSIONS

We have developed a simple model for MSO processing based on the correlation model of Jeffress. The model is capable of localizing a single shifted tone within a more complex waveform, but it does not do with a clearly-defined peak shifted to the exact value of the delay. The nature of the cochlear filters and the spike initiation process in the auditory nerve create a relatively complex MSO image which must be interpreted in order for spatial position to be ascertained. We plan to improve upon this model and perform more experiments on spatial localization and sound source segregation.

Acknowledgements. This work was supported by the Office of Naval Research, ONR grant # N00014-91-J-1333.

REFERENCES

Colburn H.S., Y. Han and C.P. Culotta (1990), Coincidence model of MSO responses. *Hear. Res.*, **49**: 335-346.

Jeffress, LA (1948), A place theory of sound localization. *J. Comp. Physiol. Psychol.*, **41**: 35-39.

Parnas, B.R., K.S. Gangnes, D.A. Feld, W.L. Lee and E.R. Lewis (1991), A parallel neural model for auditory front end processing, in **Analysis and Modeling of Neural Systems**, F.H. Eeckman, ed., Kluwer Academic Publishers, Norwell, MA, pp. 283-288.

Parnas BR and ER Lewis (1992), A computationally efficient spike initiator model that produces a wide variety of neural responses, in **Neural Systems: Analysis and Modeling**, F.H. Eeckman, ed., Kluwer Academic Publishers, Norwell, MA, pp. 67-75.

SECTION 5

OTHER SENSORY SYSTEMS

This section includes topics in chemotaxis, olfaction and electroreception. A paper on the cricket wind detection system can be found in section 1 (chapter 2). The olfactory system has always been a favorite with computational neuroscientists. Its atypical structure and connectivity have attracted theorists and modelers alike. The olfactory system has been used as a model system for the study of cortical dynamics and associative memory (See sections 9 and 10).

38

MODELING CHEMOTAXIS IN THE NEMATODE *C. elegans*

Shawn R. Lockery, Steven J. Nowlan
and Terrence J. Sejnowski

Computational Neurobiology Laboratory
Salk Institute and Howard Hughes Medical Institute
La Jolla, CA 92037

ABSTRACT

To elucidate the neural mechanisms of chemotaxis in the nematode C. elegans, we constructed a model based on the anatomically defined neural circuitry associated with identified chemosensory neurons. The model combines the temporal derivative of chemosensory input with an internal representation of behavioral state to produce a duty-cycle controller of head angle during swimming movements. The model reproduces observed chemotactic behavior and suggests that separate control circuitry is required when moving up as opposed to down the concentration gradient.

The problem of the neural basis of chemotaxis in *C. elegans* raises important issues in sensorimotor integration. The exposed tips of its chemosensory neurons are too close, and at the wrong orientation during locomotion, for the animal to take an instantaneous spatial derivative of concentration [1]. Rather, it is believed the worm computes the temporal derivative of concentration, a task containing an inherent memory component. Moreover, the behavioral significance of the temporal derivative depends upon the action performed at the time the derivative is computed. Derivative information must be integrated with behavioral state to compute the correct response.

As a first step in understanding the neural basis of chemotaxis in *C. elegans*, we analyzed the anatomical circuitry database [2] for connections that could contribute to this behavior. For simplicity, we focussed on the pathways from the chemosensory neuron ASER, whose ablation produces the greatest deficit in chemotaxis [3], to motor neurons projecting predominantly to either dorsal or ventral muscles. ASER connects directly to 11 first-order sensory and interneurons which connect in turn to

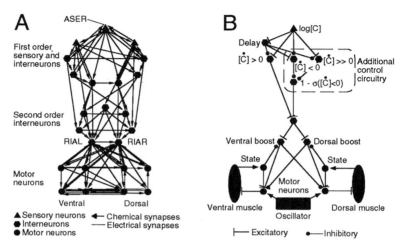

Figure 38.1. A. Anatomically defined connections from a chemosensory neuron (ASER) to motor neurons innervating dorsal or ventral muscles of the head and neck. The connections shown are those mediated by interneurons RIAL and RIAR, which uniquely connect to both dorsal and ventral motor neurons. Sensory neurons are shown as triangles, interneurons as hexagons, and motor neurons as circles. Lines with arrows indicate chemical synapses; lines without arrows indicate electrical synapses. B. Neural network model based on the anatomical connections in A. In Model 1, the derivative of concentration is computed by the unit labeled $[\dot{C}]>0$, which receives direct and delayed input from the chemosensory neuron. The derivative is combined with behavioral state by the dorsal and ventral boost neurons, which provide duty-cycle control of head movements driven by an oscillator. Model 2 is similar to Model 1 except that additional control circuitry (dashed box) has been added for moving down the gradient.

an additional 22 second-order interneurons. Two first-order and 10 second-order interneurons make direct connections to dorsal or ventral motor neurons. Chief among these are interneurons RIAL and RIAR which are presynaptic to 8 and 7 motor neurons, respectively, and are unique in contacting both dorsal and ventral motor neurons. Restricting the analysis to the motor effects of these interneurons yields a simplified circuit (Fig 38.1A) with 8 first-order neurons that have connections to RIAL, RIAR, or the 5 second-order interneurons presynaptic to them.

A simple yet plausible model for chemotaxis in *C. elegans* assumes the normal oscillatory swimming movements of the head are biased in the direction of increasing attractant concentration. To explore this possibility, we constructed a model worm having a head and tail joined by a single flexible segment. The area of the flexible segment is constant, representing the constant volume constraint imposed by the hydrostatic skeleton of the worm. Head angle is controlled by muscles, represented as spring-dash pot systems, on either side of the flexible segment. Rather than model the biomechanics of sinusoidal swimming movements in detail [4], we assume a constant and appropriately scaled tangential velocity in the direction given by the head angle. The muscles receive out-of-phase sinusoidal inputs resulting in characteristic, quasi-sinusoidal swimming movements whose amplitude matches those of previously published worm tracks [1].

The purpose of Model 1 was to determine whether swimming movements could be biased appropriately by a simple neural network (Fig. 38.1B) that computes the temporal derivative of attractant concentration and integrates this information with behavioral state. The model was constrained by the connections in the biological network (Fig. 38.1A). Neurons were represented as single electrical compartments with a sigmoidal synaptic transfer function [5] and a realistic time constant (10 ms) derived from anatomical measurements of typical *C. elegans* neurons [6] and standard values for specific membrane capacitance and resistivity [7]. The derivative was computed using a three-neuron circuit that takes the difference between the current chemosensory input and a delayed version of the same signal. Direct and delayed sensory inputs are a common feature of first-order interneurons in the biological network. Behavioral state was represented by a stretch receptor on each side of the model. However, other representations of behavioral state such as motor neuron activity would serve as well. Derivative and state information were integrated by dorsal and ventral boost neurons. RIAL and RIAR are candidate boost neurons.

Model 1 operates as a duty cycle controller of head angle. For example, if the derivative is made positive by a head sweep to the dorsal side, the ventral boost neuron is inhibited while the dorsal boost neuron is excited. The main effect of exciting the dorsal boost neuron is to inhibit the contralateral motor neuron. This delays the next contraction on the ventral side, shifting the duty cycle of head sweeps in favor of the dorsal side (Fig. 38.2A), resulting in movement biased toward higher attractant concentration. Each boost neuron also excites the ipsilateral motor neuron, but this effect is comparatively small relative to contralateral inhibition.

Model 1 orients successfully as it approaches the center of the gradient (the "x" in Fig. 38.2B), then fails in an instructive way. As it moves away from the center the amplitude of its swimming movements is reduced, and it never turns back toward the center. Analysis of the problem revealed that the primary source of derivative information is not the side-to-side head movements, but simply the forward motion through the gradient. This means that as the model worm moves away from the center of attractant the temporal derivative of concentration is strongly negative and three-neuron derivative circuit, which is specifically tuned for positive derivatives, is shut down.

The failure of Model 1 when moving away from the center of the gradient was corrected in Model 2 by additional circuitry that takes over when the overall derivative is negative. Model 2 includes a neuron that responds only when the derivative is negative ($\dot{[C]} < 0$, Fig. 38.1B). The output of this neuron is subtracted from the activity of a neuron with a strong positive bias ($1 - \sigma([C]) < 0$). The result is a pathway

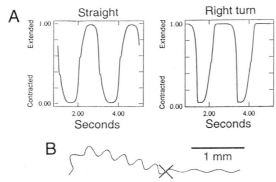

Figure 38.2. A. Duty-cycle control of orientation. The state of the left muscle in the model network is shown during straight-ahead movement (top) or a right turn (bottom). Straight-ahead movement involves a 50% duty cycle. In turns, however, the extension phase is longer on the side opposite the turn. B. Chemotaxis of Model 1 in response to a gradient centered at the "x". The model passes through the center of the gradient, but fails to turn back. This problem was solved in Model 2 by additional control circuitry.

that detects the direction in which the derivative is decreasing most slowly, the appropriate control signal when the overriding derivative is negative. Balance between the positive and negative derivative pathways is achieved by a neuron that shuts off the negative pathway when the derivative is strongly positive ($[\dot{c}] \gg 0$). Thus, chemotaxis in this system uses separate control circuitry for moving up and down the gradient, and a means of switching between these circuits.

With its additional control circuitry, plus noise added to each neuron, Model 2 successfully reproduces chemotactic behavior in a variety of conditions (Fig. 38.3). First, in the presence of a gradient model worms, like the real ones, swim toward the center of the gradient and hover there for extended periods. Second, in the absence of a gradient, the animals wander in confined regions. Finally, with the addition of a bias in head angle, the model reproduces the spiraling trajectories of a strain of worms in which the head is permanently bent to one side.

These preliminary models provide the basis for construction of more realistic models of *C. elegans* chemotaxis. The high degree of convergence and divergence of both sensory and motor information suggested by the anatomical circuitry (Fig. 38.1A) points to a distributed processing mechanism for the integration of sensory and motor state.

Using neural network training algorithms like backpropagation [8], such a model can now be constructed by optimizing a model with the anatomically correct connections to reproduce the sensory and motor neuron activity produced by Model 2 as it successfully negotiates the gradient. This procedure can be expected to reveal novel means of computing the derivative of chemosensory information. Moreover, it provides a theoretical basis for the interpretation of experiments in which identified chemosensory neurons and interneurons are ablated in living worms.

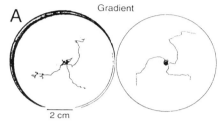

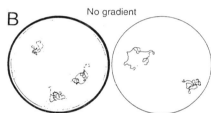

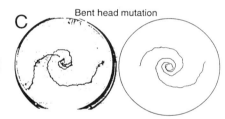

Figure 38.3. Chemotactic behavior of actual worms (left side) and of several different runs of Model 2 (right side). Performance is shown for 1-3 individuals in the presence (A) or absence (B) of a gradient, and for the case in which the worms have an inherent turning bias produced in actual worms by the bent head mutation (C). The model reproduces the behavior of actual worms in each case.

REFERENCES

1. Ward, S. 1978. Nematode chemotaxis and chemoreceptors. *Taxis and Behavior.*   G. L. Hazelbauer (eds.) Chapman and Hall, London.
2. Achacoso, T. B. and W. S. Yamamoto 1992. *AY's Neuroanatomy of* C. elegans *for Computation.* CRC Press. Boca Raton.
3. Bargmann, C. I. and H. R. Horvitz 1991. Chemosensory neurons with overlapping functions direct chemotaxis to multiple chemicals in *C. elegans.* Neuron 7:729-742.
4. Niebur, E. and P. Erdos 1992. Theory of the locomotion of nematodes: Dynamics of undulatory progression on a surface. Biophys. J. 60:1132-1146.
5. Davis, R. E. and A. O. W. Stretton 1989. Signaling properties of *Ascaris* Motorneurons: Graded active responses, graded synaptic transmission, and tonic transmitter release. J. Neurosci. 9:415-425.
6. White, J. G., E. Southgate, J. N. Thomson and S. Brenner 1986. The structure of the nervous system of the nematode *Caenorhabditis elegans.* Phil. Trans. R. Soc. Lond. 314:1-340.
7. Claiborne, B. J., A. M. Zador, Z. F. Mainen and T. H. Brown 1992. Computational models of hippocampal neurons. *Single Neuron Computation.* T. McKenna, J. Davis and S. F. Zornetzer (eds.) Academic Press, San Diego, CA.
8. Rumelhart, D. E., G. E. Hinton and R. J. Williams 1986. Learning internal representations by back-propagating errors. Nature 323:533-536.

Formal Model of the Insect Olfactory Macroglomerulus

C. Linster[*1], C. Masson[**], M. Kerszberg[***],
L. Personnaz[*], G. Dreyfus[*]

* ESPCI,Laboratoire d'Electronique, 10, rue Vauquelin,
75005 PARIS - FRANCE
** Laboratoire de Neurobiologie Comparée des Invertébrés
INRA/CNRS (URA 1190), 91140 BURES SUR YVETTE - FRANCE
***Institut Pasteur, Neurobiologie Moléculaire,
CNRS (UA 1284), 25, rue du Docteur Roux, 75015 PARIS

Abstract

We present a model of the specialist olfactory system of selected moth species and the cockroach. The model is built randomly, constrained by biological (physiological and anatomical) data. Among the observations made in our simulations a number relate to data about olfactory information processing reported in the literature, others may serve as predictions and as guidelines for further investigations. We discuss the effect of the random parameters of the model on the observed model behavior and on the ability of the model to extract features of the input stimulation.

Introduction

Related to two classes of odor signals and their associated behaviors, two olfactory sub-systems exist in insects: the specialist sub-system (devoted to the processing of sex pheromones) and the generalist sub-system (devoted to the processing of food odors) (for a review see (Masson & Mustaparta 1990)). This presentation focuses on the specialist sub-system, with a view to more general modeling of the olfactory pathway. We base our model on biological data pertaining to the specialist sub-system of several moth species and of the cockroach (for a review see (Christensen & Hildebrand 1987; Burrows et al 1982)).

The formal model

In the model, we consider two types of neurons: receptor neurons and local interneurons. Receptor neurons are *integrate and fire* neurons which receive input stimulation representing various odor components; they are characterized by their integration slope and

---

[1]Christiane Linster has been supported by a research grant (BFR91/051) from the Ministère des Affaires culturelles, Grand-duché de Luxembourg.

their firing threshold. The axons of receptor neurons project into the network of local interneurons, which are probabilistic neurons with a membrane time constant. Local interneurons may make dendro-dendritic, excitatory or inhibitory synapses with any other interneuron, thus introducing feedback into the network of local interneurons (Figure 1). The interneurons are characterized by their membrane time constant, their firing threshold and their synaptic noise. Synaptic delays associated with dendro-dendritic synapses are meant to model all sources of delay, transduction and deformation of the transmitted signal.

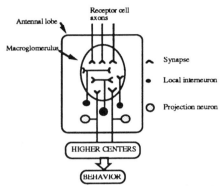

Figure 1: Schematic representation of the olfactory specialist sub-system of most insect species. Projection neurons and higher information processing centers are not considered in the model described here

The mean value of the delay distribution is longer for inhibition than for excitation: we thus take into account approximately the fact that IPSC's usually have slower decay than EPSC's, and may accumulate to act later than actually applied.

Results

In order to analyze the behavior of our network, we first introduce a classification of the possible response patterns of the neurons, based on the temporal aspects of the responses, which is inspired by the methodology used for the analysis of neurophysiological data gathered from the olfactory system (Meredith 1986).The results of the analysis for a network which exhibits a typical distribution of response patterns to stimulations with pure and mixed odors have been described in detail before (Linster et al. 1992): the neurons respond with a limited and stable number of response patterns, most of them a combination of inhibition and excitation*, which are in agreement with biological

---

* The results described in this chapter correspond to a typical network with the following parameters: 50 neurons; 30% receptor neurons, 30% excitatory interneurons, 40% inhibitory interneurons; all the neurons in the network make randomly chosen synapses with 10% of their possible target neurons; the synaptic weights are +/- 1; the sampling step for simultaneous updating (Δt) corresponds to 5ms in real time, which is enough to study the maximal physiological spiking frequencies; all interneurons have a membrane time constant of 25 ms; synaptic delays are chosen in a uniform distribution between 10 and 50 ms for excitatory synapses and between 10 and 100 ms for inhibitory synapses.

observations (Christensen and Hildebrand 1987; Olberg 1983; Burrows et al. 1982) (Figure 2).

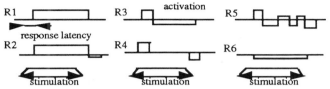

Figure 2: Schematic representation of the 6 types of activation patterns in response to receptor neuron stimulation observed in the model. Each single stimulation is characterized by its amplitude, its rise and fall times and its duration

We show how the responses of individual neurons reflect various features of the input pattern (pure odors or mixed odors) which are important for odor recognition and localization (Kaissling & Kramer 1990): amplitude, duration, frequency, rise time and composition of the odor mixture (see Figures 3, 4 and 5 as examples).

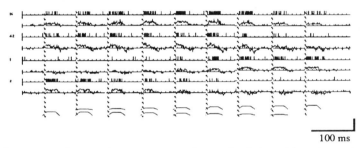

100 ms

Figure 3: Interneurons have various responses to input stimulation with various mixture ratios. The figure shows the responses (upper lines: action potentials, bottom lines: membrane potentials) of several interneurons to input stimulations which last 50 ms, and have a rise time of 10 ms. The total amplitude of the two inputs A and B is constant during the experiment, the ratio of the two inputs varies at each stimulus presentation.

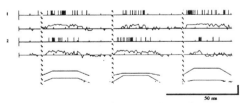

Figure 4: The latency of response to stimulation depends on the concentration ratio of odors in the stimulation mixture. The comparison of response latencies is potentially a powerful means for odor recognition in higher processing centers

The information processing that such a network can perform on the input stimulation strongly depends on the complexity of the response patterns, and therefore also on the architecture and the parameters of the network.

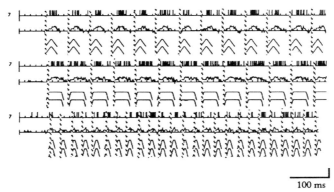

100 ms

Figure 5 shows the responses of one interneuron to stimulation with the same input ratio (A = B = 2) at varying frequencies and interstimulus-intervals. Similar behaviors in response to pulsed stimulation have been observed in antennal lobe local interneurons and projection neurons in Manduca Sexta (Christensen and Hildebrand 1988) and in Heliothis virescens (Christensen et al. 1989). In both species, some antennal lobe neurons follow pulsed input with bursts up to acut-off frequency (upper diagram: stimulation duration 30ms, interstimulus interval 20ms; middle diagram: stimulation duration 40ms; interstimulus interval 10ms; bottom diagram: stimulation duration 20ms; interstimulus interval 10ms).

When shunting the inhibition in the network, these interneurons respond continously to pulsed stimulation; this phenomena has been observed in Manduca Sexta antennal lobe neurons (Christensen & Hildebrand 1988), where blocking of GABA transmitter lead to a continous response of these neurons to pulsed stimulation. This shows that the architecture of the system, and not intrinsic neuron properties are responsible for the mixed response patterns observed in the MGC.

In order to gain more insight into the distribution of the response patterns as a function of all the network parameters, we have designed a neural classifier (a single layer perceptron trained with the pocket-algorithm), which allows us to automatically classify the response patterns observed in the networks and to perform a statistical search through parameter space. Figure 6 shows the response distribution in a network with fixed parameters, with a varying percentage of connections between interneurons.

Discussion

Some aspects of olfactory information processing predicted by our model may be considered as guidelines for further investigations: (i) both the detection of the ratio of the two components at a precise concentration by individual neurons, and the detection of ratios at varying concentrations by across-line comparison between neuronal activities, are possible in the model; (ii) the influence of the stimulus profile onto the responses can also be investigated by making use of the model.

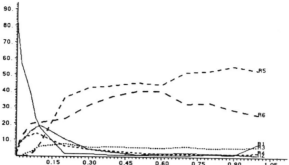

Figure 6: Mean distribution (100 trials) of the response patterns R1-R6 as a function of the connectivity in the network of interneurons (1.0 in the abscissa corresponds to full connectivity, the ordinate shows the ratio of the number of response paterns of each type to the total number of responses).

The information processing performed by the network of interneurons results in separate representations of a number of features of the input components, and in a representation of the relative concentration of the components. The model generates a diversity of cellular responses which are probably exploited by further stages (e.g. the protocerebrum) for odor identification. We thus show the importance and the role of the various response patterns observed in natural and formal networks. These results suggest that a direct relationship exists between (i) the architecture of the system, (ii) the diversity of response patterns which may be observed, and (iii) the complexity of information that may be processed by such a system.

References

Burrows, M., Boeckh, J., Esslen, J. 1982. Physiological and Morphological Properties of Interneurons in the Deutocerebrum of Male Cockroaches which respond to Female Pheromone. *J. Comp. Physiol,* 145:447-457.

Christensen, T.A., Hildebrand, J.G. 1987. Functions, Organization, and Physiology of the Olfactory Pathways in the Lepidoteran Brain. *In Arthropod Brain: its Evolution, Development, Structure and Functions,* A.P. Gupta, (ed), John Wiley & Sons: 457-484.

Christensen, T.A., Hildebrand, J.G. 1988. Frequency coding by central olfactory neurons in the spinx moth Manduca sexta. *Chemical Senses,* VOL. 13 no.1:123-130.

Christensen, T.A., Mustaparta, H., Hildebrand, J.G. 1989. Discrimination of sex pheromone blends in the olfactory system of the moth. *Chemical Senses,* Vol 14no. 3:463-477.

Kaissling, K. E., Kramer, E. 1990. Sensory basis of pheromone-mediated orientation in moths. *Verh. Dtsch. Zoolo. Ges.* 83, 109-131.

Linster, C., Masson M., Kerszberg, M., Personnaz, L., Dreyfus, G. 1992 Computational Diversity in a Formal Model of the Insect Olfactory Macroglomerulus, Neural Computation, in press.

Masson, C., Mustaparta, H. 1990. Chemical Information Processing in the Olfactory System of Insects. *Physiol. Reviews* 70(1): 199-245.

Meredith, M., O. 1986. Patterned response to odor in Mammalian olfactory bulb: the influence of intensity. *Journal of Neurophysiology* Vol 56, No 3: 572-597.

Olberg, R.M., 1983. Interneurons sensitive to female pheromone in the deutocerebrum of the male silkworm moth, Bombyx mori. *Physiological Entomology* 8: 419-428.

DYNAMIC ACTIVITY, LEARNING, AND MEMORY IN A MODEL OF THE OLFACTORY BULB

Tamás Gröbler and Péter Érdi

Biophysics Group
KFKI Res. Inst. Part. Nucl. Phys. of the Hungarian Acad. Sci.
P.O.Box 49, H-1525 Budapest, Hungary

ABSTRACT

A mathematical model to describe the associative memory character of the olfactory bulb is presented. A set of bifurcation phenomena and transition from oscillation to chaos are demonstrated. The continuous nature of the sensory input and of the learning process are explicitly taken into account. Simulation experiments demonstrate the ability of the model to recognize continuously learned patterns under the very restrictive requirement to produce physiologically relevant burst-like activity.

40.1 THE MODEL

The model of Li and Hopfield [1] is extended by taking explicitly into account the lateral interactions in the mitral layer, and the modifiability of certain synapses. The model consists of two layers of neurons: the layer of excitatory mitral cells, and the layer of inhibitory granule cells. Mitral cells receive inputs from the olfactory receptor cells. The mitral-to-granule connections are excitatory, whereas granule-to-mitral synapses are inhibitory. Lateral connections in the mitral layer can be excitatory or inhibitory. Granule cells receive modulatory input. The sigmoid transfer functions of mitral and granule cells are denoted by ϕ_m and ϕ_g. The underlying differential equations for the mitral and granule cell activities are respectively as follows:

$$\frac{dm_i(t)}{dt} = -a \cdot m_i(t) - b \sum_{j=1}^{N} G_{ij}\phi_g(g_j(t)) + c \sum_{j=1}^{N} L_{ij}(t)\phi_m(m_j(t)) + I_i^{\text{rec}}(t)$$

$$\frac{dg_i(t)}{dt} = -d \cdot g_i(t) + e \sum_{j=1}^{N} M_{ij}\phi_m(m_j(t)) + I_i^{\text{cort}} \qquad (40.1)$$

where a, b, d, e are positive constants and c can be positive or negative. The matrix L of the lateral connections may be a function of time but M and G are fixed. Odors are coded as amplitude patterns of the time varying receptor input I^{rec} that follows the sniff cycle and carries a sequence of input patterns. The linear increase during inhalation and exponential decrease during exhalation are adopted from [1]. The modulatory cortical input vector I^{cort} is constant in time.

40.2   PROBLEMS

Three classes of problems have been studied. First, the qualitative dynamical behavior of the system is studied by describing the different attractor types and bifurcation sequences in terms of the strength of lateral connections.

The second problem is how continuous synaptic modification can induce bifurcation between these regions. Learning might imply such kinds of changes between oscillatory and chaotic states (e.g. [2]). The learning rule used here for the matrix elements L_{ij} is the simple Hebb rule:

$$\frac{dL_{ij}(t)}{dt} = k \cdot \phi_m(m_i(t)) \cdot \phi_m(m_j(t)) \qquad (j = i \pm 1) \qquad (40.2)$$

where the constant k controls the rate of learning.

In the third problem, our aim is to show that incomplete input patterns can also be identified as proper stimuli if a suitable learning rule is used to modify the lateral connections between mitral cells. At the same time, only those parameters of the learning process can be accepted which provide physiologically well-defined responses (bursts).

Families of associative memory models based on dynamic systems can be classified according to the nature of input and of the learnig rule since both can be 'static' or 'dynamic'. The following examples represent the basic model classes:

■ *Constant connection strengths*

    − *Constant inputs*

      This is the classical scenario where the initial conditons are classified by fixed point attractors.

    − *Time-variable inputs*

      With nonautonomous activity dynamics, the input patterns are classified by (fixed point and/or periodic) attractors. A typical example is the Li–Hopfield model [1].

■ *Time-variable connection strengths*

    − *Constant inputs*

      Varying the connection strengths implies changes in the structure of the attractor–basin portrait.

    − *Time-variable inputs*

      The classification of time functions is rather difficult in the general case. It is possible, however, to classify parameters of the input function by transforming them into initial values [3]. It is still an easier problem to treat time-continuous inputs when the parameter space is discrete. The model given here is an example for such a simple associative memory.

The learning rule used here consists of three terms. First, a second order decay term controls the upper bound of the increase of connection strengths. Second a Hebbian term is responsible for the strengthening of connections between simultaneously firing cells. The third term is introduced to selectively decrease the synaptic strength between firing and inactive cells:

$$\frac{dL_{ij}(t)}{dt} = -k_1 \cdot L_{ij}^2(t) + k_2 \cdot \phi_m(m_i(t)) \cdot \phi_m(m_j(t)) -$$
$$-k_3 \cdot L_{ij}(t) \cdot [\phi_m(m_i(t)) - \phi_m(m_j(t))]^2 \qquad (40.3)$$

where k_1, k_2, k_3 are positive constants.

40.3 SIMULATION RESULTS

Simulation experiments show the occurrence of a series of bifurcation phenomena (Figure 40.1): Hopf bifurcation, transition from fixed point to large

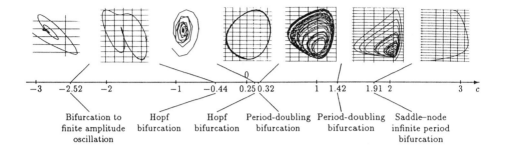

Bifurcation to   Hopf   Hopf   Period-doubling Period-doubling Saddle–node
finite amplitude bifurcation   bifurcation   bifurcation   bifurcation   infinite period
oscillation   bifurcation

Figure 40.1 Attractor regions in terms of the bifurcation parameter c. In each region, a characteristic phase diagram is shown. The activity of an arbitrary mitral–granule pair is chosen for two-dimensional representation. Bifurcation types between the regions are indicated below.

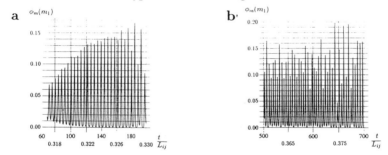

Figure 40.2 Synaptic modification induced bifurcation. Both the time scale and the average lateral synaptic strength are indicated at the horizontal axis. Oscillation and period doubling at the beginning (a) and the emergence of chaos at the end (b) of the process.

amplitude oscillation, infinite-period bifurcation, period-doubling bifurcation to chaos, coexistence of oscillation and chaos. The qualitative character of the attractors in terms of the control parameter c has been determined. The following parameter values are used in all simulations: $N=11$, $a=0.1$, $b=1.0$, $d=0.2$, $e=1.2$.

In the second series of simulation experiments, the effect of continuous-time synaptic modification on the qualitative dynamics has been studied. Adopting a moderately slow learning process ($k=0.015$), transitions between different regimes are demonstrated. In particular, a continuous time transition from oscillation to chaos is shown. Figure 40.2 shows the output of one mitral cell vs. time. In the first part of the learning process (a), period doubling can be observed. After an intermediate multiperiodic stage, the chaotic region is

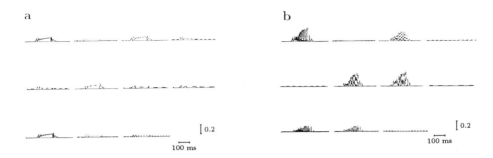

Figure 40.3 Responses of the $N=11$ mitral cells to an incomplete odor input before, and after 60 cycles of learning.

reached (b).

In the third part of the simulations the values 1 and 0 are assigned to high and low amplitudes of the receptor input, respectively. Thus, a complete odor pattern corresponds to the set of mitral cells, potentially receiving input during the sniff. In reality, however, only subsets of this set are stimulated. During learning, randomly chosen incomplete parts of the same odor pattern are presented in each sniff cycle. The values of the learning constants are $k_1=10^{-5}$, $k_2=0.1$, and $k_3=0.1$.

Before learning (Figure 40.3a), no significant response pattern can be observed. In Figure 40.3b, however, all six cells belonging to the pattern exhibit large amplitude high frequency firing bursts although only four receives the odor input. It can be concluded that the learning rule (40.3) leads to the recognition of incomplete odor patterns, simultaneously preserving physiologically justified oscillatory burst activity.

REFERENCES

[1] Li, Z. and Hopfield, J.J.: Modeling the olfactory bulb and its neural oscillatory processings. *Biological Cybernetics* **61** (379–392), 1989

[2] Skarda, Ch. and Freeman, W.J.: How brains make chaos in order to make sense of the world. *Behavioral and Brain Sciences* **10** (161–195), 1987

[3] Érdi, P., Gröbler, T., and Tóth, J.: On the classification of some classification problems. In: Tsuda, I. and Takahashi, K. (eds) *International Symposium on Information Physics* Kyushu Inst. Techn., Iizuka, Japan, pp. 110–117, 1992

41

DIFFERENTIAL EFFECTS OF NOREPINEPHRINE ON SYNAPTIC TRANSMISSION IN LAYERS 1A AND 1B OF RAT OLFACTORY CORTEX

Michael C. Vanier

James M. Bower

Department of Computation and Neural Systems,
California Institute of Technology 216-76, Pasadena, California 91125

ABSTRACT

In the rat olfactory (piriform) cortex, two distinct fiber systems exist for communicating information to pyramidal cells. Layer 1a contains afferent fibers from the olfactory bulb which terminate on the distal branches of the dendritic tree of pyramidal cells. Layer 1b contains axon collaterals from pyramidal cells which synapse onto other pyramidal cells (associational fibers). We investigated the differences in the effects of norepinephrine (NE) on synaptic transmission in layers 1a and 1b using extracellular recording techniques. NE at 25 μM causes a large but reversible decrease in field potential height in layer 1b, while causing a significant increase in field potential height in layer 1a. This effect appears to be mediated solely by α-adrenergic receptors. The functional significance of these results in the context of our ongoing efforts to build computational models of the olfactory cortex is discussed.

1   INTRODUCTION

The rat olfactory (piriform) cortex is a phylogenetically old three-layered cortex which receives input from the olfactory bulb via the lateral olfactory tract (LOT). The excitatory inputs to the intrinsic pyramidal cells of the piriform cortex are segregated into two layers: layer 1a consists of fibers originating from mitral cells of the olfactory bulb which travel through the LOT and synapse onto the distal-most branches of the dendritic tree of piriform cortex pyramidal cells, while layer 1b consists of axon collaterals from these pyramidal cells which synapse onto more proximal segments of other pyramidal cells in piriform cortex [1]. This tight segregation of inputs and the large numbers of recurrent connections have given rise to the hypothesis that the olfactory cortex may be acting as an associative memory [1, 2] Many differences exist in the anatomical and physiological properties of synapses in layers 1a and 1b. For example, layer 1a synapses show little long-term potentiation (LTP) while layer 1b synapses show significant LTP [3]. In addition, the neurotransmitter/neuromodulator acetylcholine (ACh), which is widely believed to be important in memory function [4] can reduce synaptic transmission in layer 1b by as much as 90%, while having no effect on transmission in layer 1a [5]. Both results have been interpreted in terms of the proposed function of piriform cortex as a biological associative memory [6].

The neurotransmitter/neuromodulator norepinephrine (NE), like ACh, has often been implicated in memory function [4]. In this report we demonstrate that NE also has different effects on field potentials (presumably representing summed EPSPs) in layers 1a and 1b. NE suppresses synaptic transmission (as represented by field potential height) in layer 1b (as does ACh), whereas in layer 1a NE causes a strong increase in synaptic transmission (unlike ACh, which has no effect in 1a). This effect appears to be mediated exclusively by α-adrenergic receptors. The functional implications of these results are discussed in terms of the associative-memory model of piriform cortex.

2   MATERIALS AND METHODS

Extracellular recordings were obtained from brain slices of female albino Sprague-Dawley rats cut perpendicular to the laminar organization of piriform cortex in the coronal plane with a thickness of 400 μm. Stimulating electrodes were placed in layers 1a and/or 1b, and extracellular recording electrodes were placed in the appropriate layer. Electrode position was adjusted to give the largest possible negative deflection upon stimulus. Stimuli were provided as low-voltage electric shocks lasting 0.1 msec to the appropriate layer. Paired-pulse stimuli were applied with a 50 msec interval between the first and second pulses. In experiments in which simultaneous recordings were made from layers 1a and 1b, the positions of the stimulating/recording electrode pairs were sufficiently far apart to prevent stimulation in one layer from reaching the recording electrode in the other layer. Slices were stimulated every 15 seconds, and the responses were digitized, recorded, and later analyzed by computer to give the

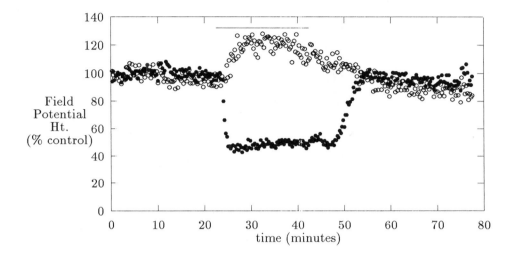

Figure 1 Time course of effect of 25 μM on field potential height in layers 1a (open circles) and 1b (filled circles) of piriform cortex. Horizontal line denotes time of NE application.

peak heights of the field potentials as well as the initial slope. For all the data presented the results from peak height measurements were essentially identical to those obtained using initial slopes. Drugs were applied by changing the perfusion medium to one containing the drug.

3 RESULTS

Figure 1 shows the results of a typical experiment using 25 μM NE. Application of NE causes a significant rise in field potential height in layer 1a but a large decrease in field potential height in layer 1b. These changes are reversible after washout of NE. In the figure the y-axis is expressed as a percentage of control (baseline) peak deflections. NE caused an increase in field potential height of ~20% in layer 1a but a decrease of ~50% in layer 1b. This effect was very rapid, especially in layer 1b where NE reaches its maximum effect in less than 2 minutes.

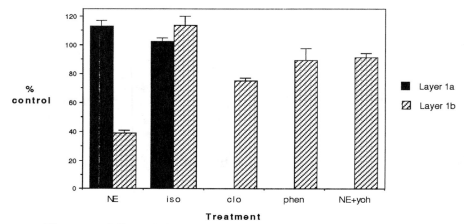

Figure 2 Effect of pharmacological treatments (25 μM) on height of field potentials in piriform cortex. Error bars are ± S.E. Abbreviations: *iso* = isoproterenol, *clo* = clonidine, *phen* = phenylephrine, *yoh* = yohimbine. 1a responses to isoproterenol, phenylephrine, and yohimbine not tested.

In figure 2 the effects of several types of pharmacological manipulations on field potential height in layers 1a and 1b are compared. As shown in figure 1, NE causes a large decrease in field potential height in layer 1b but a significant increase in 1a. The β-adrenergic agonist isoproterenol had no effect in layer 1a but produced a slight increase in field potential height in layer 1b. The α-adrenergic agonists clonidine (α-2 agonist) and phenylephrine (α-1 agonist) each produced a smaller depression of field potentials in layer 1b than NE. When NE was paired with yohimbine in layer 1b (an α-2 adrenergic antagonist), most of the effect of NE was inhibited although yohimbine alone has no effect (not shown). These results suggest that the depression of field potentials in layer 1b is mediated by α-adrenergic receptors. Although not shown, we have also demonstrated that NE at 25 μM causes a significant increase in paired-pulse facilitation in layer 1b but not in layer 1a, which is also true for ACh [5].

4 DISCUSSION

The results we have presented show that norepinephrine strongly decreases the height of extracellular field potentials, representing summed EPSPs, in layer 1b of piriform cortex while strongly increasing field potential height in layer 1a. The neurotransmitter/modulator acetylcholine has similar effects in layer 1b [5] but has no effect in layer 1a. If piriform cortex is acting as an associative memory [1, 2] then computer models suggest that the consequence of ACh's effects is to reduce interference during learning [6]. Synaptic transmission between pyramidal cells without the suppression of layer 1b synapses during learning creates interference between old memories and those being newly stored. However, suppressing 1b synapses reduces the excitatory drive on pyramidal cells, which by itself results in lower firing rates. LTP is generally believed to require high presynaptic firing rates. Therefore, one consequence of the increase in efficacy of layer 1a synapses may be to keep the total excitatory drive to the pyramidal cells constant while learning takes place. ACh has no effect on layer 1a synapses, although evidence suggests that it may compensate for the loss of excitatory drive by increasing the excitability of pyramidal cells [6]. Thus the data for both NE and ACh is consistent with the model that neuromodulation serves to switch the network between learning and recall modes, although the strategies used by the two neuromodulators are different.

REFERENCES

[1] Haberly, L. B. (1989) "Neuronal Circuitry of the Olfactory Cortex: Anatomy and Functional Implications." *Chemical Senses*, 10:219–238.

[2] Haberly, L. B. and Bower, J. M. (1989) "Olfactory Cortex: Model Circuit for Study of Associative Memory?" *Trends in Neuroscience*, 12:258–264.

[3] Kanter, E. D. and Haberly, L. B. (1990) "NMDA-dependent induction of long-term potentiation in afferent and association fiber systems of piriform cortex in vitro." *Brain Research*, 525:175-179.

[4] Squire, L. R. (1987) <u>Memory and Brain</u>. New York: Oxford University Press.

[5] Hasselmo, M. E. and Bower, J. M. (1992) "Cholinergic Suppression Specific to Intrinsic not Afferent Fiber Synapses in Rat Piriform (Olfactory) Cortex." *Journal of Neurophysiology*, in press.

[6] Hasselmo, M. E. and Anderson, B. P. and Bower, J. M. (1992) "Cholinergic Modulation of Cortical Associative Memory Function." *Journal of Neurophysiology*, in press.

42

CHOLINERGIC MODULATION OF ASSOCIATIVE MEMORY FUNCTION IN A REALISTIC COMPUTATIONAL MODEL OF PIRIFORM CORTEX

Ross E. Bergman, Michael Vanier, Gregory Horwitz, James M. Bower and Michael E. Hasselmo

M.E.H., R.E.B., G.H., Dept. of Psychology, Harvard University 33 Kirkland St., Cambridge, MA 02138 hasselmo@katla.harvard.edu

M.V. and J.M.B. Div. of Biology 216-76, Caltech, Pasadena, CA 91125

ABSTRACT

A detailed biophysical model of piriform cortex developed with the GENESIS simulation package shows associative memory properties such as completion. This model allows analysis of how the neuromodulatory effects of acetylcholine influence cortical associative memory function. Experiments in brain slice preparations demonstrate that acetylcholine causes selective suppression of synaptic transmission at excitatory intrinsic fiber synapses in piriform cortex [1]. When applied during learning in the model, this selective cholinergic suppression prevents interference between overlapping patterns of afferent input, allowing separate and distinct learning of input stimuli.

42.1 INTRODUCTION

Memory function in cortical structures has been studied using a class of abstract models termed associative memories [2, 3, 4]. These models have the capability to respond to incomplete or noisy versions of stored patterns with activity more closely resembling the originally learned pattern. Here we present a realistic biophysical model of associative memory function in the piriform cortex. This region provides an excellent model system for the study of cortical associative memory function [5, 6]. The piriform cortex is the primary olfactory cortex in all mammals, and may play an important role in recognition of complex olfactory stimuli [7]. The widely distributed excitatory intrinsic connections between pyramidal cells in the piriform cortex are believed to undergo synaptic modification with Hebbian properties consistent with those of a cortical associative memory [8].

Biophysical models are important tools for the researcher because they allow the examination of how specific physiological parameters of cortical networks impact on cortical function. In particular, the biophysical model presented here allows the study of the role of specific physiological effects of neuromodulatory agents in cortical associative memory function. Cholinergic agents have been shown to affect memory function in a wide range of tasks in humans [9] and animals [10], however these behavioral effects have not been linked to the full range of neuropharmacological evidence on acetylcholine. Results of electrophysiological experiments have shown that acetylcholine (ACh) will selectively suppress synaptic transmission at the synapses of excitatory intrinsic fibers connecting pyramidal cells in the piriform cortex without affecting synaptic transmission at synapses of afferent fibers arising from the olfactory bulb [1]. The effects of ACh on the associative memory function of a biophysical simulation of the piriform cortex are examined in this study.

42.2 METHODS

Associative memory function was analyzed in a realistic biophysical simulation of piriform cortex using the GENESIS simulation package [11, 12]. The basic elements of this model are compartmental simulations of cortical pyramidal cells, including a full range of membrane conductances. Networks of these single cell compartmental simulations were linked together with realistic axonal transmission delays and synaptic currents. We will first describe methods for single cell simulation, and then describe network interactions.

Single cell biophysics: The principal excitatory cells and inhibitory interneurons of piriform cortex were represented with compartmental simulations incorporating realistic membrane dynamics and a range of synaptic and voltage dependent currents. Passive membrane potential dynamics were based on the equivalent circuit model and cable theory, while the dynamics of action potential generation were based on numerical solutions to differential equations representing the kinetics of the fast, voltage-dependent sodium channel and the delayed rectifier potassium channel. The parameters chosen for these channels result in simulated action potentials similar to those observed in mammalian cortical intracellular recording [13].

Mitral cells were not simulated as compartmental cells, but represent the pattern presented to the network. Each cell was on or off, corresponding to whether the associated element of the input pattern was a zero or a one. Each mitral cell projected to every inhibitory interneuron. For ease of pattern representation, however, each mitral cell projected to only one pyramidal cell. Transmission delays were not included for mitral cells.

The model included two types of inhibitory interneurons, here referred to as GABAa and GABAb-activating. Both GABAa and GABAb-activating interneurons were modeled as single compartment somas, receiving excitatory afferent activation from each of the mitral cells. GABAa-activating interneurons mediated inhibition in other neurons by increasing chloride conductances, while GABAb-activating interneurons mediated inhibition in other neurons by increasing potassium conductances.

Each pyramidal cell was comprised of three compartments, two dendritic, and one somatic. The distal dendritic compartment contained chemically gated sodium channels activated by afferent axons arising from a mitral cell. The proximal dendritic compartment contained sodium channels activated by axons arising from other pyramidal cells, and potassium channels activated by GABAb-activating interneurons. The soma contained chloride channels activated by GABAa-activating interneurons.

Network simulations: The compartmental biophysical simulations of single neurons described above were combined in networks with parameters representing the transmission of action potentials along axons and the conductances mediated by synapses on the dendrites of other neurons. Transmission delays depended on distance between modeled neurons, and were derived from extensive previous simulations of piriform cortex [11, 12]. Simulations of networks containing up to 225 neurons were performed, with the examples presented here from networks with 3 and 30 neurons. Synaptic currents were represented by shifts in ionic conductances following a standard dual exponential time course with specific time constants dependent upon the receptor and channel type [11].

Learning rule: The models described in this paper used a modified Hebb rule to strengthen excitatory synapses mediating sodium currents in cortical pyramidal cells. The learning rule strengthened synaptic currents on the basis of an action potential in the pre-synaptic neuron, and the level of the post-synaptic membrane potential averaged over the previous 20 msec time period.

Acetylcholine Effect: The cholinergic suppression of synaptic transmission at intrinsic fiber synapses during learning [1] is simulated in the model by decreasing the excitatory synaptic sodium conductances activated by other pyramidal cells in the proximal dendritic compartment to 0.3 of total strength. During recall, the conductance was restored to its full strength. Sodium conductances activated by afferent input in the distal dendritic compartment were not altered.

42.3 RESULTS

The biophysical simulation of olfactory cortex showed the capacity to learn distributed patterns of afferent input from mitral cells based on Hebbian synaptic modification of the excitatory intrinsic connections between pyramidal cells. However, this learning capacity depended on the cholinergic suppression of synaptic transmission at intrinsic synapses during learning. Without suppression of synaptic transmission during learning, the depolarization induced by previously modified excitatory synapses caused undesired synaptic modification within the network, resulting in excessive spread of excitatory activity during recall.

As an example of the basic functional characteristics of the model, a very simplified network containing only three pyramidal cells and three of each type of inhibitory interneuron is shown in figure 1.

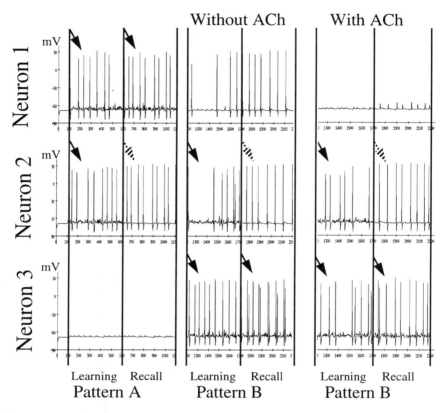

Figure 1 - Cholinergic modulation of associative memory function in a simulation of a simple network with compartmental simulations of three pyramidal cells and three of each type of inhibitory interneuron. Membrane potentials for the three pyramidal cells are shown. Black arrows represent afferent input from two active mitral cells during learning, and one active mitral cell during recall. Shaded arrows represent pattern components not presented during recall. **On the left**, afferent pattern A causes spiking in neuron 1 and 2 during learning, and the excitatory synapses between these neurons are strengthened. During recall, activity due to afferent input to neuron 1 spreads across previously strengthened synapses to activate neuron 2. Thus, the network responds to a degraded version of pattern A with spiking activity more closely resembling the response to the original learned version of pattern A. **In the center**, without cholinergic suppression, interference occurs during learning of a second overlapping pattern. During learning, afferent pattern B activates neurons 2 and 3, but activity spreads across the previously strengthened connection to cause spiking in neuron 1. All neurons are active, and all excitatory synapses in the network are strengthened. During recall, presentation of a degraded version of pattern B activates components of both pattern A and pattern B. **On the right**, cholinergic suppression of excitatory intrinsic synaptic transmission prevents the spread of activity across previously modified synapses during learning. Pattern B activates only neurons 2 and 3, and only the synapses between these neurons are strengthened. During recall, suppression is removed, and the network responds to a degraded version of pattern B with the fully learned version of pattern B, but no interference from pattern A.

As shown in figure 1, the simplified network of three pyramidal cells illustrates the basic properties of the network. Hebbian synaptic modification of intrinsic excitatory connections allows the network to respond to degraded patterns with activity more closely resembling the fully learned pattern. However, the spread of activity across previously modified synapses can interfere with learning of subsequent patterns. Cholinergic suppression of excitatory intrinsic synaptic transmission during learning prevents this interference from previously modified connections.

An important feature of the simulation concerns the timing of the inhibitory effects. Even after learning of patterns with cholinergic suppression of synaptic transmission, the network still has the capacity for excessive spread of activity during recall. For example, on the right hand side in figure 1, after presentation of afferent input to neuron 3, activity spreads to neuron 2, but could also spread across the previously strengthened synapse between neuron 2 and neuron 1. In the simulation shown here, this is prevented by the fact that at the same time as neuron 2 is activated, inhibitory interneurons are activated, which prevent spiking in neuron 1 through activation of chloride conductances. This interaction of excitation and inhibition can be seen in neuron 1 during recall of pattern B after learning with acetylcholine: the neuron shows depolarization after each spike in neuron 2, but is prevented from firing by chloride mediated inhibition.

Thus, the relative timing of inhibitory processes is very important for the associative memory function of the biophysical simulation. This suggests that recall dynamics may depend upon the timing characteristics of the gamma oscillations detected in cortical EEG recording, since these oscillations appear to be the result of fast chloride mediated inhibition [12]. The oscillatory dynamics of cortical networks also appear to be influenced by the physiological effects of cholinergic modulation. Experimental evidence suggests that cholinergic modulation induces theta rhythm oscillations in the EEG of piriform cortex and the hippocampus. This may partly result from cholinergic suppression of the neuronal adaptation of cortical pyramidal cells [14].

A simulation of 30 neurons is shown in figure 2 to further illustrate the basic properties of the network. This network was trained with five overlapping afferent input patterns, each of which contained 6 active mitral cells. The function of the network was tested after training in the presence of ACh and after training in the absence of ACh. As was the case with the 3 neuron model, without cholinergic modulation during learning, the associative memory function of the network breaks down. In this case, during recall, presentation of a degraded version of an input pattern stored in the network results in neuronal activity containing components of all patterns learned by the network. This effect resulted from the spread of activity along previously strengthened synapses causing strengthening of additional undesired connections during learning of each new pattern. Interference during learning of each new pattern caused additional interference during learning of subsequent patterns, leading to the runaway growth of a large number of synapses within the network. The result of this excessive strengthening of synaptic connections is that presentation of a degraded version of any pattern learned by the network results in pyramidal cell spiking activity containing components of all the patterns learned by the network. In contrast, cholinergic suppression of excitatory synaptic transmission between pyramidal cells during learning prevents this build-up of interference. This allows the network to show effective performance during recall. Presentation of an incomplete version of a single learned pattern during recall causes activity in a subset of neurons in the network which spreads across a particular subset of synapses into only those neurons which were components of the originally learned version of the degraded pattern. The properties of this 30 neuron biophysical simulation are similar to properties of more abstract simulations of piriform cortex [6, 15, 16].

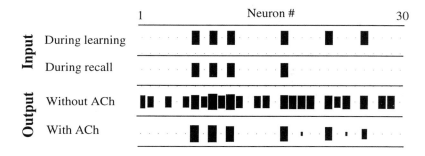

Figure 2. Cholinergic modulation of associative memory function in a 30 pyramidal cell biophysical simulation of piriform cortex. Five patterns with 6 active mitral cells each have been stored in the network. On the top, one pattern is shown, with black squares representing active afferent input to individual pyramidal cells. During learning the full pattern is presented. During recall, a version of the pattern missing two input lines is presented. On the bottom, the output of the network in response to the degraded recall pattern is shown. After learning without ACh, the network responds to the single degraded input pattern with spiking activity in pyramidal cells activated by all five patterns stored in the network. After learning with ACh, the network responds to the single degraded input pattern with spiking activity only in those neurons which were activated by the originally learned input pattern, providing completion of the missing components of the pattern presented during recall.

42.4 DISCUSSION

Simulation of associative memory function in a physiologically realistic model of cortical function supports the possibility that such associative memory function could exist in the mammalian brain. While the network presented here has considerably greater complexity than more abstract models of associative memory function, it shows the basic capability to respond to incomplete or degraded versions of a learned pattern with neuronal activity more closely resembling the activity elicited by the originally learned version of that pattern.

Demonstration of this basic capacity of associative memory function in a biophysically realistic network allows the analysis of how neuromodulatory effects influence cortical associative memory function. As can be seen from the results presented here, the absence of acetylcholine while the network is learning results in obvious failure at the time of recall; presentation of any learned pattern, or portion thereof, yields network activity containing elements of all patterns learned by the network. This effect resulted from synaptic transmission at previously modified synapses interfering with learning of each new pattern. In contrast, with cholinergic suppression of excitatory intrinsic synaptic transmission during learning, the interference due to synaptic transmission at previously modified connections is prevented. Thus, cholinergic suppression of synaptic transmission greatly enhances the associative memory function of the network.

In previous models of associative memory function [2, 3, 4], the spread of activity across modifiable synapses is ignored. Here we show how this standard technique in this class of models can be implemented using experimentally demonstrated neuromodulatory effects in a realistic biophysical simulation. In particular, it is interesting to note that prevention of interference during learning did not require complete suppression of synaptic transmission at modifiable synapses. The presence of inhibition and a threshold of synaptic modification prevented the moderate level of synaptic transmission at excitatory intrinsic synapses from causing interference during learning. This result from the biophysical simulation is in keeping with the analysis of a more abstract, linear representation of associative memory function [16].

This biophysical simulation of cortical associative memory function will allow more detailed analysis of the functional role of a range of realistic physiological parameters. In particular, this model will allow analysis of how cortical associative memory function is influenced by neuronal adaptation and the modulation of neuronal adaptation, the modulation of cortical oscillatory dynamics, the modulation of cortical inhibition, and the laminar segregation of afferent and intrinsic fiber synapses.

REFERENCES

[1]   Hasselmo M.E. and Bower J.M. (1992) Cholinergic suppression specific to intrinsic not afferent fiber synapses in rat piriform (olfactory) cortex. *J. Neurophysiol.* 67(5): 1222-1229.

[2]   Anderson, J.A. (1983) Cognitive and psychological computation with neural models. *IEEE Trans. Systems, Man, Cybern.* SMC-13: 799-815.

[3]   Kohonen T.(1984) *Self-organization and associative memory.* Berlin: Springer-Verlag.

[4]   Hopfield, J. J. (1982) Neural networks and physical systems with emergent selective computational abilities. *Proc. Natl. Acad. Sci. USA* 79: 2554-2559.

[5]   Haberly, L.B. and Bower, J.M. (1989) Olfactory cortex: Model circuit for study of associative memory? *Trends Neurosci.* 12: 258-264.

[6]   Hasselmo M.E., Wilson M.A., Anderson B. and Bower J.M. (1991) Associative memory function in piriform (olfactory) cortex: Computational modeling and neuropharmacology. In *Cold Spring Harbor Symposium on Quantitative Biology: The Brain.* Cold Spring Harbor Laboratory, Cold Spring Harbor, N.Y., p. 599- 610.

[7]   Staubli, U., Schottler, F. and Nejat-Bina, D. (1987) Role of dorsomedial thalamic nucleus and piriform cortex in processing olfactory information. *Behav. Brain Res.* 25: 117-129.

[8]   Kanter, E.D. and Haberly, L.B. (1990) NMDA-dependent induction of long-term potentiation in afferent and association fiber systems of piriform cortex in vitro. *Brain Res.* 525: 175-179.

[9]   Kopelman, M.D. (1986) The cholinergic neurotransmitter system in human memory and dementia: A review. *Quart. J. Exp. Psychol.* 38: 535-573.

[10] Hagan, J.J. and Morris, R.G.M. (1989) The cholinergic hypothesis of memory: A review of animal experiments. In *Psychopharmacology of the Aging Nervous System*, L.L. Iversen, S.D. Iversen and S.H. Snyder, eds. New York: Plenum Press, p. 237-324.

[11] Wilson, M.A. and Bower, J.M. (1989) The simulation of large-scale neuronal networks. In: *Methods in neuronal modeling: From synapses to networks.*, edited by C. Koch and I. Segev. Cambridge, MA: MIT Press, p. 291-334.

[12] Wilson, M.A. and Bower, J.M. (1992) Cortical oscillations and temporal interactions in a computer simulation of piriform cortex. *J. Neurophysiol.*, 67: 981-995.

[13] Hasselmo M.E. and Barkai E. (1992) Cholinergic modulation of the input/output function of rat piriform cortex pyramidal cells. *Soc. Neurosci. Abstr.* 18: 521.

[14] Liljenstrom, H. and Hasselmo, M.E. (1993) Acetylcholine and cortical oscillatory dynamics. *Computation and Neural Systems.*

[15] Hasselmo M.E., Anderson, B.P. and Bower, J.M. (1992) Cholinergic modulation of cortical associative memory function. J. Neurophysiol. 67(5): 1230-1246.

[16] Hasselmo, M.E. (1993) Acetylcholine and learning in a cortical associative memory. *Neural Computation* 5: 22-34.

43

NUMERICAL SIMULATIONS OF THE ELECTRIC ORGAN DISCHARGE OF WEAKLY ELECTRIC FISH

Christopher Assad
Brian Rasnow*
James M. Bower†

Divisions of Electrical Engineering, Physics, and Biology†
California Institute of Technology, Pasadena, CA 91125*

ABSTRACT

A model of a weakly electric fish was constructed with data taken from *Apteronotus leptorhynchus*, and the electric organ discharge was simulated using boundary element and finite element methods. Maps of the electric potential measured around a live fish were used to calibrate the model parameters and test the results. Our goal is to quantitatively simulate various sequences of electrosensory input to the central nervous system resulting from the fish's exploratory behavior.

INTRODUCTION

Weakly electric fish have evolved specialized active electrosensory systems for detecting nearby objects in their environment [1]. An organ in the fish's body generates electric discharges (EOD), causing weak electric currents to flow in the water surrounding the fish. The currents are detected by an array of electroreceptors distributed throughout the fish's skin, which are sensitive to local transepidermal potential. Electrolocation is possible because objects that differ in their conductivity and dielectric properties from the surrounding water can be detected by the perturbations they cause in the electric field sensed by the fish.

This electric sense presents a unique opportunity to study sensory-motor integration in the brain. Weakly electric fish exhibit a rich repertoire of characteristic behaviors while exploring their environment, including both body movements and modulations of the EOD (e.g., see [2]). The sensory consequences of these behaviors can be reconstructed by simulating the electric fields in the water surrounding the fish and predicting the receptor responses. Sensory input coded by the electroreceptors projects directly to the electrosensory lateral line lobe (ELL) in the brain, where it terminates in somatotopically organized maps [3]. The ELL in turn interacts strongly with cerebellum and other higher processing centers. We are investigating electroreception in weakly electric fish to study the role of the cerebellum in exploratory behavior. Our initial goal is to quantitatively simulate various sequences of input to the ELL. This will facilitate the analysis of behavioral consequences, and provide a basis for computational modeling of electrosensory processing in the ELL and higher brain areas.

FIELD MEASUREMENTS

As a basis for these studies, we have measured the potential on the skin and in the midplane of paralyzed specimens of *Apteronotus leptorhynchus*, a "wave-type" species with a continuous, periodic discharge [4]. Several potential waveforms measured at different points on the body are illustrated in figure 1A. It can be seen that the waveform varies considerably with different positions on the body surface, and differs significantly from a sinusoid over at least one third of the trunk and tail. The complicated EOD patterns for each specimen were found to be uniquely identifiable, composing an "electric signature". With the use of a phase reference, we have combined waveforms measured at different locations into snapshots of the EOD at each point in time. Smooth surface maps were constructed by interpolation between sample points. Our field mappings have sufficient resolution in both space and time to enable the quantification of sensory input (i.e., the electrosensory "images"). One such map is presented in figure 2A, showing isopotential contours in the midplane of a fish at a single phase of the EOD.

The field mappings clearly illustrate the spatiotemporal structure of the EOD in the water about the fish. However, these high resolution measurements can only be recorded using paralyzed fish, oriented in stationary configurations. To study exploratory strategies in a more natural setting, the fish must be allowed to swim freely. Accordingly, we have developed an electric fish simulator that will enable us to recreate the electrosensory input from recorded body movements of freely behaving fish.

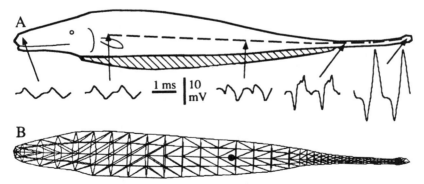

Fig 1. A. Outline of *A. leptorhynchus,* showing the extent of the electric organ (dashed line), and potential waveforms measured at different points on the body relative to an electrode far from the fish. The EOD is periodic in this species, classified as a "wave-type" fish (two periods are shown). **B.** BEM mesh on surface of 3-d fish model. 256 nodes and 508 planar triangular elements define the body. The electric organ is modeled by two current monopoles: a source in the body and a sink in the tail tip.

SIMULATIONS OF THE EOD

The electric potential $\Phi(x)$ generated by an electric fish is a solution of Poisson's equation. Multiplying both sides by the conductivity $\sigma(x)$ results in an equation for current source densities (ρ = charge density, ε = dielectric constant):

$$-\nabla^2\Phi = \frac{\rho}{\varepsilon} \quad \text{---------->} \quad -\sigma\nabla^2\Phi = \sigma\frac{\rho}{\varepsilon} = I$$

Function $I(x)$ then corresponds to the current source density generated by the electric organ. Outside the fish this reduces to Laplace's equation since there are no current sources ($I = 0$). This is a boundary value problem to solve for the potential and electric field, given the geometry of the surfaces, the conductivities corresponding to the fish, water and external objects, and the current source density generated by the electric organ. The fish model must also be flexible to allow simulation under varying configurations for a fish that is changing position.

Finite Element Method

In the finite element method (FEM), the domain is broken up into different sized elements with homogeneous conductivities, which facilitates treatment of irregular shapes and boundaries. Nodes defining the element vertices can be efficiently spaced to give higher spatial resolution in the areas of interest, close to the fish's body where the potential gradient is largest and where errors due to discretization would be most severe (figure 2C; also see [5]). $\Phi(x)$ over each element is approximated as a linear combination of nodal values Φ_i, using linear interpolation functions. A variational form of Poisson's equation is then applied to each element separately; it is equivalent to minimizing a quadratic function that corresponds to the energy in the electric field by varying the potential on each node. The resulting equations, one for each nodal potential, are combined to form a linear system with scalar coefficients defining local interactions between adjacent nodes. This large system of equations can then be solved using a variety of standard iterative techniques (the matrix is symmetric, banded and very sparse).

The FEM simulator produces high resolution solutions in 2-dimensional domains, such as the midplane of the fish (figure 2C), and we have used it to simulate the effects of body bending and nearby objects. However, 2-d solutions assume no variation in the third dimension, and hence are inherently skewed from the actual 3-d fields. A major disadvantage of FEM is in constructing 3-d fish models. To achieve the desired resolution the elements, now polyhedra, have to be very small and dense around the body, especially in the skin layer. Since the nodes and elements must completely fill the problem domain, their numbers increase drastically for 3-d problems, and the efficiency of this method is greatly reduced.

Boundary Element Method

In the boundary element method (BEM), Poisson's equation is reformulated so that the nodes are only placed on discontinuous boundaries. This greatly decreases the number of nodes required for 3-dimensional problems, and results in a much smaller but full matrix. Several simplifications were applied that are typically assumed for electric fish, including the internal body conductivity is uniform and higher than that of the water, and the skin conductivity is much lower and can vary depending on its location on the body. More importantly, the skin is very thin and has very low conductivity, so that the current density is assumed to only have a component normal to the surface. Therefore the skin is approximated as a single surface composed of planar triangular elements, and appropriate boundary conditions are applied for compensation. This results in two unknowns to solve for at each node: the potential and the normal electric field component. Once the potential and current density are solved on the fish surface, the fields in the water about the fish can also be calculated.

Surface data taken from *A. leptorhynchus* specimens were incorporated into a computer model of a fish body for the simulation (figure 1B). Other input parameters include the orientation of the fish body, the conductivities of the body, skin, and water, and the strength and location of current sources comprising the electric organ. Comparisons between simulation results and experimental mappings are being used to optimize the parameters and ensure the accuracy of the model. For example, to balance the electrical load on the organ and fit the contour lines of the field mappings, the skin conductivity at the tail was set 10 times higher than along the trunk and head.

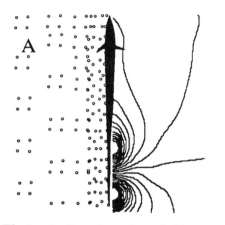
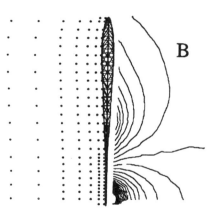

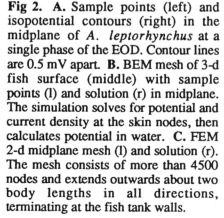
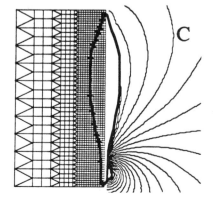

Fig 2. A. Sample points (left) and isopotential contours (right) in the midplane of *A. leptorhynchus* at a single phase of the EOD. Contour lines are 0.5 mV apart. **B.** BEM mesh of 3-d fish surface (middle) with sample points (l) and solution (r) in midplane. The simulation solves for potential and current density at the skin nodes, then calculates potential in water. **C.** FEM 2-d midplane mesh (l) and solution (r). The mesh consists of more than 4500 nodes and extends outwards about two body lengths in all directions, terminating at the fish tank walls.

The electric organ has usually been treated as an oscillating dipole, since much of the current is channeled out of its endpoints, resulting in a dipolar far field [6]. We are iteratively refining an electric organ model by varying simple source distributions and comparing the results to the field measurements on the skin. Dipole model solutions, such as in figure 2B, are similar to certain phases during the period of the EOD. However, they are not sufficient to recreate the overall EOD pattern, which becomes very complex along the tail and requires higher order multipole moments. For example, simulations using four current sources were used to better approximate phases of the EOD in which the potential is trimodal along the fish axis. Successive phases can also be simulated by sequentially shifting the source configurations and strengths, with a periodic movement phaselocked to the overall EOD.

CONCLUSIONS

We have developed a flexible, 3-dimensional EOD simulator that will solve electric field boundary value problems with complicated geometries and high resolution. Comparisons between simulations and field measurements are being used to improve the fish model, and allow us to test the assumptions and simplifications usually applied to electric fish. Initial results have already led to a better understanding of the fish's electromotor system, as stated above.

These methods will enable us to quantitatively predict the sensory consequences of electric fish behavior. In these experiments the fish will swim freely and be trained to discriminate between different objects based on their electrical properties. Characteristic exploratory behaviors will be classified and recorded using both video and electrodes fixed in the tank. The fish model described here will be used to simulate the fields, with field recordings used for calibration. The transdermal potential, comprising the image that the fish senses, will then be extracted from the results. The final step will be to convolve the images derived from the simulations with the transfer function of the electroreceptor array, resulting in an accurate representation of the information available to the ELL brain maps.

REFERENCES

[1] C. E. Carr. Neuroethology of electric fish. *Bioscience* 40:259-267, 1990.
[2] M.J. Toerring, P. Belbenoit. Motor programmes and electroreception in mormyrid fish. *Behav. Ecol. Sociobiology* 4:369-379, 1979.
[3] W. Heiligenberg. Electrosensory maps form a substrate for the distributed and parallel control of behavioral responses in weakly electric fish. *Brain Behav. Evol.* 31:6-16, 1988.
[4] B. Rasnow, C. Assad, J.M. Bower. Phase and amplitude maps of the electric organ discharge of the weakly electric fish, *Apteronotus leptorhynchus. J. Comp. Physiology A.* (in press).
[5] N. Hoshimiya, K. Shogen, T. Matsuo, S. Chichibu. The *Apteronotus* EOD field: waveform and EOD simulation. *J. Comp. Physiol.* 135:283-290, 1980.
[6] E. Knudsen. Spatial aspects of the electric fields generated by weakly electric fish. *J Comp Physiol* 99:103-118, 1975.

SECTION 6

PATTERN GENERATORS

Central pattern generators (CPG's) is one area where mathematicians and neural modelers have always interacted closely with experimentalists. Interested readers may also want to check out chapter 10 in section 1, and chapters 13 and 14 in section 2. Presented here are "traditional preparations" such as leech and lamprey and not-so-traditional ones as respiration and peristalsis.

IDENTIFICATION OF LEECH SWIM NEURONS USING A RESONANCE TECHNIQUE

Richard A. Gray

W. Otto Friesen

Center for Biological Timing
University of Virginia,
Charlottesville, VA 22901

ABSTRACT

We have developed a technique to identify neuronal oscillators and to predict the period in the oscillations. The technique involves injecting subthreshold sinusoidal current over a range of frequencies into the neuron of interest and recording the resulting membrane potential oscillations. The analysis requires the decomposition of these oscillations into their harmonic components and the determination of the amplitude and phase of the first harmonic for each frequency. We have shown that neuronal oscillators exhibit a resonance peak in the membrane potential amplitude as well as positive and negative phase angles. In addition, the frequency at which the resonance peak occurs predicts the normal oscillating frequency. In applying this technique to the mathematical model of the squid axon (Hodgkin and Huxley 1952) we predicted accurately the range of firing frequencies for this model. Our preliminary results suggest that this technique can be applied to neuronal circuit oscillators as well, for we found a resonance peak at approximately 1 Hz in the membrane potential of a leech motor neuron (part of the leech swim circuit), which agrees well with the normal swimming frequency of the leech (0.5 to 2.0 Hz).

44.1 INTRODUCTION/BACKGROUND

Two processes are required for oscillations to occur: 1) a restorative process that returns the system to the equilibrium state, and 2) a process that leads to overshoot of the steady-state value. We have developed a method to identify these two processes in biological systems.

In order to understand the elements that are necessary for oscillations in biological, non-linear systems, we first consider the requirements for oscillations to occur in linear physical systems. In the two linear systems often used to analyze oscillatory behavior, the mass-spring physical system and an electric circuit made up of a resistor, capacitor and inductor, the restorative elements are the spring and the capacitor, respectively. The elements that lead to overshoot are inertia in the physical system and inductance in the electric circuit. For oscillations to occur in these linear systems the roots of the characteristic equations must contain complex eigenvalues.

The input-output relationship for a linear system is completely determined by the transfer function, which is described by two graphs. These graphs are the ratio of output and input amplitudes plotted against frequency and the phase angle of the output minus the phase angle of the input plotted against frequency. Linear systems are completely described by the transfer function because of frequency independence, i.e. for sinusoidal input the output is also sinusoidal. The output frequency is identical to the input frequency, but the output has different amplitude and phase. Two distinguishing features that characterize the transfer function of linear oscillating systems are 1) a resonance peak near the natural frequency of the system in the graph of amplitude versus frequency, and 2) both positive and negative phase angles in the graph of phase versus frequency. We now propose that these two features also identify neuronal oscillator circuits in non-linear systems, in particular neuronal circuits that generate oscillations.

44.2 METHODS

Our technique involves injecting sinusoidal current into identified neurons and recording the resulting membrane potential oscillations. The sinusoidal current is superimposed on a constant injected current and the frequency of the sinusoidal current is varied through a specified range. For non-linear systems the resulting output is not necessarily be sinusoidal, however, if the output is stationary (i.e. periodic), the output signal may be decomposed into its harmonic components. These components may then be analyzed as described above. Currently, we have applied our analysis to the first harmonic component of the output membrane potential signal.

44.3 RESULTS

To test our hypothesis that neuronal oscillators exhibit resonance behavior at their physiological oscillating frequency we performed two computer experiments on the Hodgkin-Huxley (HH) mathematical model of the squid axon (Hodgkin and Huxley 1952) using the *NeuroDynamix* software (Friesen and Friesen 1991). For a constant

(DC) current applied to this model, the simulated axon produces impulses repeatedly (i.e. the membrane potential oscillates). The lower and upper limits for the range of input currents that generate these endogenous oscillations are called the *lower* and *upper* thresholds for impulse activity. As the amplitude of the input current is increased from the lower to the upper threshold the impulse rate also increases. We tested the HH model with sinusoidal currents for which the *maximum* current excursion was less than the *minimum* DC current needed to elicit spiking behavior (subthreshold current). As we varied the frequency of the injected current from 1 Hz to 500 Hz, we found a unique maximum in the amplitude (resonance peak) of the output oscillations (see Figure 1).

We then repeated the experiment with a greater depolarizing DC current so that the *minimum* current excursion was less than the *maximum* DC current (suprathreshold current exceeding the upper threshold) needed to elicit endogenous membrane potential oscillations. We again found a resonance peak in the output oscillation amplitude. The resonant frequencies for these two experiments were 68 Hz and 174 Hz, respectively. These values compare favorably to the minimum and maximum firing frequencies (68 Hz and 169 Hz) of the HH model response to constant current injection. Therefore, we demonstrated that by injecting sinusoidal current below the lower threshold and above the upper threshold (into the Hodgkin-Huxley mathematical model) we could predict the range of endogenous oscillation frequencies in the membrane potential (impulse rates) of the squid giant axon without forcing the axon to generate impulses (i.e., by exploring the resonant properties of this nonlinear system outside its oscillating range).

For both experiments, graphs of the phase angles (of the first harmonic of the resulting membrane potential oscillations) as a function of frequency displayed positive and negative phase angles. This result shows that the first harmonic component in the output of a non-linear neuronal oscillator displays resonance and phase properties similar to those found in linear physical oscillators.

We are applying this "resonance" technique to the leech nervous system to identify candidate neurons of the circuit that generates oscillating swimming activity (see Friesen 1989 for a review of the neuronal circuitry involved in leech swimming). We propose that the cells that contribute significantly to generating the swimming oscillations will show a resonance peak near the normal swimming frequency. Our preliminary experiments involved a motor neuron, cell 3, in the isolated leech ventral nerve cord perfused with saline containing serotonin. We used a discontinuous current clamp (DCC) apparatus (AXOCLAMP) to inject both a constant (DC) current to hold the resting membrane potential at -40 mV and a superimposed sinusoidal current. When the frequency of the sinusoidal current was swept from 0.08 Hz to 5.0 Hz we observed a resonance peak at approximately 1 Hz, which agrees well with the normal swimming rate of 0.5 Hz to 2 Hz (Kristan et al. 1974). We believe that this resonance peak is due to properties in this cell relevant to swimming, e.g. post-inhibitory rebound.

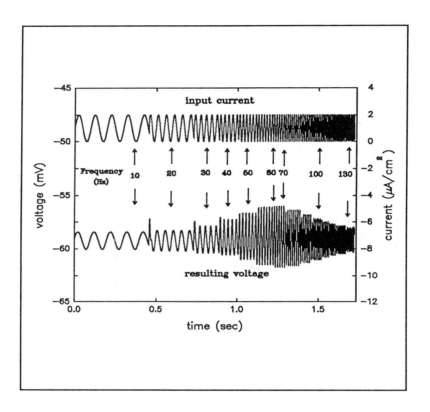

Figure 1. Hodgkin-Huxley Computer Experiment The membrane voltage response to a subthreshold sinusoidal input current from a computer experiment on the HH model of the squid axon. The amplitude of the voltage response (output) is a function of input current frequency, with the maximum amplitude near 70 Hz.

44.4 DISCUSSION

We have developed a simple technique to identify neuronal oscillators. There are two features that indicate that neurons are members of an oscillating system: a resonant peak near the normal oscillating frequency, and the occurrence of both positive and negative phase angles.

As Cole (1972) pointed out, there are electric circuit analogs (resistors, inductors and capacitors) that can be used to model the ion channels of nerve membranes. Cole suggested that the restorative processes in the HH model are due to the capacitive properties of the cell membrane and to the activation of sodium channels. He also proposed that the inertia-like processes result from the inactivation of sodium channels and the activation of potassium channels. It is well known that the voltage sensitivity of ion channels can induce membrane potential oscillations; the advantage of our method is that the tendency of neurons to generate oscillations can be detected without the need for identifying the restorative and inertia-like processes explicitly.

We have shown that our "resonance" technique identifies the two functionally significant processes necessary for oscillations (restorative and inertia elements) and predicts the range of endogenous oscillating frequency (impulse rate) for the non-linear HH mathematical model. Moreover, our technique can be used to predict the natural frequency of a neuronal circuit in the absence of the expression of endogenous rhythmicity. Following our success in applying this technique to a leech motor neuron we are now extending our experiments to the oscillator interneurons of the leech swim circuit.

(Research supported by the Center for Biological Timing)

REFERENCES

Cole KS (1972) Membranes, Ions, and Impulses, University of California Press, Berkeley.

Friesen WO (1989) Neuronal control of leech swimming movements. In Cellular and Neuronal Oscillators (JW Jacklet ed). Marcel Dekker, New York pp, 269-316.

Friesen WO, and Friesen JA (1991) Analysis of neuronal function in the leech with NeuroDynamix, a computer-based system for simulating neuronal properties. Soc Neurosci Abstr 17:522.

Hodgkin AL, Huxley AF (1952) A quantitative description of membrane current and its application to conduction and excitation in nerve. J. Physiol. 117:500-544.

Kristan WB Jr, Stent GS, and Ort CA (1974) Neuronal control of swimming in the medicinal leech. I. Dynamics of the swimming rhythm. J Comp Physiol 94:91-119.

45

THE USE OF GENETIC ALGORITHMS TO EXPLORE NEURAL MECHANISMS THAT OPTIMIZE RHYTHMIC BEHAVIORS: QUASI-REALISTIC MODELS OF FEEDING BEHAVIOR IN *APLYSIA*.

I. Kupfermann, D. Deodhar, S.R. Rosen, and K.R. Weiss[*]
Center for Neurobiology and Behavior, Columbia University, 722 W. 168 St. New York, NY 10032; [*]Dept. of Physiol. & Biophysics, Mt. Sinai Medical Center; One Gustave Levy Pl.; New York, NY 10029

ABSTRACT

One important function of the modulatory mechanisms present in the feeding circuits of *Aplysia* and other animals is to dynamically adjust the parameters of the neurons and of muscles, so as to optimize the functioning of the system. One way in which we have been exploring this hypothesis is to design elementary neural networks that perform feeding-like functions, and then investigate how the performance of the circuits varies when the system is exposed to various "challenges". We used a genetic algorithm to "evolve" the synaptic parameters that permit a two neuron system to generate rhythmic behavior that results in a net gain of energy. The system performs poorly and unpredictably when challenged with internal or external variability, perhaps because it lacks modulatory mechanisms of the type that occur in the real system.

45.1 INTRODUCTION AND RESULTS

Studies of feeding in the marine mollusc *Aplysia* have proven useful for exploring how motivational states affect the nervous system and behavior (1,2). The relevant neural machinery is organized into two main parts: one which controls the appetitive responses (orienting), and one which controls the consummatory behavior, which consists of rhythmic protraction and retraction movements of the radula. Rhythmic bite-swallow movements can be driven by individual command-like cells, which activate a rhythmic central pattern generating circuit (3). This basic neural machinery is regulated by two forms of modulation: extrinsic, and intrinsic. Extrinsic modulation is typified by the serotonergic MCC (metacerebral cell). Intrinsic modulation is mediated, in part, by neuropeptides which exist as cotransmitters in motor neurons.

We have postulated that one important function of the modulatory mechanisms present in the feeding circuits of *Aplysia* is to dynamically adjust the parameters of the neurons and of muscles, so as to optimize the functioning of the system (1). One approach toward exploring this hypothesis is to design elementary neural networks that perform feeding-like functions, and then investigate how the performance of the circuits varies when the system is exposed to various "challenges" e.g. food of different hardness, or variations in environmental temperature. To start, we attempted to determine the parameters of the minimum circuits capable of generating rhythmic outputs similar to those observed during feeding. Our basic circuit consists of just two single-compartment neurons. Each neuron makes two synaptic connections to the other (Fig. 1). The dual synapses provide for multiphasic synaptic potentials, whose components (excitatory or inhibitory) can have different onset and decay time constants (4,5). The continuous state of the system was determined by stepwise integration based on an integrate and fire model (6,7) in which dV/dt is calculated as a function of the sum of the various inward and outward currents in the neuron (leakage, externally applied currents, synaptic currents and active currents). Afterpotentials following a spike were simulated by including synapses from each neuron back on itself (not shown in Fig. 1).

To complete the elementary feeding network we added two antagonistic muscles whose contraction move the radula forward or backward. The muscles are excited by a protractor (P) or retractor (R) neuron. Similar to *Aplysia* buccal muscles (8), the model muscles are non-spiking, and their contraction is proportional to their membrane potential as determined by the synaptic input. Since at this early stage of model development, the prime purpose was to explore the oscillatory properties of the neural circuit, the muscle properties and energetics (9) were kept very simple (most likely a gross oversimplification of the actual properties, which currently are only poorly understood). The system was additionally provided with a fixed external current into the protractor neuron to represent the presence of an excitatory drive provided by food stimuli that contact the lips of the animal.

We also calculated the energy used (set proportional to the area of the muscle forces), and the energy gained (set proportional to the difference between the most forward and most backward position of the radula, within defined limits, for each bite, as the radula pulls in a length of seaweed). The fitness of a circuit was defined as the difference between the energy gained and the energy used during a 10 sec simulation in which food was continuously present.

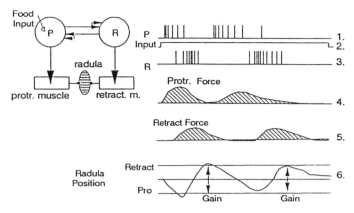

Fig. 1. Upper left. Two neurons interconnected with dual synapses (double arrows), and also making synaptic connections to a protractor or retractor muscle (large arrows). The protractor neuron is provided with a constant current input, representing afferent activity excited by food stimuli. Right side. Diagrammatic representation of the output of the system as calculated by a detailed fire and integrate model (see text).

We wished to determine the parameters of the networks, which could produce an effective output, i.e. could gain a maximum amount of energy, using a minimum amount of energy. Based on preliminary studies, we fixed the non-synaptic cell parameters (e.g. capacitance, input resistance, resting potential, threshold) at values similar to those described for molluscan neurons involved in rhythmic behaviors. First we wished to determine if a genetic algorithm could be used to effectively explore the solution space of the possible parameters (synaptic sign, conductance and onset/decay time constants) of the four synapses between the neurons. Genetic algorithms are powerful search techniques that have been used to optimize a variety of functions including those that describe the operation of neural networks (10-12), including adaptive oscillatory networks (13), but to our knowledge this method has not been used to determine the optimal parameters of realistic, continuous-time recurrent networks which operate with detailed dynamics of inhibitory and excitatory synaptic currents. A genetic algorithm, rather than alternative optimization methods such as back propagation, was selected as a method of choice because: a) it is easily adapted to any simulation model of the circuit, including the detailed continuous time model we used, and there are no constraints on the types of functions that can be optimized. b) Its performance does not depend on precise specification of an output vector (which often can not be specified a priori), but simply on any overall measure of adequacy of performance. c) The procedure somewhat mimics a natural process and its operation is easily understood.

The procedure we used was: 1. Generate a population of circuits, with randomly chosen parameters encoded as a linear string of numbers that represent the modifiable genes of each member. 2. Compute a fitness for each member, by simulating the network for 10 sec. The following is then repeated until a given level of fitness is reached. a) Select a number of mating pairs as a probability proportional to their fitness. b) Generate children in which the circuit parameters are modified by operations that mimic the genetic processes of crossover and mutation. c) Replace some members of the population with children.

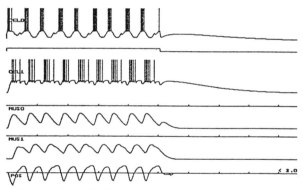

Fig. 2. Example of the functioning of one of the best fit circuits that evolved after the generation of approximately 15,000 children. Note that this circuit stopped feeding when the food (trace 2) was removed after 10 sec. The six traces represent: 1 & 3. spiking and membrane potential of the P and R neurons. 2. external current (feeding stimulus) to P neuron. 4 & 5. forces generated by the protractor and retractor muscles. 6 position of the radula (function of the difference between the protraction and retraction forces).

With successive generations the average and maximum fitness of the population increased, and a maximum fitness (Fig. 2) of approximately 25 was attained (i.e., the "organism" gained a net of 25 units of energy above the energy it used).

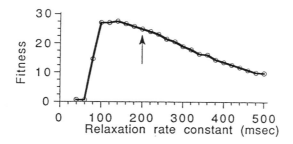

Fig. 3. Example of an external "temperature" variable that results in a change of a decay constant. Muscle relaxation decay constant was varied from 80 msec to 500 msec. Circuit was evolved at a decay constant of 200 msec and had a fitness of 25, arrow.

We found however that when selected parameters of the external environment or of internal parameters were varied over a relatively narrow range, the fitness of the networks often dramatically decreased, sometimes in unpredictable ways. In other words, rather than graceful degradation the circuits that evolved exhibited klutzy degradation when the "optimal" parameters were varied. One variable explored was food abundance. Because of excessive positive feedback in some of the highly fit circuits, they failed to stop feeding when the food was removed, resulting in a degradation of their fitness in environments in which food was not continuously present. Other variables that had large effects of the fitness of the systems were: a) food attractiveness (magnitude of the input current); b) environmental temperature, with concomitant alterations of the onset and decay time-constant of various parameters (Fig. 3); and c) "normal" ranges of variation of the genetically specified parameters due to noise or developmental factors.

45.2 CONCLUSIONS

1) A genetic algorithm can evolve the synaptic parameters that permit a two neuron system to generate rhythmic behavior that can result in a net gain of energy. 2) The system performs poorly and unpredictably when challenged with internal or external variability. 3) Current work is extending this approach, using genetic algorithms to increase the number of variables that can be explored, and to

study how the systems can be improved by the addition of mechanisms that can dynamically alter relevant parameters according to system demands (as we have postulated to occur in the real system). This may provide insights into the function of modulatory systems, and may offer leads for experimental tests in the *Aplysia* feeding system.

Bibliography

1. Weiss, K. R., Brezina, V., Cropper, E. C., Hooper, S. L., Miller, M. W., Probst, W. C., Vilim, F. S., and Kupfermann, I. (1992) *Experientia* **48**, 456-463

2. Kupfermann, I., Teyke, T., Rosen, S. C., and Weiss, K. R. (1991) *Biol. Bull.* **180**, 262-268

3. Rosen, S. C., Teyke, T., Miller, M. W., Weiss, K. R., and Kupfermann, I. (1991) *J. Neurosci.* **11**, 3630-3655

4. Jan, Y. N. and Jan, L. Y. (1983) *Federation Proceedings* **42**, 2929-2933

5. Kupfermann, I. (1991) *Physiol. Rev.* **71**, 683-732

6. Perkel, D. H., Mulloney, B., and Budelli, R. W. (1981) *Neuroscience* **6**, 823-837

7. Getting, P. A. (1989) in *Methods in neuronal modeling: From synapses to networks* (Koch, C. and Segev, I., eds) pp. 171-194, The MIT press, Cambridge

8. Cohen, J. L., Weiss, K. R., and Kupfermann, I. (1978) *J. Neurophysiol.* **41**, 157-180

9. Woledge, R. C., Curtin, N. A., and Homsher, E. (1985) *Energetic aspects of muscle contraction*, Adademic Press, Orlando

10. Stork, D. G., Jackson, B., and Walker, S. (1991) in *Artificial life, II* (Farmer, C., Langton, C., Rasmussen, S., and Taylor, C., eds) Addison,

11. Goldberg, D. E. (1989) *Genetic algorithms*, Addison-Wesley Pub. Co. Inc., New York

12. Bornholdt, S. and Graudenz, D. (1992) *Neural Netwks.* **5**, 327-334

13. Beer, R. D. and Gallagher, J. C. (1992) *Adapt. Behav.* **1**, 91-122

46

UNDULATORY LOCOMOTION — SIMULATIONS WITH REALISTIC SEGMENTAL OSCILLATOR

T. Wadden, S. Grillner*, T. Matsushima* & Anders Lansner

Department of Numerical Analysis and Computing Science Royal Institute of Technology, S-100 44 Stockholm, Sweden

*Nobel Institute for Neurophysiology Karolinska Institute, S-104 01 Stockholm, Sweden

ABSTRACT

Vertebrates like eel and lamprey move about by means of a undulatory locomotion. The neural mechanisms which underly this movement have been investigated in terms of circuitry, types of synaptic transmission, and general membrane properties. Intersegmental coordination is characterized by a phase lag (fixed proportion of cycle duration) between each consecutive spinal segment. In simulation of 20 segments, using realistic cell models, in which each segment mutually excites its' closest neighbors, it is shown that a phase lag can be produced by increasing the excitability of only one segment. It will thus become the "leader" and consecutively entrain the remaining segments. This very simple neural organization can produce a flexible (forward and backward coordination), complex, and yet easily controlled motor behavior.

46.1 INTRODUCTION

In the lamprey, a simple vertebrate, the left and right sides of each spinal segment contract in an alternating fashion at a frequency varying from 0.25–10 Hz. This produces a laterally directed wave which is propagated down the body, propelling the animal forwards through the water. The speed of the travelling wave is proportional to the forward speed of swimming and the lag between the onset of contraction in consecutive spinal segments along the body is, as a rule, a fixed proportion of the cycle duration (approximately +1%; [3, 4]). Forward swimming is achieved by a lag between the onset of contraction in

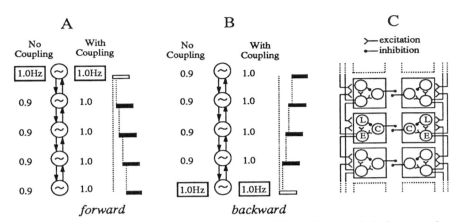

Figure 1   A, Local difference of excitability induces global changes of intersegmental phase lag. During forward swimming **A**, muscular contractions propagate rostro–caudally along the body, resulting in a phase lag between consecutive segments (see text). A reversed direction of the propagation brings about backward swimming (**B**). **C**, Neuronal network coordinating locomotion in the lamprey. The segmental circuitry of each hemisegment (box) is based on experimentally identified connections ([1, 2]; see details in text). In each segment the two symmetric halves are coupled by reciprocal inhibition from crossed interneurons (CC). In each hemisegment excitatory interneurons (E) activate all neurons of the same side including CC interneurons, lateral interneurons (L) and motorneurons (not shown here). The intersegmental connections are formed by excitatory interneurons (contained within each box) projecting to all ipsilateral interneurons in the adjacent rostral and caudal segments.

consecutive segments rostro-caudally along the spinal cord, while backward swimming is achieved by a reversed direction of lag, i.e. in the caudo–rostral direction. In the isolated spinal cord the intersegmental phase lag is commonly around 1% of the cycle duration, yet values of $+2\%$ to -2% can be produced [5]. It has been shown that the intersegmental phase lag along the spinal cord can be influenced by local increases of excitability [5]. The segments with increased excitability will, regardless of their location in the spinal cord, become the leading segments from which a lag will be produced between each segment in the rostral or caudal direction. Computer simulations have been performed to see whether the phase lag can be produced by the established neural circuitry.

46.2   METHODS AND RESULTS

The components of the segmental oscillatory network are shown in Fig. 1C. This network has been identified experimentally by paired intracellular record-

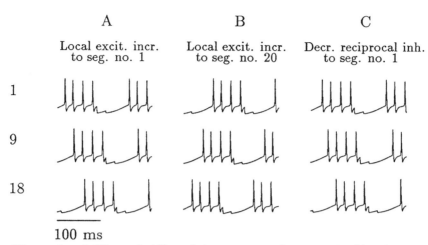

Figure 2 **A** The excitability of the most rostral segment is 6% higher than that of all other oscillators, resulting in a rostro–caudal phase lag (forward swimming) of approximately 2%/seg. When in **B** the most caudal segment is the given the extra excitability the result is an equivalent caudo–rostral phase lag (backward swimming). **C** Crossed reciprocal inhibition reduced to 85% of that of the other segments results in higher excitability which leads to a stable phase lag.

ings [2, 1], and simulated with realistic neuronal models [1, 6]. The neuron model is of intermediate complexity, and consists of a soma and three dendritic compartments, equipped with Na^+, K^+, Ca^{2+}, and Ca^{2+} dependent K^+ channels [7, 1]. Each segmental network, simulated with one to three cells of each type, can be made to cover the normal physiological range of alternation [1, 6]. Each segment is also known to be interconnected with neighboring segments by means of excitatory synapses ([1]; Fig. 1C). In order to test whether this connectivity is sufficient to account for the intersegmental phase coupling, we have connected 20 segmental networks with excitatory interneurons projecting to all ipsilateral neurons in the adjacent rostral and caudal segments (Fig. 1C). Model intersegmental synaptic strengths are made equal to 30% of the corresponding intrasegmental connections previously reported [6]. This intersegmental connectivity is simplified by letting the excitatory intersegmental connections extend only to the adjacent segment, although they actually extend over a few segments. The long (greater than 15 segment), and mainly descending inhibitory fibers from CC and L interneurons have not been included at this point. The reason being that these long fibers have a fixed fast conduction time (approximately 2ms/segment), whereas the time lag corresponding to the phase lag (1%) will vary in a wide range from 1 to 40ms (see [1]) within the physiological range of cycle durations. We are, however, presently studying what affect these inhibitory fibers actually have on phase lag. Twenty such segments were interconnected and each simulation was run for 10 seconds using the "SWIM" simulator [7] on a workstation and a Connection Machine (CM-200). The simulated network is turned on by increasing the excitability of all neurons [1, 6].

When the most rostral segment (no. 1 in Fig. 2A) of 20 interconnected segmental networks is given a small amount of extra excitatory drive (6%), it will attain a slightly higher bursting frequency than before. This will result in a certain lag between the onset of activity in segments 1 and 2, and between consecutive segments. The first segment with higher excitability will entrain the second segment to start its bursting cycle earlier than it would have started otherwise, but still with a delay. In the same manner there will be a lag as the second segment entrains the third segment and so forth, resulting in a forward intersegmental lag. When the same amount of excitation is instead added to the most caudal segment, it will become the leading oscillator and the segments will now be entrained in a caudo–rostral (backward) direction (Fig. 2B). Therefore the "inherent" frequency, which is the frequency of the single segment had it been uncoupled from other segments, is critical in determining which segment shall become the leading one. An effect analogous to adding extra excitation (Fig. 1A–B, 2A–B) is to increase the "inherent" frequency by reducing the reciprocal inhibition between the two halves of one segment [8]. Fig. 2C shows that a phase lag is similarly produced if the reciprocal inhibition in segment no. 1 is reduced by 15%. As with an excitability increase a phase lag will then occur between each consecutive segment along the chain of 20 segments. This condition with a reduced reciprocal inhibition in the rostral segments is similar to that occurring in the isolated spinal cord (phase lag around 1%) in which the inhibitory current in the rostral segment may be only 50-60% of that in the more caudal segments [9].

46.3   DISCUSSION

Stable forward and backward intersegmental lags can be generated by the simulated multisegmental network of Fig. 1C by adding extra excitation only to one segment in the rostral, or caudal part of the oscillator chain. The connectivity and cellular properties of the segmental network model correspond to those experimentally established [1]. What then is the physiological mechanism modifying selectively the excitability in different portions of the spinal cord? Indeed, a number of reflex and descending mechanisms provide such an effect [3, 1, 5]. The fact that there is usually a rostro–caudal lag also in the isolated spinal cord [5, 4] can most likely be ascribed to the finding that there is less reciprocal inhibition in the most rostral segments (see above Fig 2C). The current network model is considerably simplified with only a minimal number of neurons of each type and a simplified intersegmental coupling, yet it can still account for the experimental findings. The effects of the inhibitory interneurons as well as the use of a model with populations of neurons is now under investigation. Several other investigators have used mathematical models describing chains of oscillators, with these models representing populations of neurons [10, 11, 12]. When using these models to describe intersegmental coupling in the lamprey they have assumed that no excitability difference occurs along the spinal cord, based on the experiments of Cohen (as cited in [12]) and have relied on asymmetric connections to produce a constant phase lag. Such models produce a fixed phase lag, but are not well suited to provide for the flexibility of a changing phase lag as is apparent under physiological conditions. Cohen (as cited in [12]) stated that there is no overall excitability gradient along the spinal cord after subdividing it into smaller parts, each of which could display a similar burst rate. After transection the most rostral segments in each part will have less recipricol inhibition compared to the other segments (as demonstrated experimentally; Wallén *et al.* 1993). This results in a higher inherent frequency and therefore the rostral segments will tend to form a leading segment in relation to the remaining segments resulting in a rostro–caudal lag. The smaller reciprocal inhibition in the rostral segment is due to the mainly caudal arrangement of crossed inhibitory CC interneurons [9]. The intersegmental network model here (Fig. 1C) uses a very simple neuronal organization with segmental oscillators connected by mutual excitation and generates a flexible intersegmental motor pattern. A local "extra" excitation of the most rostral or caudal segments (or a reduced reciprocal inhibition) will thus suffice to produce an intersegmental phase lag throughout the entire spinal cord.

Acknowledgement

This study was supported by the Swedish Medical research Council (3026), Swedish Natural Research Council (B-TU-3531) and the Agency for Technical Developement.

REFERENCES

[1] Grillner, S., Wallén, P., Brodin, L., and Lansner, A., "Neuronal network generating locomotor behavior in lamprey: Circuitry, transmitters, membrane properties and simulations," *Ann. Rev. Neurosci.*, vol. 14, 1991, pp. 169–199.

[2] Buchanan, J. and Grillner, S., "Newly identified 'glutamate interneurons' and their role in locomotion in the lamprey spinal cord," *Science*, vol. 236, 1987, pp. 312–314.

[3] Grillner, S., "On the generation of locomotion in the spinal dogfish," *Exp. Brain Res.*, vol. 20, 1974, pp. 459–470.

[4] Wallén, P. and Williams, T., "Fictive locomotion in the lamprey spinal cord in vitro compared with swimming in the intact and spinal animal," *J. Physiol.*, vol. 347, 1984, pp. 225–239.

[5] Matsushima, T. and Grillner, S., "Neural mechanisims of intersegmental coordination in lamprey - local excitability changes modify the phase coupling along the entire spinal cord.," *J. Neurophysiol.*, vol. 67, 1992, pp. 373–388.

[6] Wallén, P., Ekeberg, Ö., Lansner, A., Brodin, L., Tråvén, H., and Grillner, S., "A computer based model for realistic simulations of neural networks: II. the segmental network generating locomotor rythmicity in the lamprey," *J. Neurophysiol.*, vol. 68, 1992, pp. 000–000.

[7] Ekeberg, Ö., Stensmo, M., and Lansner, A., "SWIM – a simulator for real neural networks," Technical Report No. TRITA-NA-P9014, Dept. of Numerical Analysis and Computing Science, Royal Institute of Technology, Stockholm, Sweden, 1990.

[8] Grillner, S. and Wallén, P., "Does the central pattern generation for locomotion in the lamprey depend on glycine inhibition?," *Acta Physiol Scand*, vol. 110, 1980, pp. 103–105.

[9] Wallén, P., Shupliakov, O., and Hill, R., "Origin of phasic synaptic inhibition in motor neurons during fictive locomotion in the lamprey," *Exp. Brain Res.*, 1993, in press.

[10] Williams, T. L., Sigvardt, K., Kopell, N., Ermentrout, G., and Remler, M., "Forcing of coupled nonlinear oscillators: studies of intersegmental coordination in the lamprey locomotor central pattern generator," *J. Neurophysiol.*, vol. 64, 1990, pp. 862–871.

[11] Kopell, N. and Ermentrout, G., "Coupled oscillators and the design of central pattern generators," *Math. Biosci.*, vol. 89, 1988, pp. 14–23.

[12] Rand, R., Cohen, A., and Holmes, P. "Systems of coupled oscillators as models of central pattern generators,". in *Neural Control of Rythmic Movements in Vertebrates* (Cohen, A., Rossignol, S., and Grillner, S., eds.). John Wiley & Sons., New York, 1988, pp. 333–367.

NETWORK MODEL OF THE RESPIRATORY RHYTHM

Eliza B. Graves
William C. Rose
*Diethelm W. Richter
James S. Schwaber

Neural Computation Program, DuPont Company, Wilmington, DE 19880-0323
*Physiology Institute, Georg-August-Universät Göttingen, D-3400 Göttingen, FRG

ABSTRACT

We propose a model for the generation of the respiratory rhythm. The model is composed of four distinct neuron types recorded in vivo in the cat, defined by their phase relationship to the respiratory cycle as seen in the phrenic neurogram. It is proposed that the respiratory rhythm emerges from the connectivity and membrane dynamics of these neurons. Computational models are created of the four cell types using Hodgkin-Huxley form kinetics to describe active membrane properties that are then tuned to reproduce the recorded patterns of neuronal behavior. The neurons are reciprocally connected with inhibitory synapses. Simulation results accurately reproduce the three-phased respiratory cycle, and demonstrate the way in which this activity may arise from intrinsic membrane, as well as network, properties.

47.1   INTRODUCTION

The nervous command to various respiratory muscles is controlled by a neuronal network which is localized in the lower brainstem. In the adult mammal the origin of the rhythmic respiratory commands has been hypothesized to be organized by neuronal interaction within a central respiratory network [1]. In the present paper we explore this hypothesis in simulation. Our goal is to constrain the model based on known properties of the neurons, connections and synapses of the network and determine under what conditions the respiratory rhythm can plausibly emerge.

The rhythm generator normally produces a three-phased respiratory cycle: inspiration, post-inspiration and active expiration. For the present purposes we propose that a reduced network model consisting of early-inspiratory (e-I), post-inspiratory (post-I) and stage-2 expiratory (E2) neurons with inputs from pre-inspiratory neurons (pre-I) might produce a three-phased respiratory cycle (see Fig. 47.1). The first three

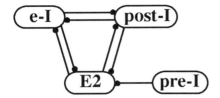

Figure 47.1. Four neuron network used to model respiratory rhythm generation. Early inspiratory (e-I), post-inspiratory (post-I), and stage-2 expiratory (E2) neurons are reciprocally connected with inhibitory synapses; pre-inspiratory neuron (pre-I) provides inhibitory synaptic input to E2.

of these neuron types are reciprocally interconnected by inhibitory synapses. The neuronal activity patterns arise from synaptic activity and from voltage- and calcium-dependent active membrane channels, for example, from "rebound excitation" that occurs from a low voltage activated calcium current, Ca(T) which is re-activated when neurons are released from postsynaptic) inhibition.We have translated the hypothesized biological model into computational models by first creating neuron models in the Hodgkin-Huxley form. In the models reported here the pre-I neuron contains Na(HH), K(DR), and K(A) [2] and K(Ca) [3]. The E2 neuron also contains Ca(T) [4], and the e-I and post-I neurons include Ca(L) [3] as well. It should be noted that the present K(Ca) kinetics from Yamada et al. (3) contributed to accommodation in the neuron models. Inhibitory synapses are modeled as ligand-gated ($t \cdot e^{-\alpha t}$) conductances, with reversal potentials of -90 mV and time constants matched to *in vivo* measured ipsp's.

Using these neurons, the goals for network simulation are (a) to achieve the observed patterns of neuronal activity, including e-I and post-I neurons with a gradually accommodating pattern and E2 activity which ramps up and then stops; (b) to extend the accommodating patterns of activity for each neuron for seconds as needed for the length of the respiratory period, and; (c) to maintain the proper sequence of activity through the 3-neuron chain.

47.2 RESULTS

In order to create a three-phased oscillator of the desired frequency in which the E2 neuron firing pattern was realistic (ramping up) an additional source of inhibition was needed to terminate simulated E2 activity, and to do this we added the pre-I neuron. The pre-I neuron model is simply one with a linear I/F range so that it can be driven to match its observed biological firing pattern [5,6] by current injection. The biological basis of the pre-I activity pattern is presently unknown.

Initial simulations using neuron models and connections appropriately constrained by the known biology were unsuccessful in meeting the above goals, indeed could not be made to sustain activity. This is not surprising since much remains unknown about the network, and since the system would have to be exceptionally robust to "work" start-

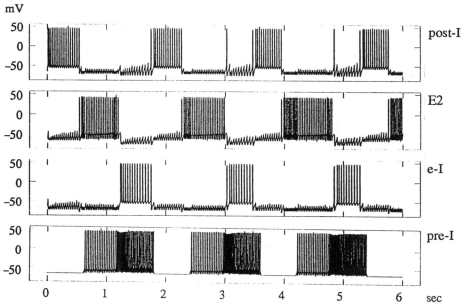

Figure 47.2. Network model I. post-I, E2, and e-I neurons all receive tonic excitation. Pre-I neuron receives a variable stimulus that is needed to generate the activity observed *in vivo*.

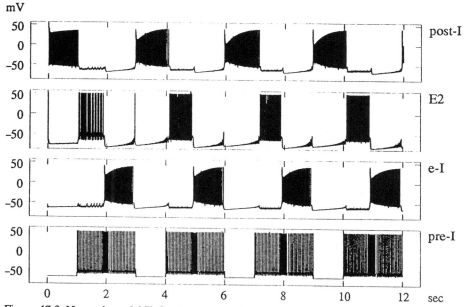

Figure 47.3. Network model II. See text for details.

ing from initial guesses within the huge parameter space defined by the model. Fortunately, the neuron models were based on earlier work that defined a "kinetics library" [7] which allows model tuning in a biophysically predictable fashion. Thus, neuron models were tuned, primarily by adjusting the maximum channel densities and time-dependence of kinetics.

In order to achieve the sustained patterns of accommodation needed in e-I and post-I neurons we have used two different tuning strategies, involving alternatively either largely synaptic (Model I; Fig. 47.2) or intrinsic neuronal (Model II; Fig. 47.3) properties. The Model I cells are highly sensitive to current injection and their activity pattern sensitively depends on patterns of synaptic input and synaptic strengths. The post-I and e-I neurons achieve accommodation by a build up of calcium in the cell via the high threshold calcium current, Ca(L), and subsequent activation of the calcium-dependent potassium conductance, K(Ca). This model was successful in achieving many of the goals for respiratory network activity. However, the simulated activity pattern was difficult to achieve and highly sensitive to small changes in a number of parameters. In order to sustain activity over time all the neurons (except the pre-I) are receiving high levels of tonic excitation. Sensitive tuning of calcium concentrations and pumping, of calcium and potassium kinetics and of synaptic weights were required to achieve plausible neuronal firing frequencies, adequate rebound excitation and long-lasting accommodation. In addition, with the present Model I dynamics it was not possible to make the respiratory period longer than 1.8 seconds (vs. 3 seconds observed *in vivo*)

In order to create a more robust network, in Model II we explored the use of state-dependent neurons having sustained, nonlinear currents that could produce a long-lasting, stable period of accommodation of firing rate within bursts. Since we did not have kinetics data for the conductances that could perform this role, we created a conductance for this function by tuning the kinetics of the low threshold calcium conductance, Ca(T) [4] to a high channel density as well as to an increased time constant (by a factor of 60). The new kinetics proved to alter the neuronal intrinsic response properties as needed to provide longer accommodation in the post-I and e-I neurons, as well as still providing rebound excitation. Activity in these neurons adapts over time due to activity-dependent activation of K(Ca). The neurons do not receive tonic excitation as in Model I. With the longer accommodation the periodicity of the rhythm is 3 seconds.

47.3 CONCLUSIONS

The simulation results with the present models indicate that our modeling goals for respiratory rhythm generation are most robustly achieved if the network contains cells with intrinsic membrane properties producing a slow decrease in excitability that contributes to the patterning of neuronal activity. Neuronal dynamics of this kind could be produced by any of several conductances, such as an additional TTX sensitive Na, Ca conductance [8] and a calcium-dependent activation of cation conductances [9] which are additive with Ca(T). Although present Ca(T) levels are unrealistic, it still could be involved to some degree, as it has been shown that there is a large variation in the density and time dependence of inactivation of Ca(T) [10].

However, the present model contains only one representative of each neuron class, which can not be the case in life. Thus, an alternative means of achieving robust respiratory rhythm generation might be to change the model by increasing the numbers of neurons. For example, creating populations of each neuron type with wide-spread interconnectivity might achieve stability by spreading inputs in time.

The simulation results suggest two lines of future investigation: whole-cell patch studies of the present neuron types to look for a sustained transient current with duration in the second range, and simulation studies using a model of the respiratory network with populations of model neurons.

Acknowledgments: This work was supported by the DuPont Company and the Deutsche Forschungsgemeinschaft.

REFERENCES

[1] Richter, D.W., Champagnat, J., Mifflin, S. In *Neurobiology of the Control of Breathing.* Raven Press, 1986.

[2] Conner, J.A., Walter, D., McKown, R. *Biophys. J.* **18**: 81-102, 1977.

[3] Yamada, W.M., Koch, C. Adams, P. R. In *Methods in Neuronal Modeling.* MIT Press, 1989.

[4] Coulter, D.A., Huguenard, J.R., Prince, D.A. *J. Physiol.* **414**: 587-604, 1989.

[5] Cohen, M.I. *J. Neurophysiol.* **31**:142-165, 1969.

[6] Smith, J.C., Greer, J., Liu, G., Feldman, J.L. *J. Neurophysiol.* **64**:1149-1169, 1990.

[7] Schwaber, J.S., Paton, J.F.R.P., Graves, E.B. *Brain Res.,* in press.

[8] Meves, H., Vogel, W. *J. Physiol.* **235**: 225-264, 1973.

[9] Partridge, L.D., Swandulla, D. *Trends Neurosci.* **11**: 69-72, 1988.

[10] Lytton, W.W., Sejnowski, T.J. Personal communication.

A MODEL OF THE PERISTALTIC REFLEX

A. D. COOP and S. J. REDMAN

Division of Neuroscience, The John Curtin School of Medical Research,
G.P.O. Box 334, Canberra, A.C.T. 2601, Australia.
Email: allan.coop@anu.edu.au

1 INTRODUCTION

The first descriptions of the peristaltic reflex are attributed to Bayless and Starling [1]. They observed that excitation at any point of the gut induced contraction above and inhibition below, and that reactivation of the reflex response induced a peristaltic wave that acted to move the intestinal contents aborally. Cannon [3] later distinguished between peristalsis (short movements of intestinal contents over a few millimeters) and the peristaltic rush (movements over tens of millimeters). Peristalsis has since been reported to occur in all chordates possessing a muscular gastrointestinal tract. A biologically plausible model based on a distension sensitive component of the peristaltic reflex in the guinea-pig has been developed. It has been used as an aid in understanding the neural mechanisms underlying peristalsis.

2 COMPONENTS OF THE PERISTALTIC REFLEX

The myenteric plexus is a part of the enteric nervous system. It lies between the longitudinal and circular muscle (CM) within the wall of the intestinal tract and is considered responsible for the neuronal control of the peristaltic reflex. It is composed of ganglia (containing the majority of the neuronal cell bodies) and interganglionic neuronal projections. Together these give the myenteric plexus its net-like appearence along and around the intestine.

Two classes of neuron have been identified electrophysiologically within the myenteric plexus. Putative sensory neurons (SN) typically generate from 1–

3 action potentials before being inactived by a characteristically prolonged afterhyperpolarization (< 20 s) [4]. They have been shown to project 2–4 mm circumferentially and in some cases < 0.5 mm anally. It has been estimated that each of these neurons contacts ≈ 15–20 other neurons, but it is not known to which pathways they belong. The second class of neuron includes interneurons and motoneurons (MNs). These fast firing neurons generate ≈ 200 action potentials before inactivating for several minutes [4]. At least four monosynaptic projections for MNs have been identified, two orally directed and excitatory [5], and two anally directed and inhibitory [2]. The majority of ascending projections extend up to 1 mm orally, with a small proportion extending up to 11 mm orally. Similarly, the majority of descending projections extend up to 2 mm anally, with the longest projecting up to 30 mm anally. There are also thought to be ascending and descending interneurons, due to reflex responses being observed over distances greater than the known lengths of possible participating MNs. The CM is the reflex effector and may be thought of as a series of longitudinally stacked smooth muscle rings, each receiving input from excitatory and inhibitory MNs.

3   THE MODEL

No functional significance is attached to the location of ganglia within the myenteric plexus. Consequently, in the model the two dimensional structure of the plexus was reduced to a single array of ganglia that represented the longitudinal axis of the intestinal wall. The model also assumed that menteric ganglia were ≈ 0.5 mm apart and that the inter-ganglionic propagation delay for the action potential was ≈ 1 ms. Further, it was assumed that each ganglion contained at least one SN and the origin of at least one excitatory and one inhibitory pathway. These assumptions allowed the construction of a reflex "unit" that consisted of a ganglionic "node" (GN) plus an associated "patch" of CM. The model was constructed of 200 of these reflex units which represented a 10 cm length of the small intestine.

The connectivity of neurons within and between GNs depended upon where reflex paths were assumed to converge. The simplest circuit that allowed for convergence of inputs onto the CM is shown in Fig. 1(top). Within each GN, the SN activated the ascending and descending MN projections commencing at that node. The ascending excitatory MNs projected monosynaptically to the CM patch associated with each of 24 more orally located GNs. Similarly, projections of the descending inhibitory MNs at a GN monosynaptically projected to the CM patch associated with each of 60 more anally located GNs. An alternative circuit allowed convergence onto the MNs. Fig. 1(bottom) shows that in this case the SN within a GN must synapse with ascending and descending interneurons to reach target MNs. In the simplest case these MNs only project between their GN of origin and its associated CM patch.

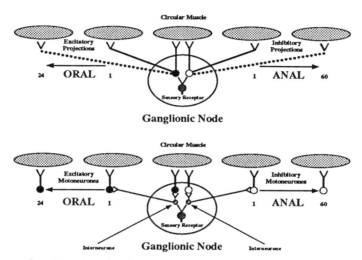

Figure 1 The reflex circuits modeled depended upon where convergence of reflex paths occurred. In both circuits, convergence at the CM (top) and convergence at the MNs (bottom), ascending projections connected 24 more orally located and descending projections 60 more anally located CM patches.

Sensory neurons were modelled as binary units activated by radial distension of the CM. Once activated they fired with a specific frequency (10 Hz, 20 Hz or 30 Hz) which represented the degree of radial distension caused by the presence of a bolus. Interneurons were not explicitly modelled as it was assumed that there was sufficient convergence of SNs onto interneurons for them to reliably generate an action potential in response to each SN event. No distinction was made between the electrophysiological properties of the excitatory and inhibitory MNs. Both types of neuron exhibited a resting membrane potential, threshold for impulse generation, membrane time constant, and absolute and relative refractory periods.

The spatial activity of the CM was determined at each timestep by summing the total supra-threshold activity of the inhibitory and excitatory MNs at each of the 200 CM patches in the model. The temporal activity of the CM was determined by summing the total supra-threshold excitatory and inhibitory activity at each point in the model for the previous 900 timesteps. At any timestep, if the CM activity at the oral end of the bolus was greater than resting tension, and the CM activity at the anal end of the bolus was less than that of the oral end, the bolus moved anally one GN.

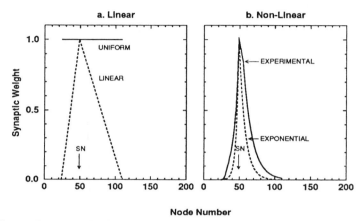

Figure 2   The distributions of synaptic weights used by simulations. The absolute magnitude of synaptic weights in the inhibitory path (right of peak) is plotted to allow comparison with the magnitude of synaptic weights in the excitatory path (left of peak). a. UNIFORM and LIN-EAR. b. EXPONENTIAL and EXPERIMENTAL. SN: sensory neuron. Orientation: oral: left, anal: right.

Four different distributions of synaptic weights across the ascending and descending pathways were tested to determine their effect on bolus movement. Figure 2 shows the normalized absolute values used. One distribution (UNIFORM) used the same magnitude for each synaptic weight in a path. The other distributions decremented with distance from the SN originating the path. They included, linear (LINEAR) and exponential (EXPONENTIAL) decrement, and a distribution adapted from experimental data (EXPERIMENTAL). In all simulations the magnitude of the first synaptic weight in each pathway was identical. For any simulation the distribution of synaptic weights was normalized to one of three different magnitudes; 0.25, 0.5, or 0.75.

4   RESULTS

Activation of sensory receptors due to the presence of a bolus induced activity in the ascending excitatory and descending inhibitory reflex pathways. If the bolus was of a sufficient length, and the synaptic weights were of sufficient magnitude, suprathreshold activity of ascending and descending MNs induced an orally directed contraction and an anally directed relaxation of the CM (Fig. 3a). At any one GN, the passage of the bolus was seen as a wave of CM relaxation followed by a wave of CM contraction (Fig. 3b).

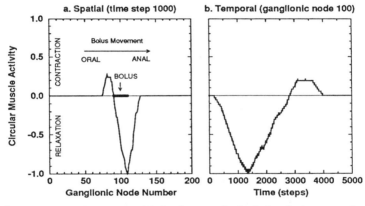

Figure 3   Spatio-temporal behaviour typical of simulations. a. Spatial activity of the CM after bolus moved from GN 20 to GN 100. b. Temporal activity of CM at SN 100.

4.1   Bolus movement depends on the weight distribution

When reflex paths converged at the MNs, only the EXPERIMENTAL distribution gave an increase in the maximum distance a bolus travelled for both an increase in the magnitude of the synaptic weights, and an increase in the number of suprathreshold SN responses (Fig. 4a). Consequently, subsequent simulations using convergence at the MNs used only the EXPERIMENTAL distribution.

When convergence occurred at the CM, the maximum distance a bolus moved depended upon which distribution of synaptic weights was used. For any one distribution (EXPERIMENTAL, LINEAR or UNIFORM), the maximum distance a bolus moved was independent of the magnitude of the synaptic weights, and each distribution produced a different maximum bolus movement for a specific number of suprathreshold SN events (Fig. 4b). In contrast to convergence at the MNs, only the LINEAR distribution gave an increase in the maximum distance a bolus moved for an increase in the number of suprathreshold SN responses from 3 to 5.

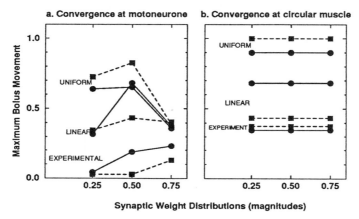

Figure 4   The effect of the type of synaptic weight distribution (EXPER-IMENTAL, LINEAR or UNIFORM) at three different magnitudes (0.25, 0.5, 0.75), and SN sensitivities (3 (dashed), or 5 (solid) suprathreshold events), on the maximum distance moved by a bolus (normalization: the maximum distance moved by the bolus under convergence at the CM, UNIFORM weight distribution, and SN sensitivity of 3 action potentials). a. Convergence at the MN. b. Convergence at the CM.

4.2   Sensory neuron discharge affects bolus movement

When a SN was allowed to generate an unlimited number of action potentials in response to reflex activation, periodic movements of the bolus were observed (whether convergence occurred at either the MNs or the CM) for as long as simulations were run (< 30,000 time steps) (Fig. 5a). However, in the case of convergence at the MNs, an increase in the magnitude of the synaptic weight distribution increased the rate of bolus movement compared with convergence at the CM.

Limiting the number of action potentials that could be generated by the SNs to a maximum of either 3 or 5 induced movement of the bolus that eventually ceased (Fig. 5b). This was independent of whether convergence occurred at the MNs or the CM. Two types of bolus movement typically observed during simulations are shown by the 0.50 record (Fig. 5b). The first was a small bolus movement of one GN, that occurred at ≈ 50 and ≈ 1100 timesteps. The other was a larger movement of 20 GNs, initiated at ≈ 950 timesteps. All subsequent simulations were run with a limit of 5 SN action potentials.

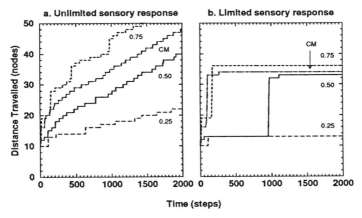

Figure 5 The effect of limiting sensory receptor activity on bolus movement in the presence of an EXPERIMENTAL distribution of synaptic weights. a. Unlimited SN activity. b. SN activity limited to 5 suprathreshold events.

4.3 Bolus length alters bolus movement

The movement of a bolus was sensitive to its length whether convergence occurred at the MNs or the CM. When convergence occurred at the MNs, increasing the length of the bolus from 21 to 31 GNs (10.5 to 15.5 mm respectively) substantially reduced the total distance a bolus moved. However, decreasing the size of a bolus from the equivalent of 21 to 11 GNs had no effect on the total distance the bolus moved. Making the bolus smaller (11 nodes: 5.5 mm) increased the number of times it paused before finally ceasing to move when compared with the behaviour of a physiologically sized control bolus (21 nodes). When convergence occurred at the CM, increasing the length of a bolus from 11 GNs to 31 GNs increased the total distance the bolus moved.

4.4 Sensory neuron firing frequency affects bolus movement

When convergence occurred at the MNs, bolus movement was most sensitive to a SN firing frequency of 20 Hz. Lower (10 Hz) or higher (30 Hz) SN activity almost eliminated bolus movement. When convergence occurred at the CM, altering the SN frequency of firing had little effect on the total distance the bolus moved. However, a SN response frequency of 20 Hz considerably delayed the time at which the bolus finally stopped moving.

5 CONCLUSIONS

The model predicted that the activation of a simple reflex circuit consisting of a sensory receptor, interneurons, and excitatory and inhibitory MNs, was sufficient to replicate much of the known behaviour of a bolus within the small intestine, including both small (peristaltic) and large (peristaltic rush) bolus movements. Simulations also suggested that a smaller bolus should induce greater reflex activity. The results supported the conjecture that the peristaltic response may be due to the longitudinal replication of a simple reflex circuit and that feedback provided by bolus movement coordinates the activity of the ascending and descending components of the reflex circuits.

REFERENCES

[1] Bayless, W. M. and E. H. Starling. (1899). The movements and innervation of the small intestine. *J. Physiol.* (Lond). **24**. pp. 100–143.

[2] Bornstein, J. C., M. Costa, J. B. Furness, and R. J. Lang. (1986). Electro-physiological analysis of projections of enteric inhibitory motor neurones in the guinea-pig small intestine. *J. Physiol.* (Lond). **370**. pp. 61–74.

[3] Cannon, W. B. (1902). The movements of the intestines studied by means of Rontgen rays. *Am. J. Physiol.* **6**: 251–277.

[4] Hirst, G. D. S., M. E. Holman, and I. Spence. (1974). Two types of neurones in the myenteric plexus of duodenum in the guinea-pig. *J. Physiol.* (Lond). **361**. pp. 297–314.

[5] Smith, T. K., J. B. Furness, M. Costa, and J. C. Bornstein. (1988). An electro-physiological study of the projections of motor neurons that mediate non-cholinergic excitation in the circular muscle of the guinea-pig small intestine. *Auton. Nerv. Sys.* **22**. pp. 115–128.

SECTION 7

CORTEX, CEREBELLUM AND SPINAL CORD

The chapters in this section include studies on reflexes such as the stretch reflex, the eyeblink reflex, and the vestibulo-ocular reflex, as well as more general papers on information processing in the cerebellum and the brain stem. The last two chapters present high level models of speech production and basal ganglia information processing during movement.

49

SIMULATION OF THE MUSCLE STRETCH REFLEX BY A NEURONAL NETWORK

Bruce P. GRAHAM[1] and Stephen J. REDMAN

*[1]Centre for Information Science Research and
Division of Neuroscience, The John Curtin School of Medical Research,
The Australian National University,
P.O. Box 334, Canberra, A.C.T., 2600, AUSTRALIA
Email: Bruce.Graham@anu.edu.au*

1   INTRODUCTION

A dynamic model of the muscle stretch reflex has been developed to allow the investigation of the effects of the different neuronal types in the reflex circuit on the reflex response. The situation being modelled is shown in Figure 1. Two

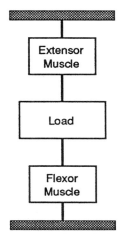

Figure 1   Physical layout - two antagonistic muscles acting on a load. The extensor muscle acts against gravity, and the flexor muscle acts with gravity. The load moves in a straight line.

antagonistic muscles act on a load to control its position. The load can move in a straight line, which, for small movements, is a reasonable approximation to the real situation in which the muscles act to rotate a limb around a joint. The muscle forces, and hence load movement, are controlled by the stretch reflex circuit, which is shown in Figure 2.

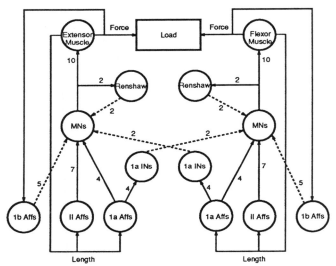

Figure 2 Stretch reflex circuit diagram (solid lines are excitatory connections, dashed lines are inhibitory connections, numbers are conduction delays in msecs).

While many of the neuronal types in this circuit receive signals from higher levels of the nervous system, we consider the situation in which such signals are constant. In this case, if the load position is disturbed, the reflex circuit must act to return the load to its initial equilibrium position. The roles of the different neuronal types in the circuit are evaluated using the simulation by selectively including and excluding the different types from the circuit and comparing the reflex responses that result.

The simulation has been implemented using the neural network simulation package GENESIS from the California Institute of Technology, running on a Sun 4 workstation.

2   STRETCH REFLEX CIRCUIT

The load is simply modelled as an inertial mass with a frictional damping factor.

The muscles are modelled using data from the cat medial gastrocnemius (MG) muscle [1]. For simplicity, both muscles are identical. Each muscle is modelled as a population of 30 motor units, with each motor unit consisting of a motoneuron, and the bundle of muscle fibres it stimulates. Slow (S), fatigue resistant (FR) and fast fatiguing (FF) motor unit types are represented in the same proportion they are found in the cat MG muscle. The muscle is approximately one-tenth the size of the real MG muscle, which contains some 300 motor units. The overall muscle force is given by the sum of the forces from the individual motor units. Each bundle of muscle fibres is modelled as a linear, critically-damped, second-order system [6]. The inputs to the system are impulses from the motoneuron. The output is the force generated. The motoneurons fire in response to excitation from the Group Ia and Group II afferents. The firing rate is a thresholded, linear function of the average excitatory current induced by the arrival of impulses from the afferents. This is a model of the primary firing range for the motoneurons [5]. The current threshold, and minimum and maximum firing rates, vary for each motoneuron, with a general increase in the values as the motor unit type changes from S to FR to FF. The parameter values are derived from cat MG data.

The primary sensors in the stretch reflex are the Group Ia and Group II afferents. The Group Ia afferents signal the muscle length and the rate of change of muscle length. The Group II afferents signal muscle length only. A simple model of afferent firing rate during ramp stretch of a muscle [3] has been modified for use when the velocity is not constant. The firing rates of these afferents are converted to excitatory current at the motoneurons by multiplying the average firing rate by a weight, which represents the number of synapses and the charge transferred to the soma from a synapse due to the arrival of an impulse. Each afferent type is represented by a single cell which excites all of the motoneurons equally.

Another type of sensor in the circuit is the Group Ib afferents, which respond to force in the muscle. They feed back inhibition onto the motoneurons of the homonymous muscle. Again, a single cell is used to represent the pool of afferents. The firing rate of the cell is a linear function of the muscle force [2].

There is a class of interneurons that are excited by the Group Ia afferents of one muscle, and inhibit the motoneurons of the opposing muscle. A single cell is used to represent a pool of these interneurons excited by the Group Ia afferents of a particular muscle. Its firing rate is taken to be identical to the firing rate of the Group Ia afferent that excites it.

Another class of interneurons provides recurrent inhibition onto motoneurons, as compared to the reciprocal inhibition from the Group Ia inhibitory interneurons. The Renshaw cells are excited by the motoneurons of a muscle, and inhibit those same motoneurons. Again, a single cell represents the pool of Renshaw cells associated with a particular muscle. Its firing rate is the weighted average of the motoneuron firing rates, with the motoneurons of fast motor units providing approximately four times the excitation to the Renshaw cell as the slow motor units [4]. The Renshaw cell inhibits all of the motoneurons equally.

3   SIMULATION RESULTS

We will present the results for a single situation in which the load is disturbed from its resting position, and the reflex circuit must act to return it to rest. This is analagous to the standard tendon jerk reflex test, and a typical simulation is summarised in Figure 3. The performance of a particular reflex response was

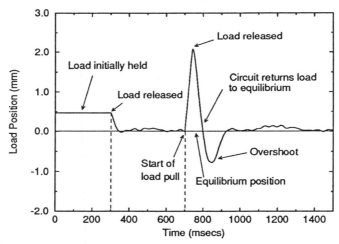

Figure 3   Example of a simulation of the tendon jerk reflex test. Load position is shown relative to the equilibrium position (0mm).

graded by how fast it returned the load to equilibrium (rest), and how well it maintained it there.

The first circuits trialled contained only the Group Ia and Group II afferents, with none of the inhibitory loops. The initial results, shown in Figure 4, are for circuits containing either Group Ia or Group II afferents, but not both. With Group Ia afferents only, the response is highly oscillatory, at a frequency of about 10Hz. This is comparable to the condition known as clonus. Group II afferents provide low frequency, damped oscillations. The load is returned to rest much slower than by the Group Ia afferents. By trial and error an optimum combination of Group Ia and Group II afferents was found. A combination of one sixth Group Ia afferents to five sixths Group II afferents provided a quick return of the load to equilibrium, and good control of the load at equilibrium. This optimum response is shown in Figure 5.

These responses are for a load of 100g. The response of the circuit containing the optimum ratio of Group Ia and Group II afferents for this load is qualitatively good, and not much improvement upon it could be expected. However, if the load is decreased to 50g, which is much easier to move, the response of this circuit deteriorates markedly. This is shown in Figure 6. Trial and error again confirmed, though, that one sixth Group Ia to five sixths Group II was still the optimum ratio for the different afferents. So the response shown in the Figure 6 is the best that can be obtained with only excitatory loops in the circuit. A load of 50g was taken as the test case for investigating the effects of the inhibitory loops.

The addition of reciprocal inhibition from the Group Ia inhibitory interneurons to the circuit containing the optimum ratio of excitatory afferents improved the response by curtailing the overshoot in load position past its rest position. The best response was obtained with the strength of inhibition at 3% of the strength of afferent excitation. This is shown in Figure 7.

Including the inhibitory loops of the Group Ib afferents and the Renshaw cells led to a further improvement in performance. The best performance with both these loops added to the circuit already containing the excitatory afferents and the Group Ia inhibitory interneurons is shown in Figure 8. This was obtained with the strength of Group Ib inhibition set at 14%, and the Renshaw cell inhibition at 28%, of the strength of afferent excitation.

REFERENCES

[1] Burke, R.E. (1981). Motor units: anatomy, physiology and functional organization. In Brooks, V.B. (Ed.), *Handbook of Physiology, Sect. 1:*

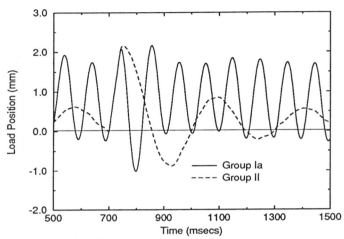

Figure 4   Reflex response with a load of 100g for a circuit containing only Group Ia afferents (solid line), or Group II afferents (dashed line).

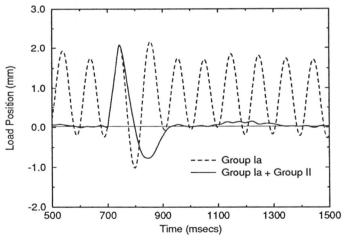

Figure 5   Reflex response with a load of 100g for a circuit containing both Group Ia and Group II afferents, with afferent excitation of the motoneurones being five times stronger from the Group II afferents than from the Group Ia afferents (solid line), compared with Group Ia afferents only (dashed line).

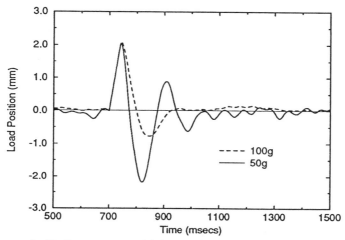

Figure 6   Reflex response with loads of 50g (solid line) and 100g (dashed line) for a circuit containing the optimum connection strengths of Group Ia and Group II afferents i.e. Group II five times stronger than Group Ia.

The Nervous System, Volume II: Motor Control, Part 1, (pp. 345-422). Bethesda: American Physiological Society.

[2] Crago, P.E., Houk, J.C. & Rymer, W.Z. (1982). Sampling of total muscle force by tendon organs. *Journal of Neurophysiology, 47,* 1069-1083.

[3] Houk, J.C., Rymer, W.Z. & Crago, P.E. (1981). Dependence of dynamic response of spindle receptors on muscle length and velocity. *Journal of Neurophysiology, 46,* 143-166.

[4] Hultborn, H., Katz, R. & Mackel, R. (1988). Distribution of recurrent inhibition within a motor nucleus. II. Amount of recurrent inhibition in motoneurones to fast and slow units. *Acta Physiologica Scandinavica,* **134,** 363-374.

[5] Kernell, D. (1984). The meaning of discharge rate: excitation-to-frequency transduction as studied in spinal motoneurones. *Archives Italiennes de Biologie, 122,* 5-15.

[6] Rymer, W.Z. (1984). Spinal mechanisms for control of muscle length and tension. In Davidoff, R.A. (Ed.), *Handbook of the Spinal Cord: Volumes 2 & 3 Anatomy & Physiology,* (pp. 609-646). New York: Marcel Dekker.

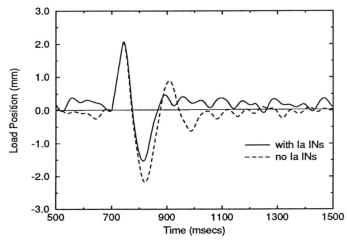

Figure 7   Reflex response with a load of 50g for a circuit containing the optimum strengths of afferent excitation, and inhibition from Group Ia interneurones (3%) (solid lines - basic circuit), compared with the circuit without the Ia interneurones (dashed lines).

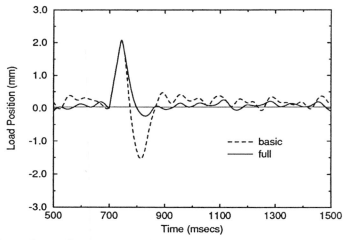

Figure 8   Reflex response with a load of 50g for the basic circuit plus the optimum strength of inhibition from Group Ib afferents (14%) and Renshaw cells (28%) (solid lines - full circuit), compared with the basic circuit only (dashed lines).

50

Modeling and Simulation of Compartmental Cerebellar Networks for Conditioning of Rabbit Eyeblink Response

P.M. Khademi, E.K. Blum[*], P.K. Leung, D.G. Lavond,
R.F. Thompson, D.J. Krupa and J. Tracy
Neural, Informational and Behavioral Sciences Program
University of Southern California, Los Angeles, CA 90089
[*] Corresponding author's address: Mathematics Dept., U.S.C.

ABSTRACT

Continuing previous research, compartmental models of cerebellar networks were developed based on new data from extracellular recordings of cerebellar cortex and interpositus nucleus in rabbit during conditioning of the eyeblink response. Sites of synaptic plasticity were postulated, as suggested by these recordings and by recent cooling probe experiments, and these were incorporated into the models. The models were tested by simulation using the CAJAL simulator, which produces action potential spike trains that can be compared against the experimental time-trace data. Conditioning is simulated by varying parameters in the model synapses at the implicated sites. Results provide further evidence that these cerebellar circuits may be a main neural substrate of this conditioning paradigm.

50.1 INTRODUCTION

In a previous paper [BKT92], we introduced a simplified cerebellar network of leaky integrator neurons, and showed how unconditioned and conditioned behavior, in terms of frequency of firing histograms of implicated cells, could be reproduced in the model by means of a least squares algorithm which identified the set of synaptic weights. Comparison of the model weights showed significant changes in the synapses of parallel fibre to Purkinje and basket cells, and mossy fibre collaterals to the interpositus. (Due to lack of space, the reader is referred to [BKT92] for further background on our current model, the underlying hypotheses and a full set of references.)

Recent cooling probe experiments [LKT93] show that the interpositus is a necessary site of plasticity for conditioning (learning) and that the red nucleus is necessary for the expression of conditioned response (performance). Given this result and our earlier modeling results, we have modeled the implicated circuitry with compartmental neurons using the CAJAL simulator [BK92]. There are two main advantages to compartmental modeling of the problem in contrast to modeling the network with leaky integrator neurons. First, the compartmental model is a truer model, a closer abstraction of real neurons, where model parameters represent measurable electrical properties of neurons. Therefore a compartmental model of the cerebellar network, though still limited by substantial simplifying assumptions, can lay claim to a much closer

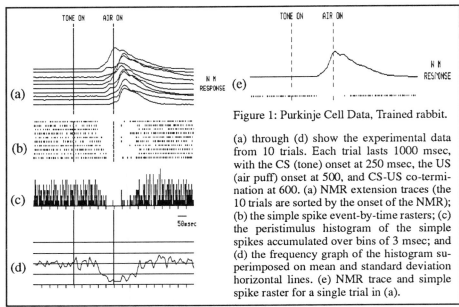

Figure 1: Purkinje Cell Data, Trained rabbit.

(a) through (d) show the experimental data from 10 trials. Each trial lasts 1000 msec, with the CS (tone) onset at 250 msec, the US (air puff) onset at 500, and CS-US co-termination at 600. (a) NMR extension traces (the 10 trials are sorted by the onset of the NMR); (b) the simple spike event-by-time rasters; (c) the peristimulus histogram of the simple spikes accumulated over bins of 3 msec; and (d) the frequency graph of the histogram superimposed on mean and standard deviation horizontal lines. (e) NMR trace and simple spike raster for a single trial in (a).

approximation to the real rabbit than the leaky integrator model. Second, the variable model parameters are the maximum synaptic conductances, which can be restricted to values within experimentally obtained ranges. Subsequently, a modeling scenario in which changing certain conductance parameter values is sufficient to transform the before conditioning model behavior to after conditioning model behavior, can provide motivation for experimental verification.

50.2 METHODS

The experimental data we use is from single cell recordings made at the Behavioral Neurosciences Laboratory at U.S.C. Figure 1 depicts the experimental data from a Purkinje cell from a trained rabbit. (For lack of space, we have not shown here, the other cell data which are Purkinje cell data from naive rabbit, and interpositus data from naive and trained rabbits. See [Kha92].) Extracellular single units are recorded from cerebellar cortex (simple spikes) and from interpositus using stainless steel microelectrodes. The neural signals are amplified with a gain of about 10,000 and bandpass filtered from 500 to 5,000 Hz. These records are then amplitude-discriminated for action potentials of the cells. The spikes are counted in successive 3 msec time bins for each training trial. The unit activity of each trial can be displayed as an event-by-time raster (Figure 1b), and the trials can be accumulated into a histogram (Figure 1c). Simultaneously, nictitating membrane position is monitored during an eyeblink by a mini-torque potentiometer attached by a nylon loop sutured in the margin of the nictitating membrane (third eyelid). The movement is transduced by the potentiometer into a voltage which is proportional to the movement (Figure 1a).

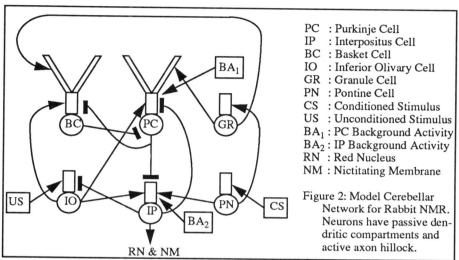

PC : Purkinje Cell
IP : Interpositus Cell
BC : Basket Cell
IO : Inferior Olivary Cell
GR : Granule Cell
PN : Pontine Cell
CS : Conditioned Stimulus
US : Unconditioned Stimulus
BA$_1$: PC Background Activity
BA$_2$: IP Background Activity
RN : Red Nucleus
NM : Nictitating Membrane

Figure 2: Model Cerebellar
Network for Rabbit NMR.
Neurons have passive den-
dritic compartments and
active axon hillock.

Figure 2 depicts a compartmental model of the cerebellar circuit implicated in NMR conditioning. In comparison with the network architecture in [BKT92], the only significant change is the addition of an inhibitory collateral feedback from the interpositus to the cerebellar cortex [BBC89]. It is more likely that this pathway is comprised of mossy fibres terminating in glomeruli. However, there is no evidence that the feedback pathway would terminate at the same site as the CS pathway. Thus the IP-PC synapse in the model is regarded as a shorthand for an IP-GR-PC pathway, through another GR.

50.3 RESULTS AND DISCUSSION

The model in Figure 2 is capable of displaying a wide range of behavior depending on the values of the large number of model parameters. We set the values of many of the individual cell parameters, propagation delays, and certain synaptic parameters based on known neurophysiology of the circuit and previous simulation results, but also based on our overarching modeling philosophy which is: given that using reasonable parameters we are able to reproduce the before-conditioning behavior in the Purkinje and interpositus cells (Figure 3), what modifications (Table 1), if any, to the set of maximum synaptic conductance parameters, G_i, will result in reproducing the after conditioning scenario (Figure 4). This approach, while not without quantitative significance, allows us to make strong qualitative statements about sites of synaptic plasticity in the circuit, and the signs of the ΔG_i. Simulations subsequent to the results shown here of a more comprehensive model than in Figure 2 which addresses through normalization the discrepancy between pre-stimuli firing frequencies of same-type cells from naive and trained rabbits, and also targets IO, establish more conclusively that the key synaptic sites are PN-IP and GR-PC [Kha92].

In order to establish an effective measure of closeness of model results to target data

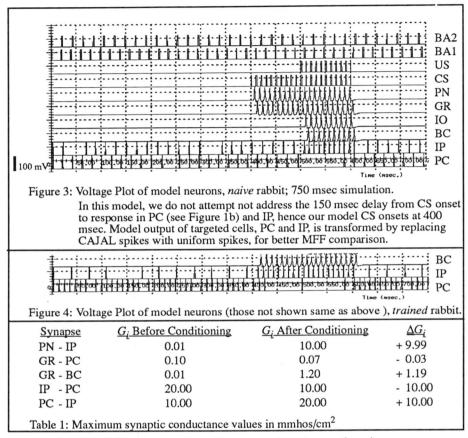

Figure 3: Voltage Plot of model neurons, *naive* rabbit; 750 msec simulation.
In this model, we do not attempt not address the 150 msec delay from CS onset to response in PC (see Figure 1b) and IP, hence our model CS onsets at 400 msec. Model output of targeted cells, PC and IP, is transformed by replacing CAJAL spikes with uniform spikes, for better MFF comparison.

Figure 4: Voltage Plot of model neurons (those not shown same as above), *trained* rabbit.

Synapse	G_i Before Conditioning	G_i After Conditioning	ΔG_i
PN - IP	0.01	10.00	+ 9.99
GR - PC	0.10	0.07	- 0.03
GR - BC	0.01	1.20	+ 1.19
IP - PC	20.00	10.00	- 10.00
PC - IP	10.00	20.00	+ 10.00

Table 1: Maximum synaptic conductance values in mmhos/cm^2

we have developed the Moving Firing Frequency (MFF) error function:

$$E(G_S) = \sum_{t_m} \sum_i \left(\int_{t_m - WIN}^{t_m} V_i(\tau)\ d\tau - \int_{t_m - WIN}^{t_m} T_i(\tau)\ d\tau \right)^2$$

where $0 \le t_m \le t_{\text{end-of-trial}}$, T_i are the target voltage traces constructed based on single trial data (e.g. Figure 1e) with uniform (elongated sine wave) spikes, V_i are the corresponding model voltage traces, and *WIN* is constant value of the moving window. The MFF function is a measure of average local-time firing frequency, in that it penalizes discrepancies in frequency of spikes, rather than point-wise discrepancies. It proves to be a good approximation of the histogram even when based on data from single trials (e.g., compare target MFF integral of trained Purkinje cell in Figure 5c, which is based on raster in Figure 1e, with the histogram in Figure 1c). Figure 5 shows how

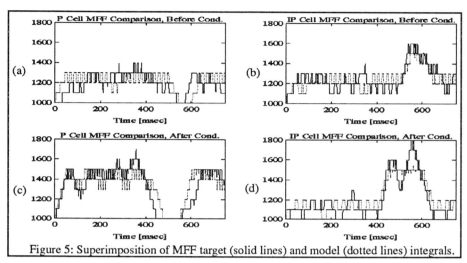

Figure 5: Superimposition of MFF target (solid lines) and model (dotted lines) integrals.

MFF integrals of the model results (from Figures 3 and 4) compare against those of the targets based on experimental data. (See [Kha92] for further elaboration of the MFF error function, and methods of applying it for fitting model behavior to data.)

REFERENCES

[BBC89]   Batini, C., Buisseret-Delmas, C., Compoint, C. and Daniel, H. "The GABAergic neurones of the cerebellar nuclei in the rat: projections to the cerebellar cortex." *Neuroscience Letters*, 99, 251-256, 1989.

[BK92]   Blum, E. K. and Khademi, P. M. "CAJAL-91: A biological neural network simulator." To appear in *Analysis and Modeling of Neural Systems II*, Ed. Eeckman, F.H. Kluwer Academic Publishers, 1992.

[BKT92]   Blum, E. K., Khademi, P. M. and Thompson, R.F. "Model and simulation of a simplified cerebellar neural network for classical conditioning of the rabbit eyeblink response." To appear in *Analysis and Modeling of Neural Systems II*, Ed. Eeckman, F.H. Kluwer Academic Publishers, 1992.

[Kha92]   Khademi, P. M. "Simulation and Modeling of Neurobiological Networks with Applications to Cerebellar Networks Implicated in Classical Conditioning of the Rabbit Eyeblink Response." *Ph.D. Thesis*, U.S.C., 1992.

[LKT93]   Lavond, D.G., Kim, J.J. and Thompson, R.F. "Mammalian brain substrates of aversive classical conditioning." To appear in *Annual Review of Psychology*, 44, 317-342, 1993.

RECURRENT BACKPROPAGATION MODELS OF THE VESTIBULO-OCULAR REFLEX PROVIDE EXPERIMENTALLY TESTABLE PREDICTIONS

Thomas J. Anastasio

University of Illinois, Beckman Institute
405 North Mathews Avenue, Urbana, Illinois 61801

ABSTRACT

Recurrent backpropagation models have been used to study the dynamic features of the vestibulo-ocular reflex (VOR). They demonstrate how the VOR could be brought about by nonlinear and distributed processing mechanisms operating at the neural level. The models are based on actual VOR anatomy, and reproduce the salient response properties of VOR neurons. The models predict that nonlinearity and dynamic variability in the responses of VOR neurons should decrease following sectioning of specific connections. The testability of the recurrent models encourages a continued dialog between theory and experiment.

51.1 INTRODUCTION

The vestibulo-ocular reflex (VOR) is a sensorimotor transformation which stabilizes vision by producing eye rotations that counterbalance head rotations. It is mediated by neurons in the vestibular nuclei (VN) that relay head rotational velocity signals from vestibular afferent sensory neurons to the motoneurons of the eye muscles (Wilson and Melvill Jones 1979). The time constant of the VOR is longer than that of the sensory signal that drives it, indicating that some form of signal processing must be occurring along the VOR pathway. This processing, designated as velocity storage (Raphan et al. 1979), is known to occur at the VN level (Buettner et al. 1978).

To gain insight into the neural mechanisms of velocity storage, recurrent backpropagation is used to construct dynamic neural network models of VOR. These models are based on actual VOR anatomy, and reproduce the salient response properties of VN neurons (Anastasio 1991). Testable predictions from the models are generated by making parameter adjustments that correspond to experimentally realizable manipulations, and determining what effects these manipulations should have on the observable response properties of VN neurons.

CHAPTER 51

51.2   CONSTRUCTING AND TRAINING THE NETWORKS

The structure of the recurrent VOR neural network models is shown in Figure 1. To reflect actual VOR organization, the models represent a bilaterally symmetric sensory-to-motor path through a series of three neural types. Modeling focuses on the horizontal VOR, which operates in the horizontal plane. The sensory neurons (sensoneurons) represent afferents from the left and right horizontal canal vestibular receptors (*lhc* and *rhc*). The motoneurons are those of the lateral and medial rectus eye muscles (*lr* and *mr*). The interneurons correspond to VN neurons on the left and right sides of the brainstem (*lvn* and *rvn*). All models had two each of sensoneurons and motoneurons, but the number of interneurons could be varied for different simulations. All units compute the weighted sum of their inputs and pass the result through the sigmoidal squashing function.

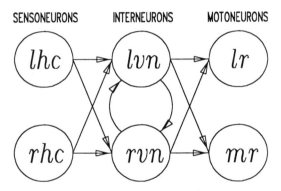

Figure 1.   Architecture of the recurrent VOR neural network models.

To represent the VOR relay, sensoneurons project to interneurons (VN neurons), which in turn project to motoneurons. Inhibitory vestibular commissures are represented by crossing connections to the interneurons. The recurrent connections between the VN neurons represent closed-loop vestibular commissures.

The models are trained to reproduce horizontal VOR dynamics using recurrent backpropagation (Williams and Zipser 1989). Sensoneuron and motoneuron responses to two oppositely directed head rotations are shown in Figure 2. In fully trained networks, the motoneuron eye velocity command (solid line) is equal but opposite in magnitude to the sensoneuron head velocity signal (dashed line). This ensures that eye velocity will counterbalance head velocity. Additionally, the motoneuron command time constant is longer than that of the sensoneuron signal, reflecting velocity storage. In any layer, responses of units on either side are mirror symmetric. Thus, when the sensoneuron on one side is activated, the sensoneuron on the opposite side is deactivated, and vice-versa. Interneurons and motoneurons operate in a similar, reciprocal fashion.

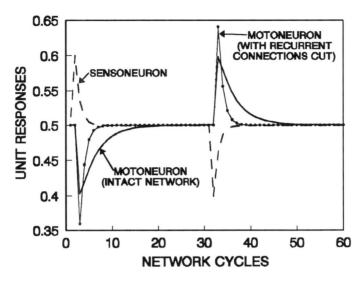

Figure 2. Responses of sensoneurons and motoneurons.

The networks learn to produce velocity storage by using the inhibitory, recurrent connections between interneurons.   Each interneuron inhibits its contralateral counterparts and is inhibited by them in turn, producing positive feedback that allows the interneurons to integrate (imperfectly, with loss) the sensoneuron head velocity signals.   Cutting the recurrent connections decreases the VOR time constant, but increases its sensitivity (dotted line). These effects are also observed for the real VOR following vestibular commissurotomy (Blair and Gavin 1981).

51.3 NETWORK MECHANISMS AND PREDICTIONS

To reproduce the VOR, the network has to learn to pass the sharp, initial peak of the sensoneuron signal, but prolong its rapidly decaying tail.   The networks develop two mechanisms to realize this transformation.   The first mechanism is nonlinear enhancement (Figure 3).   Interneurons (solid line) develop lower baseline firing rates and higher sensitivities than sensoneurons (dashed line).   This causes the deactivating interneuron responses to rectify at peak.   Real VN neurons also have these features (Buettner et al. 1978).

Nonlinear enhancement works like this: Rectification of interneurons on the deactivated side opens the recurrent loops. This temporarily switches-off velocity storage integration and allows the activated interneurons to pass the sharp, initial peak of the sensoneuron signal. After the peak, when deactivated interneurons

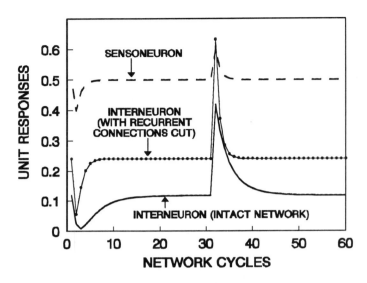

Figure 3. Nonlinear properties of interneuron responses.

come out of rectification, velocity storage is switched-on again and prolongs the rapidly decaying tail of the sensoneuron signal.

Cutting the recurrent connections removes a source of inhibition from the interneurons and increases their background firing rates (dotted line). This decreases rectification and makes interneuron responses more symmetrical. This in turn increases their ability to reciprocally drive the motoneurons, and accounts for the increase in VOR gain after commissurotomy.

In order to realize the VOR transformation, the second mechanism networks develop is distributed processing. In networks with many interneurons, the sensoneuron signal, and the velocity storage integrated version of it, become nonuniformly distributed over the population of interneurons (Figure 4). This causes the model VN neurons to vary continuously in their dynamic properties (solid lines). Real VN neurons also vary continuously in their dynamics (Buettner et al. 1978). Cutting the recurrent connections removes the velocity storage integrated component from the network, and so reduces the dynamic variability of the interneurons (dotted lines).

Thus, the recurrent VOR neural network models predict that cutting vestibular commissures should reduce the nonlinearity and dynamic variability of VN neurons. The predictions are readily testable, because commissurotomy by midline section, and single unit recording from VN neurons, are experimentally feasible.

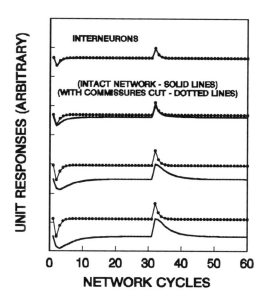

Figure 4. Distributed properties of interneuron responses.

REFERENCES

Anastasio TJ (1991) Neural network models of velocity storage in the horizontal vestibulo-ocular reflex. Biol Cybern 64:187-196

Blair SM, Gavin M (1981) Brainstem commissures and control of time constant of vestibular nystagmus. Acta Otolaryngol 91: 1-8

Buettner UW, Büttner U, Henn V (1978) Transfer characteristics of neurons in vestibular nuclei of the alert monkey. J Neurophysiol 41: 1614-1628

Raphan Th, Matsuo V, Cohen B (1979) Velocity storage in the vestibulo-ocular reflex arc (VOR). Exp Brain Res 35: 229-248

Williams RJ, Zipser D (1989) A learning algorithm for continually running fully recurrent neural networks. Neural Comp 1: 270-280

Wilson VJ, Melvill Jones G (1979) Mammalian Vestibular Physiology. Plenum Press, New York

52

PROLONGED ACTIVATION WITH BRIEF SYNAPTIC INPUTS IN THE PURKINJE CELL: INTRACELLULAR RECORDING AND COMPARTMENTAL MODELING

Dieter Jaeger
Erik De Schutter
James M. Bower

Division of Biology, California Institute of Technology, Pasadena, CA 91125

ABSTRACT

Purkinje cells in crus IIa of rat cerebellar cortex often respond to brief tactile stimulation of the face with long lasting increases in activity. We examined the properties of single Purkinje cells leading to such prolonged responses using the methods of intracellular recording *in vitro* and computer simulation of a detailed Purkinje cell model. We found that brief electrical stimulation of the granule cell layer directly underlying recorded Purkinje cells is sufficient to trigger long lasting depolarizations and increases of spike rate. In the model, the simulation of a synchronous burst of excitatory synaptic input led to a very similar effect, which could be traced to subthreshold activation of dendritic voltage sensitive calcium currents. Consistent with this mechanism, the *in vitro* preparation as well as the model also showed a prolonged activation following brief depolarizing current injections. These results show that voltage sensitive dendritic currents are sufficient to explain long lasting responses to brief synaptic input and that this mechanism is likely to be important in the temporal integration of input information performed by cerebellar Purkinje cells.

INTRODUCTION

A strong activating influence of granule cells directly underlying extracellularly recorded Purkinje cells has been proposed [1]. In crus IIa of cerebellar cortex this input is likely to be responsible for the pronounced responses of Purkinje cells to brief (10 msec) tactile stimulation of facial areas, which are organized in a fractured somatotopic pattern in register with underlying granule cell responses [2,3]. A common property of these responses is a prolonged increase of spike rate that can outlast the activation of the underlying granule cell layer by more than 150 msec (Thompson and Bower, this volume). We have shown with intracellular recordings *in vivo* that, following brief electrical stimulation of the lip, Purkinje cells at resting potential show a marked depolarization far outlasting the underlying granule cell activation. This situation is depicted in Figure 1, in which the trace C shows multi-unit activity in the granule cell layer of the upper lip patch in response to 0.1 msec electrical lip stimulation and the upper traces show a dendritic Purkinje cell

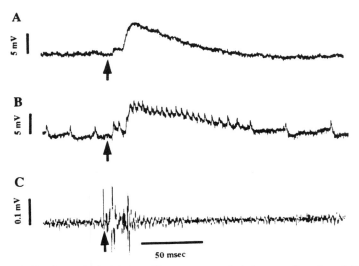

Figure 1. Prolonged Purkinje cell response to brief upper lip stimulation *in vivo*

recordings shown in this figure, bicuculline methiodide was applied locally to block inhibition.

In the present report, we examined the activating influence of the ascending granule cell projection in isolation by using an *in vitro* preparation, in which slices were cut sagittally to transect the parallel fiber input to recorded Purkinje cells. We found that brief electrical stimulation in the granule cell layer is sufficient to generate long lasting depolarizations and increases of spike rate in overlying Purkinje cells. In conjunction with the intracellular physiology we used a detailed compartmental computer simulation of the cerebellar Purkinje cell (Erik De Schutter and Bower, this volume) to test the proposition that single cell properties alone can explain a prolonged activation following brief synaptic input. The simulation of synchronous synaptic input and current injection pulses closely matched the results obtained *in vitro* and furthermore provided a functional hypothesis concerning the mechanism of response generation.

METHODS

In vitro recordings were obtained from 350 μm thick slices of guinea pig cerebellum cut sagitally. The slice medium contained (in mM): $NaHCO_3$ (26), NaCl (124), KCl (5), K_2HPO_4 (1), $CaCl_2$ (2.4), $MgSO_4$ (1.3) and glucose (10). This solution was continuously gassed with a mixture of 95% O_2 and 5% CO_2. Recordings were obtained at 36^0 Celsius. Electrodes of 70 to 150 MegΩ impedance were advanced manually in steps of 3-10 μm until a Purkinje cell was encountered as indicated by a burst of fast somatic and/or Ca spikes as described by Llinás and Sugimori [4]. Recordings that stabilized at a resting membrane potential below -50 mV without current injection were used to examine the results of synaptic stimulation.

The details of the realistic compartmental model of a Purkinje cell used here have been described elsewhere (De Schutter and Bower, this volume). Basically the simulation assumed slice-like conditions, in which no tonic synaptic input was present. Periods of spontaneous activity were simulated with current injection into the soma. For the simulation of bursts of synchronous synaptic input during spontaneous activity, changes in the kinetics of the modeled dendritic calcium currents were found necessary, since the existing model would show a dendritic calcium spike under these stimulation conditions. These changes consisted of a shift of the activation and inactivation curves towards more depolarized membrane potential and an increase in the inactivation time constants. All simulations described here were performed with a single set of channel parameters.

RESULTS

In 21 recordings from Purkinje cell somata and 23 recordings from dendrites we invariably found that a brief (0.1 msec) electrical stimulation in the underlying granule cell layer led to a depolarization of 100 to 300 msec duration when the cell previously had been in a non-spiking state (Figure 2A). This depolarization sometimes triggered somatic spikes. During periods of spontaneous spiking, the response consisted of an increase of spike rate of identical duration as the depolarization seen in the absence of spiking (Figure 2B). The increase of spike rate was accompanied by a depolarization in dendritic recordings (Figure 2C) but not in the soma (Figure 2B), where active currents apparently shunted the depolarizing currents. This finding strongly suggests a dendritic source for prolonged depolarizing currents following synaptic stimulation. A simulation of a synchronous synaptic input to the spiny dendrites in the model (170 synapses stimulated at the same time) led to results closely matching those seen *in vitro* (Figure 2, right side).

Short injections of depolarizing current into the soma or main dendrite (40 msec duration, 0.4 nA amplitude) led to a depolarization and/or increase of spike rate very similar to that seen following synaptic stimulation (Figure 3A), indicating that specialized synaptic currents are not needed to account for prolonged responses. Again this result was paralleled in the simulation (Figure 3B).

One of the attractive features of computer simulations is that the time course of all variables in the model is readily available to the researcher. In the present study, the time courses of all ionic currents were plotted for various locations in the Purkinje cell to establish the cause of prolonged responses in the model (Figure 4). We found that a combination of depolarizing Ca currents and hyperpolarizing K currents of smaller amplitude lead to the observed prolonged activation. The close match between the observed responses in the model and *in vitro* therefore leads us to predict that a similar combination of dendritic calcium and potassium currents is responsible for the responses in the real cell.

CONCLUSIONS

The results indicate that intrinsic properties of Purkinje cells are responsible for prolonged activation following brief stimulation in the underlying granule cell layer.

Dendritic voltage sensitive Ca channels (T and P type) are likely to account for this activation. The activation is kept in balance by calcium sensitive K channels (see De Schutter and Bower, this volume).

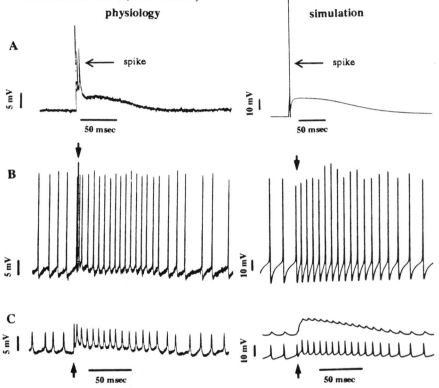

Figure 2. Prolonged activation following brief stimulation *in vitro* and in the computer simulation

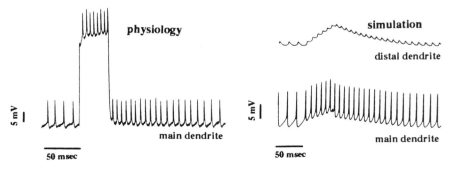

Figure 3. Prolonged activation following brief current injection *in vitro* and in the computer simulation

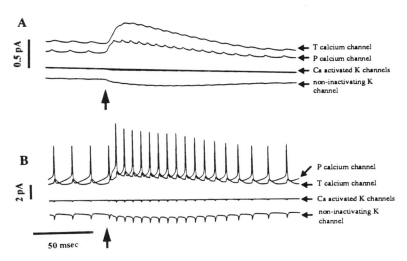

Figure 4. Ionic currents during prolonged activation in the computer simulation. A. Spiny dendrite. B. Main dendrite.

Therefore, Purkinje cell properties have a significant influence on the time course of observed responses to peripheral stimulation. This influence is likely to be important in the temporal integration of input information under more natural conditions of continuous peripheral stimulation.

Present theoretical models of cerebellar function do not take into account the details of temporal response properties of Purkinje cells observed here. We suggest that incorporation of these findings will significantly change our understanding of the dynamics and function of cerebellar activity.

REFERENCES

[1] Llinás, R. Radial Connectivity in the cerebellar cortex: a novel view regarding the functional organization of the molecular layer. *Exp. Brain Res. Suppl.* 6: 189-194, 1982

[2] Shambes, G.M., Gibson, J.M. and Welker, W. Fractured somatotopy in granule cell tactile areas of rat cerebellar hemispheres revealed by micromapping. *Brain Behav. Evol.* 15: 94-140, 1978

[3] Bower, J.M. and Woolston, D.C. Congruence of spatial organization of tactile projections to granule cell and Purkinje cell layers of cerebellar hemispheres of the albino rat: vertical organization of cerebellar cortex. *J. Neurophysiol.* 49: 745-765, 1983

[4] LLinás, R. and Sugimori, M. Electrophysiological properties of in vitro Purkinje cell somata in mammalian cerebellar slices. *J. Physiol.* 305: 171-195, 1980

ELECTROPHYSIOLOGICAL DISSECTION OF THE EXCITATORY INPUTS TO PURKINJE CELLS

John H. Thompson
James M. Bower

Division of Biology, California Institute of Technology, Pasadena, CA 91125

ABSTRACT

We have used our knowledge of the detailed pattern of tactile mossy fiber projections to the granule cell layers of crus IIa to look at the effects of the three major inputs on Purkinje cells. Our results show that Purkinje cell responses to brief peripheral stimuli consist of a short latency (5-8 ms), short duration (15 ms) component and a longer latency prolonged (100-400 ms) increase in simple spike activity. Our data suggest a role for the ascending portion of the granule cell axon in the generation of each. Also, we have used spike separation techniques to separate these granule cell effects from those of the climbing fiber system.

INTRODUCTION

Experiments in our and other laboratories have described the spatial organization of tactile mossy fiber projections to the granule cell layer of crus IIa of rat cerebellar cortex [4]. Using high density micro-mapping techniques, it has been shown that the perioral regions are represented by a fractured somatotopic map. That is, there are patches within whose boundaries a region of the rat's face is somatotopically mapped. At patch boundaries, however, there are discontinuities after which a new, not necessarily peripherally adjoining region is mapped. The granule cells which receive these inputs project to the molecular layer which contains the dendrites of the Purkinje cells. In this layer the granule cell axons bifurcate forming the parallel fibers which run parallel to each other for several millimeters across patch boundaries contacting numerous Purkinje cells along their length. Bower and Woolston (1983) found that the shortest latency tactile projections to the Purkinje cells (Fig.1) in this region were spatially congruent to those in the granule cell layer. That is, activated Purkinje cells were directly over activated granule cells. This suggested that the ascending portion of the granule cell axon (which synapses on Purkinje cells on the way up) was responsible for this short latency response.

The afferent organization of cerebellar cortex shows a great degree of convergence from numerous sensory pathways. In particular, crus IIa receives tactile input from perioral regions directly via the trigeminal sensory nuclei and indirectly from somatosensory cortex via the pons. Bower *et al.* (1981) showed that these two pathways have similar topography, that is, they converge to a single fractured somatotopic map in the granule cell layer.

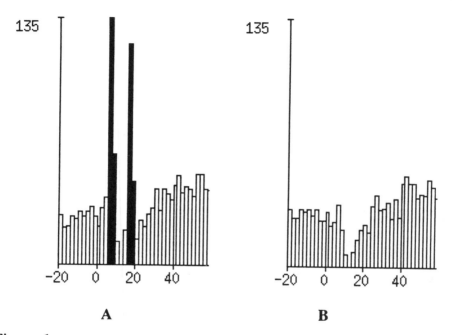

A   **B**

Figure 1. PSTs of spikes occurring from 20 ms prestimulus to 60 ms poststimulus. A. Short latency responses to tactile stimulation are highlighted. B. Short latency excitatory responses are absent when stimulus activates an adjacent patch only.

We designed experiments which would allow us to view Purkinje cell responses to peripheral stimulation with respect to the activity in the underlying granule cell layer. The design allows us to separate out the climbing fiber pathway from the mossy fiber/granule cell pathways. We were also able to see effects which can be ascribed to the ascending branch of the granule cell axon versus the parallel fibers.

METHODS

In these experiments, we anesthetize the animals using standard procedures[2]. Using two electrodes, we record from a Purkinje cell and the granule cell layer directly underlying it while tactilely stimulating perioral regions projecting to the granule cells. We construct a peri-stimulus time histogram (PST) from the spikes recorded from 300 trials. Then, the peripheral stimulus position is changed to activate adjacent patches.

A window discriminator is used to separate the complex spikes generated by climbing fiber input from the simple spikes generated by granule cell input. The trials containing complex spikes can be subtracted from the main body of data, revealing Purkinje cell responses generated by the mossy fiber/granule cell pathway.

RESULTS

The typical field potentials recorded in the granule cell layer reveal that the short latency response is spatially restricted but the longer latency one is more extensive(Figures 2 and 3).

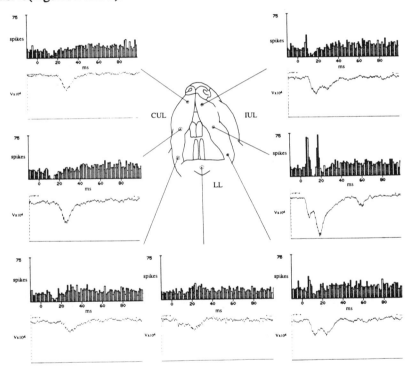

Figure 2. PSTs of spikes occurring from 10 ms prestimulus to 100 ms poststimulus from one Purkinje cell (PC) overlying a region of the granule cell layer activated by tactile stimulation of the ipsilateral upper lip (iul). Lines from the PSTs point to the perioral region stimulated while the data was collected. Below the PSTs are typical field potentials recorded from the granule cell layer.

There are two shorter latency peaks present, corresponding to the two afferent pathways [3]. The shortest latency peak, caused by the direct pathway from the trigeminal sensory nuclei, is strongly present only when the ipsilateral upper lip (iul) is stimulated. The second peak, due to the pathway through somatosensory cortex, is evoked by stimuli on either side of the face.

The PSTs of Purkinje cell spike activity are analyzed in terms of short (Fig.2) and long (Fig.3) latency responses. In figure 2, the short latency response clearly reveals a pattern that is spatially congruent to that of the granule cell field potentials. This PC overlies a granule cell layer iul patch. Tactile stimulation of the iul yields a short latency (6-10 ms) increase in spike activity which is followed by a 5 ms reduction and another 4 ms increase in spike number. As the stimulator is moved further away from this point, the most prominent change is the loss of the peaks of higher spike activity. When stimulating the contralateral upper lip (cul), the peaks are absent, but the short latency reduction of activity is clearly present. Stimulation of the lower lip (ll) produces neither of these patterns.

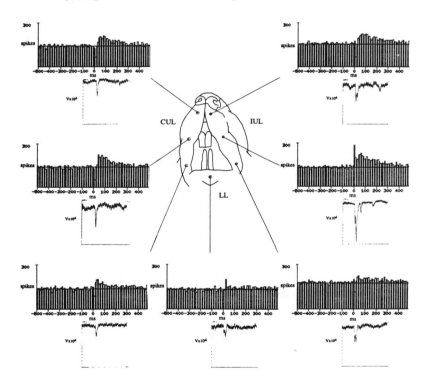

Figure 3. PSTs with 10 ms bins of spikes occurring from 500 ms prestimulus to 500 ms poststimulus from the same PC as in fig. 2. Below them are the granule cell layer field potentials.

The longer latency pattern in figure 3 reveals a "plateau" of enhanced activity starting at roughly 40 ms and lasting as long as 400 ms which is clearly visible in all records except those for lower lip stimulation (fig. 3). The plateau occurs as strongly on the contralateral side as on the ipsilateral. Finally, when all stimulus trials which produced a complex spike in the first 100 ms are removed from the PSTs, the plateau remains a feature of the PST.

DISCUSSION

Comparing the PSTs and the field potentials from the various trials, it is clear that Purkinje cells alter their firing characteristics shortly after underlying granule cell activation. The short latency responses to stimulation are spatially restricted. That this peak in activity is not present, even with some delay, in PSTs generated by stimulation outside the receptive field but along the parallel fiber "beam" strongly suggests that the ascending branch and not the parallel fibers mediates this effect. The prolonged, longer latency "plateau" response begins shortly after the second waveform in the granule cell layer field potential, again suggesting ascending branch mediation. The duration of the response, however, can't be attributed to underlying activation. Nor can it be attributed to the climbing fiber input as demonstrated by the presence of the response in the absence of climbing fiber generated activity. Jaeger *et al.* (this volume) suggest that ascending branch activation leaves the Purkinje cell in a depolarized state which enhances its response to excitatory synaptic input.

REFERENCES

[1] Bower, J.M., Beermann, D.H., Gibson, J.M., Shambes, G.M. and Welker, W.I. (1981) Principles of organization of a cerebro-cerebellar circuit. Micromapping the projections from cerebral (SI) to cerebellar (granule cell layer) tactile areas of rats. Brain Behav. Evol. 18: 1-18

[2] Bower, J.M. and Woolston, D.C. (1983) Congruence of the spatial organization of tactile projections to the granule cell and Purkinje cell layers of the cerebellar hemispheres of the albino rat: The vertical organization of the cerebellar cortex. Journal of Neurophysiology. 49: 745-766.

[3] Morissette, J.M.L.C., Lee, M. and Bower, J.M. (1991) Temporal relationships between cerebral cortical and cerebellar responses to tactile stimulation in the rat. Soc. Neurosci. Abstr. 17: 138.

[4] Welker, W.I. (1987) Spatial organization of somatosensory projections to granule cell cerebellar cortex: Functional and connectional implications of fractured somatotopy (summary of Wisconsin studies). In: *New Concepts in Cerebellar Neurobiology.* J. S. King (ed), New York, Alan R. Liss, Inc., pp. 239-280.

INTEGRATION OF SYNAPTIC INPUTS IN A MODEL OF THE PURKINJE CELL

Erik De Schutter
James M. Bower

Division of Biology, California Institute of Technology, Pasadena, CA 91125

ABSTRACT

The response of the cerebellar Purkinje cell to granule cell synaptic inputs was examined with a compartmental model that included active dendritic conductances. The model was optimized to replicate the Purkinje cell response to current injection in *in vitro* experiments. Synaptic inputs were applied to the same model to examine the effect of synchronous excitatory inputs during asynchronous stimulation. The active conductances in the dendritic membrane amplified post-synaptic responses. Small synchronous inputs applied distally resulted in somatic postsynaptic potentials of similar size to those generated by more proximal inputs. In contrast, in a purely passive model the peak amplitude of the somatic response to distal inputs was 76 % smaller than the response to proximal synaptic inputs. The model predicts that the calcium channels in the Purkinje cell dendrite make this cell relatively insensitive to the exact dendritic location of its synaptic inputs. This result has important consequences for the function of the cerebellar cortex.

54.1 INTRODUCTION

The interaction between the geometry of a particular neuron's dendrite and the way in which that neuron processes synaptic information is a central question in neurobiology [1]. With the introduction of more sophisticated experimental techniques, evidence has accumulated that the complex distribution of diverse ionic channels in dendritic membranes can have a profound effect on synaptic integration. Ultimately, efforts to understand single neuron computation in such cells will have to take into account both dendritic morphology and membrane biophysics.

In this paper we describe our use of a computer model to explore synaptic integration in the cerebellar Purkinje cell. While this cell's enormous [2, 3] and electrically active [4] dendrite have made it the subject of several modeling efforts [2, 5], most of these models have explored only its passive electrical properties [6, 7]. The few Purkinje cell models which have included active voltage dependent conductances have not studied the effects of these active properties on synaptic integration [8].

The 150,000 to 175,000 granule cell inputs received by each Purkinje cell [9] is believed to constitute the most massive synaptic convergence found on any neuron in the brain [1]. Further, recent experiments have shown that these granule cell synapses can activate Ca^{2+}-conducances in Purkinje cell dendrites [10]. It is clear that understanding information processing in the Purkinje cell and thus the cerebellar cortex requires a knowledge of how this massive synaptic input is integrated.

54.2 MODEL AND RESULTS

The model was implemented with the GENESIS software [11]. The dendritic geometry was based on morphological data provided by Rapp, Yarom and Segev [8]. The modeled cell was represented by 4588 electrically distinct compartments [12], including 1475 passive spines. Subsets of these compartments contained various combinations of 10 different types of voltage dependent channels (8021 channels in total) previously shown to be present in Purkinje cells. Channel kinetics were modeled using Hodgkin-Huxley like equations based on Purkinje cell specific voltage clamp data [13, 14] or, when necessary, on data from other mammalian cells [15, 16]. As suggested by experimental data [4], channel distributions in the model were not uniform but were distributed with the same density in each of 3 domains (the soma, the main dendrite and the spiny dendrites). The soma had fast and persistent Na^+-channels, low threshold (T-type) Ca^{2+}-channels, a delayed rectifier, an A-current, non-inactivating K^+-channels [13] and an anomalous rectifier [15]. The dendritic membrane included P-type and T-type Ca^{2+}-channels [14], two different Ca^{2+}-activated K^+-channels (BK and K2) [13, 16] and a non-inactivating K^+-channel. The P-type Ca^{2+}-channel is a high-threshold, very slowly inactivating channel, first described in the Purkinje cell [17]. In the model, the P-channel constituted about 90% of the total Ca^{2+}-conductance [18]. While there is experimental evidence that Ca^{2+} release from internal Ca^{2+}-stores may play a role in Purkinje cell responsiveness [19], such effects have not been incorporated into the model.

Once constructed, model parameters (e.g. channel densities and Ca^{2+}-removal kinetics)

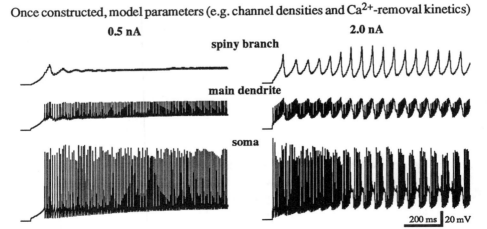

Figure 54.1 Response of the modeled Purkinje cell to current injection in the soma.

were adjusted until the model replicated published *in vitro* physiological responses to current injections in the soma and dendrites [4]. Fig. 1 briefly summarizes these results. The model shows the typical firing of somatic sodium spikes during low amplitude current injection (0.5 nA), while dendritic calcium spikes appear with higher intensity currents (2.0 nA). Note also the delay in onset of firing during the 0.5 nA current injection. In some experiments all active dendritic conductances were removed to contrast the electrical responses of a passive dendrite with an active one.

To explore the effects of synaptic activation, 3189 synaptic channels were added to the Purkinje cell model, without changing any of the other parameters. Granule cell excitatory synaptic inputs were modeled as a 0.7 nS AMPA type conductance [20], applied on one passive spine [21] located on each spiny branch compartments. The 1475 spines represented approximately 1% of the number of spines found on real Purkinje cells [9. For random, asynchronous inputs the missing spines can be compensated for by increasing the firing rate of each synapse [7]. Assuming a linear scaling, an asynchronous firing rate of 10 Hz in the model would thus correspond to an average firing rate of about 0.1 Hz for real parallel fibers. Measured by the firing frequency of the Purkinje cell model, simulations showed that the scaling was linear in the range 1 to 10% of the spines (results not shown). All synaptic input firing rates mentioned are unscaled. Each spiny branch compartment also included one to two inhibitory GABA$_A$-type synapses [22].

Recent physiological and anatomical results suggest that the effects of granule cell synapses can be divided into those associated with the ascending branch of the granule cell axon and those associated with the parallel fiber branch [5, 23]. The ascending branch synapses contact the smallest diameter spiny dendrites and are probably the main source of synchronous excitatory inputs to the Purkinje cell [23]. Excitatory synaptic input was provided to the model in two forms: asynchronous and synchronous signals. Asynchronous synaptic activation represented background parallel fiber firing due to spontaneous mossy fiber input [2] and initially synchronous signals, dispersed by the variable propagation speeds along the parallel fibers [24]. Synchronous inputs corre-

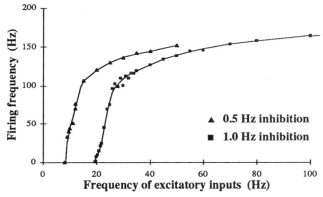

Figure 54.2   Firing frequency of the model during asynchronous inputs.

sponded to activation of the ascending branch synapses by bursts of granule cell activity in localized regions of the granular layer caused by, for example, peripheral tactile stimulation [25]. Synchronous stimuli were always given in the presence of both asynchronous excitatory and inhibitory inputs.

In the presence of only asynchronous input, the modeled cell fired somatic action potentials at rates from 1 Hz to 200 Hz (Fig. 2). Note that the model is quite sensitive to the frequency of excitatory inputs in the normal range of spontaneous firing of Purkinje cells *in vivo*, i.e. 30 to 100 Hz [3]. Inhibition seems to change the baseline of the response curve without affecting its shape (Fig. 2). The modeled neuron also closely replicated the peristimulus histograms (PST) and interspike interval distributions seen in real neurons (c.f. Fig. 5). In most experiments asynchronous excitatory inputs were provided at 28 Hz and random inhibitory synaptic inputs at a low rate of 1 Hz, with no relation between the two types of asynchronous input.

Having established that the modeled Purkinje cell generated responses quite similar to those seen *in vivo*, we were interested in exploring the importance of the dendritic position of synchronous synaptic inputs. Such inputs were provided by including an additional set of spines distributed over various branchlets of the dendritic tree. These spines were put on thin branches, corresponding to the location of ascending fiber inputs [23]. We restricted synchronous inputs to single branchlets to make comparisons between different locations easier, though actual synchronous inputs are probably more dispersed. The synapses were intended to represent 200 inputs with a 0.7 nS peak conductance, a small fraction of the 150,000 excitatory granule cell synaptic contacts on the Purkinje cell. For computational efficiency, in some cases only 20 spines with a peak synaptic conductance of 7 nS were added. Control simulations with all 200 spines showed that this simplification had no effect on the results.

Fig 3 presents data contrasting the amplitudes of excitatory postsynaptic potentials (EPSPs) generated in the active membrane model with a passive soma, to the same model without active conductances, i.e. completely passive. Peak EPSP-amplitudes in the

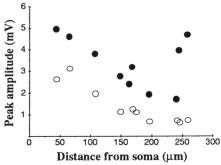

Figure 54.3 Peak amplitude of the somatic EPSP in the passive membrane (O) and active membrane model (●) versus distance of the synaptic input. Average of 40 traces.

soma were always greater in the active than in the passive model. Thus, inclusion of active properties had the effect of amplifying all synaptic inputs. Further, EPSPs in the active model showed less variation in amplitude (S.D. of 35% versus 62%). When the EPSP-amplitudes were evaluated with respect to the position of the input, additional differences became apparent. In the passive model, peak EPSP-amplitude decreased with distance of the synaptic input. This was expected from the classical passive cable properties of the dendrite [26]. As Rapp et al. [7] also reported, this attenuation was augmented by the background synaptic activity, because the increased membrane conductance distended the cable length of the dendrite. In the active model, however, near and far inputs produced similar EPSP amplitudes (Fig. 4). When compared to the passive model, synaptic inputs in the active model were amplified differentially, with distal synaptic inputs more affected than those on proximal dendrites.

Variable amplification of synaptic inputs in the active model resulted from an interaction between the electrical structure of the neuron and the active dendritic channels, especially the P-type Ca^{2+}-channels in the dendritic shafts. P-channels were activated a few milliseconds after the input, provoking a pronounced local depolarization. This P-channel activation following modest synaptic input was possible because the asynchronous background excitation kept the dendritic membrane potential slightly depolarized, close to the activation threshold of the P-channel. Depolarization then spread easily to adjoining regions of the dendritic tree, where additional P-channels activated. The P-channel mediated depolarization of the surrounding dendrite prolonged the local EPSP (Fig. 4), so that the half-width of the EPSP at the spine head increased significantly. The long local depolarization of the dendritic tree subsequently spread more effectively to the soma, producing a larger somatic EPSP.

The same amplification mechanism was found to occur in all regions of the dendrite where synchronous inputs were applied. However, the geometry of the Purkinje cell

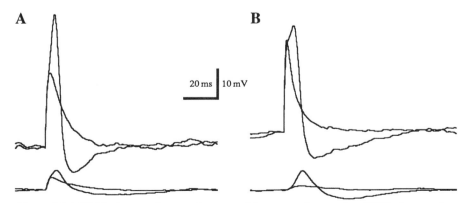

Figure 54.4 EPSPs in an active membrane model are superimposed on EPSPs in a passive model. **A** EPSPs in a spine head (upper traces) and the soma (bottom) after a proximal input (branchlet 4). Average of 8 traces. **B** Same after a distal input (branchlet 44).

itself made amplification less effective in proximal dendrites. In proximal dendrites, the depolarization due to P-channel activation was prevented from spreading to adjacent branchlets by the current sink caused by the soma. This current drain effectively decreased the amplification of proximal dendritic input in the soma. Similar effects were also seen in the passive model lacking the P-channel amplification, where the somatic current sink diminished EPSP-amplitudes in spine heads on proximal dendrites (Fig. 4A), as compared to those on distal dendrites (Fig. 4B).

Remaining differences in amplification associated with particular distal branches (Fig. 3) were caused by intrinsic variations in the membrane surface area of the branchlets over which the synaptic inputs were distributed and thus the number of P-channels in each branchlet. Normalizing the amplification of the somatic EPSP for the total membrane surface area of the input branchlet, produced a linear relation between distance and degree of amplification for distances beyond 100 μm. In other words, synaptic inputs distributed over a similar sized area anywhere in the dendritic tree would cause the same size of somatic EPSP.

When considered in the context of synaptic integration in the entire Purkinje cell, the primary consequence of the progressive amplification with distance is that all regions of this large dendrite effectively have equal access to the soma of the cell. This is seen clearly in Fig. 5 in which PSTs generated in response to synchronous input applied to corresponding dendritic regions are compared. As expected from the modeled EPSPs, at each stimulus location the synchronous input produced a similar short latency excitatory peak in firing. This peak occurred 2 to 3 ms after the stimulus for proximal inputs and slightly later [6 to 7 ms] for distal inputs, where the response was also broadened. Nevertheless, the cell always fired a somatic spike within 10 ms after the synchronous input, regardless of the dendritic location of the input (Table 1). These responses did not depend on the restricted location of the synchronous inputs, as similar responses were obtained when the inputs were distributed over the whole dendrite. The response was also relatively independent of the average firing frequency of the cell. Thus, the amplitude of the response to inputs in different dendritic positions showed little variation. There was a difference in the timing of the response, however such a difference would not be noticed in most recordings of extracellular *in vivo* responses of Purkinje cells, because usually 10 ms to 100 ms bins are used to average the recorded

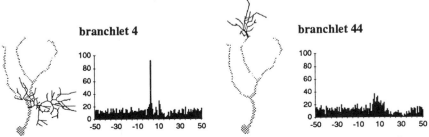

Figure 54.5 Simulated PSTs (200 events) in response to localized synchronous inputs.

Table 1 Somatic firing response within the first 10 ms after a synchronous synaptic input on different dendritic branchlets during 2 rates of asynchronous excitation and 1 Hz inhibition (average over 200 events).

Excitation (Hz)	Firing frequency (Hz)	All inputs on branchlet #				
		4	10	19	29	44
28	66.6	1.16	1.10	1.07	1.12	1.21
32	90.7	1.34	1.19	1.17	1.23	1.33

data [27]. Whether the 3 ms difference in response times between proximal and distal inputs could have a functional repercussion [28] thus remains an open question.

54.3 DISCUSSION

In summary, our results suggest that the active properties of the Purkinje cell may make this cell relatively insensitive to the precise dendritic location of its synaptic inputs. This does not mean that the pattern of dendritic activation is insignificant, but that where that activation occurs might not matter. This result has several important implications. First, this property might provide Purkinje cells with the capacity to detect common features of synaptic input arising from different patterns of granule cell layer activity. Such patterns could, for example, arise from variations in the activation of the complex fractured maps of the body surface, as found in the tactile regions of the cerebellum [25, 29]. Second, this property may greatly simplify the problem of cerebellar development. Establishing 150,000 connections to a single Purkinje cell is a formidable task if the precise position of each synapse is critical for function. The fact that hundreds of Purkinje cells share inputs from parallel fibers coursing through the same plane of the molecular layer [2, 3] makes this an even more difficult developmental feat. However, if the detailed pattern of connections onto individual Purkinje cells makes little difference, requirements for precise connectivity might be greatly reduced. Finally, when the number of synaptic inputs received by the Purkinje cell is combined with the size and electrical complexities of its dendritic tree, it seems reasonable to call this neuron the most complex one in the nervous system. The results presented here suggest that at least some of this cell's dendritic complexity may actually serve to simplify the development of its connections and the processing of its synaptic inputs. In this way, biophysical and anatomical complexity may subserve functional simplicity.

54.4 ACKNOWLEDGEMENTS

We thank M. Rapp for providing us the morphological reconstruction and D. Jaeger and B. Mel for fruitful discussions. This research used the Intel Touchstone Delta System operated by Caltech on behalf of the Concurrent Supercomputing Consortium. U.S. Bhalla, D. Bilitch and M.D. Speight helped us in porting GENESIS to the Delta System. EDS was supported by Fogarty Fellowship F05 TW04368 and JMB by NSF grant DIR-9017153.

54.5 REFERENCES

[1]   G. M. Shepherd, *The Synaptic Organization of the Brain*. Oxford University Press, New York, 1990.

[2]   M. Ito, *The Cerebellum and Neural Control*. Raven Press, New York, 1984.

[3]   S. L. Palay and V. Chan-Palay, *Cerebellar Cortex*. Springer-Verlag, New York, 1974

[4]   R. Llinas and M. Sugimori, *J. Physiol. London* **305**, 171 and 197 1980; W. N. Ross, N. Lasser-Ross, R. Werman, R, *Proc. Roy. Soc. London Ser. B* **240**, 173 1990

[5]   R. Llinas, in *Handbook of Physiology. The nervous system II. Motor control*, V. B. Brooks, Ed. Am. Physiol. Soc., Bethesda, MD, 1981, pp. 831- 876.

[6]   R. Llinas and C. Nicholson, *J. Neurophysiol.* **39**, 311 1976; D. P. Shelton, *Neuroscience* **14**, 111 1985.

[7]   M. Rapp, Y. Yarom, I. Segev, *Neural Comput.* , 1992.

[8]   A. Pellionisz and R. Llinas, *Neuroscience* **2**, 37 1977; P. C. Bush and T. J. Sejnowski, *Neural Comput.* **3**, 321 1991.

[9]   R. J. Harvey and R. M. A. Napper, *Prog. Neurobiol.* **36**, 437 1991.

[10] P. E. Hockberger, H. Y. Tseng, J. A. Connor, *J. Neurosci.* **9**, 2272 1989; M. Sugimori and R. R. Llinas, *Proc. Natl. Acad. Sci. USA* **87**, 5084 1990.

[11] M. A. Wilson, U.S. Bhalla, J.D. Uhley, J.M. Bower, in *Advances in neural information processing systems*, D. Touretzky, Ed. Morgan Kaufman, San Mateo, CA, 1989, pp. 485-492.

[12] W. Rall, *Ann. N.Y. Acad. Sci.* **96**, 1071 1962; R. W. Joyner, M. Westerfield, J. W. Moore, N. Stockbridge, *Biophys. J.* **22**, 155 1978.

[13] T. Hirano, S. Hagiwara, *Pflügers Arch.* **413**, 463 1989; B. H. Gähwiler, I. Llano, *J. Physiol. London* **417**, 105 1989; D. L. Gruol, T. Jacquin, A. J. Yool, *J. Neurosci.* **11**, 1002 1991

[14] M. Kaneda, M. Wakamori, M. Ito, N. Akaike, *J. Neurophysiol.* **63**, 1046 1990; L. J. Regan, *J. Neurosci.* **11**, 2259 1991.

[15] C. R. French, P. Sah, K. J. Buckett, P. W. Gage, *J. Gen. Physiol.* **95**, 1139 1990.

[16] J. Farley and B. Rudy, *J. Biophys.* **53**, 919 1988; F. Franciolini, *Biochim. Biophys. Acta* **93**, 419 1988.

[17] R. R. Llinas, M. Sugimori, B. Cherksey, *Ann. N.Y. Acad. Sci.* **560**, 103 1989.

[18] I. M. Mintz *et al.*, *Nature* **355**, 827 1992.

[19] C. A. Ross *et. al.*, *Nature* **339**, 468 1989; I. Llano, J. Dreessen, M. Kano, A. Konnerth, *Neuron* **7**, 577 1991

[20] W. R. Holmes and W. B. Levy, *J. Neurophysiol.* **63**, 1148 1990.

[21] K. M. Harris and J. K. Stevens, *J. Neurosci.* **8**, 4455 1988.

[22] N. Ropert, R. Miles H. Korn, *J. Physiol. London* **428**, 707 1990.

[23] J. M. Bower and D. C. Woolston, *J. Neurophysiol.* **49**, 745 1983; G. Gundappa-Sulur and J. M. Bower, *Abstr. Soc. Neurosci.* **16**, 896 1990.

[24] C. Bernard and H. Axelrad, *Brain Res.* **565**, 195 1991.

[25] G. Shambes, J. M. Gibson, W. Welker, *Brain Behav. Evol.* **15**, 94 1978;

[26] W. Rall, in *Excitatory Synaptic Mechanisms*, P. Anderson and J. K. S. Jansen, Eds. Universiteto Forlaget, Oslo, 1970, pp. 175-187; J. J. Jack, D. Noble, R. W. Tsien, *Electric Current Flow in Excitable Cells*. Clarendon Press, Oxford, 1975.

[27] K. Sasaki, J. M. Bower, R. Llinas, *Eur. J. Neurosci.* **1**, 572 1989; L. S. Stone and S. G. Lisberger, *J. Neurophysiol.* **63**, 1241 1990; D. E. Marple-Horvat and J. F. Stein, *J. Physiol. London* **428**, 595 1990.

[28] W. Bialek, F. Rieke, R. R. de Ruyter van Steveninck, D. Warland, *Science* **252**, 1854 1991.

[29] J. M. Bower and J. Kassel, *J. Comp. Neurol.* **302**, 768 1990.

UNSUPERVISED LEARNING OF SIMPLE SPEECH PRODUCTION BASED ON SOFT COMPETITIVE LEARNING

Georg Dorffner, Thomas Schönauer

Dept. of Medical Cybernetics and Artificial Intelligence
University of Vienna, Freyung 6/2, A-1010 Vienna, Austria
and Austrian Research Institute for Artificial Intelligence
georg@ai.univie.ac.at

55.1 INTRODUCTION

In this paper we present a simple connectionist model for the adaptive sensory-motor loop involved in perceiving and producing speech. At the heart of the production part lies an articulatory model which approximates the human vocal tract through polygons and splines. Output of this model is the envelope of the acoustic filter function, realized by this vocal tract, which is comparable to the spectrum of real speech segments. The goal of this research was to find a learning method to train a multi-layer neural network to produce the correct set of twelve articulatory parameters when given the spectrum of recorded real speech (stationary vowels). The method introduced in this paper explicitly makes use of a neural network categorization component. Through so-called *soft competitive learning* it learns to gradually compress the responses to more and more unitized categorical patterns. After a precategorization phase, during which presented real speech patterns are classified, the model starts to randomly produce output signals. A goodness-of-fit measure, which can be computed easily, is taken as the criterion whether the self-produced signal is close enough to any of the known categories, and as the learning rate to adapt the weights between the categorization layer and the output units.

55.2 THE ARTICULATORY MODEL

The used articulatory model implements a geometrical approximation of the midsagittal image of the human vocal tract (Fig. 1.1). The tongue body is approximated by a circle with variable center and radius. The other parts are

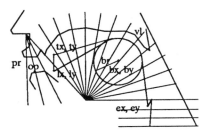

Figure 55.1   The articulatory model

seen as straight lines or polygons. Each configuration of the vocal tract can be
uniquely described with a set of twelve parameters. A grid of 25 intersecting
lines is used to compute the distances between upper and lower articulators.
From this distance profile the cross-sectional areas are computed. Then the
vocal tract is viewed as consisting of 24 tube segments which together behave
like a filter. By computing the reflection coefficients at each border between
two segments (based on the areas), and LPC coefficents, one can finally arrive
at a set of frequency values representing the envelope of this filter. These values
can be taken as corresponding to the speech signal produced by this model.

55.3   THE NEURAL NETWORK MODEL

The problem of teaching a network to produce the parameters of the articulatory model is an instance of the classical reinforcement learning problem
in motor control (see e.g. [5]). The idea behind our solution is based on the
follwing observations.

- Learning should mainly be based on receiving external examples and on
  monitoring the system's own performance (self-supervised learning).

- The criterion for evaluating the error in order to make the system adapt
  should not be what is similar in the physical world, but instead, what the
  system *perceives* as similar.

- The acquisition of sounds should be guided by recognizing that they can
  be divided into several categories (the phonemes).

The chosen connectionist approach is depicted in fig. 1.2. The core part of the

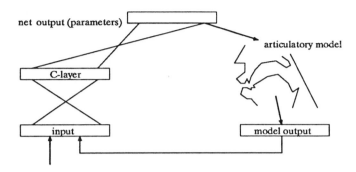

Figure 55.2 The overall connectionist architecture

network is a layer that performs *soft competitive learning* to categorize station-ary speech signals in the frequency space presented at the input layer (25 units) into distinct categories. The output layer consists of 12 units corresponding to the input parameters of the articulatory model (appropriately scaled into values in the range [0..1]). The next two subsections describe how learning is done.

55.3.1   Soft competitive learning

What distinguishes *soft* from regular competitive learning (such as [6, 2]) is that the responses in this layer (called *C-layer*) initially are distributed. For this a C-layer has full intralayer connections and operates with the interactive activation rule ([4]). These connections are initialized with zero or small neg-ative (inhibitory) weights. This achieves the well-known *rich-get-richer* effect between units in this layer, which tends to strengthen highly active units and weaken others. As in hard competitive learning, the winning (i.e. the most highly active) unit is chosen and its connections are strengthened. In addition, however, the other connections are also adapted. In mathematical terms, the learning rules are implemented the following way.

$$\Delta w_{ij} = \begin{cases} \eta(x_i - w_{ij}) & \text{if unit } i \text{ is the winner} \\ \eta(\mu_j x_i + (\lambda_j - 1)w_{ij}) & \text{otherwise} \end{cases}$$

$$\mu_j = -\frac{net_j}{\sqrt{1-net_j^2}} \quad \lambda_j = \frac{1}{\sqrt{1-net_j^2}}$$

where net_j is the net input (the weighted sum) of the j-th unit in the C-layer. The weight vector of the winner is moved toward the input vector (see [2]). The two factors μ and λ are derived from the assumption that the non-winning weight vectors should approach a vector orthogonal to the input. Both weight and input vectors are normalized to length 1. Finally, intra-layer connections increase their inhibitive effects through inverse Hebbian learning. This assures that the final responses (after competition) are compressed.

55.3.2   Goodness-of-fit

The above learning rule assures that only input vectors close enough to the learned prototype lead to a quasi-unitized response. Thus by quantifying how unitized a pattern is we can easily derive a goodness-of-fit value g^p indicating how well a given input pattern p belongs to the class.

$$g^p = \frac{1}{2}((x_w^p - \frac{1}{n-1}\sum_{i\neq w} x_i^p) + (x_w^p - x_r^p))$$

where x_w is the activation of the winner, x_r that of the runner-up (the second-highest activation); n is the number of units in the C-layer. The first part in this formula reflects the difference between the winner and the average of the other activations in the C-layer. The second part ensures that indeed only a single highly active unit leads to a large goodness-of-fit. Whenever g^p reaches a value above a threshold θ the weights between the winner of the C-layer and the output layer (denoted by v) are adapted according to the outstar rule ([3]), using the difference between g^p and θ as an additional learning rate.

$$\Delta v_{wj} = \eta(g^p - \theta)(x_j - v_{wj}) \quad \text{for all } j$$

The value of g^p also controls the random component producing the articulatory parameters during the exploratory learning phase. After each successfully categorized sound a few cycles are added where Gaussian noise is added to the previous set of parameters, with a standard deviation indirectly proportional to *g*. The function of this learning scheme can be described as follows. First randomly, then more and more guided by previously recognized phonemes, the system continually produces speech signals. These signals are categorized by the C-layer, which was trained on real speech. Whenever it hits one of the categories, expressed by a large *g*, the weights between the winner and the output are adapted, gradually associating it with the set of parameters which produced the signal.

55.4 RESULTS

During categorization of external input the network was trained with 4 recorded and Fourier transformed instances of each of 5 distinct German vowels. Depending on the number of units in the C-layer and the weight initialization between 3 and 15 classes were learned. Of course, ideally five classes should be learned. But as it turned out, these classes cannot be defined naturally through Euclidean distance, pointing to the need of more complex preprocessing of the speech data. For the sake of the experiments this was of no importance. In the subsequent phase of exploratory self-supervised learning between 100 and 500 cycles were performed in each training task. The random generator was controled as described above. After each trial with one random sound, five more trials in the "neighborhood" (defined through the Gaussian noise) were performed in each cycle. The parameters varied between learning tasks were the threshold θ and the number of pre-learned classes.

The network successfully learned to reproduce the prototypes of between 20 and 70 % of the classes it had recognized in the first phase. The higher θ the closer the reproduced signals were, due to the fact that only very good fits lead to adaptation. Also, the results were better when the number of pre-learned classes was large, although this came with a lower overall percentage of learned sounds. Two main reasons can be identified why the results were not better in these experiments. First, the articulatory model appears to be still incapable of producing some of the desired sounds. Secondly, the problem of non-uniqueness (two different sets of parameters leading to the same signal) cannot be solved in this simple network, leading to bad solutions where this is the case. Nevertheless, the results demonstrate the validity of the core approach of reinforcement learning based on soft categorization.

55.5 DISCUSSION

Among others, the approach is interesting for the following reasons. First, the criterion for goodness-of-fit is moved *inside* the model. In other words, the network learns whenever it recognizes something as similar, and not when a subjective distance measure in the physical signal is low. Secondly, the model explains some psychological phenomena, such as the loss of sensitivity toward subtle variations in sounds, once categorization has gone beyond a certain level. In [1] a similar approach is described. There, also the importance of partially compressed responses is stressed. Our approach differs from theirs in that it is simpler and achieves a goodness-of-fit in a strictly feedforward manner, by-passing the feedback necessary for resonance in ART. This might have negative implications what stability of the categories is concerned, which however was not our major concern in this work.

55.6   ACKNOWLEDGMENTS

The Austrian Research Institute for Artificial Intelligence is supported by the Austrian Federal Ministry for Science and Research.

REFERENCES

[1] Cohen M.A., Grossberg S., Stork D.G.: Speech Perception and Production by a Self-organizing Neural Network, in Lee Y.C.(ed.), *Evolution, Learning and Cognition*, World Scientific Publishing Co., London, p. 217-231, 1989.

[2] Grossberg S.: Adaptive pattern classification and universal recoding, I: Parallel development and coding of neural feature detectors, *Biological Cybernetics* 21, 145-159, 1976.

[3] Grossberg S.: *Studies of mind and brain*, Reidel Press, Boston, 1982.

[4] McClelland J.L., Rumelhart D.E.: An Interactive Activation Model of Context Effects in Letter Perception: Part 1. An Account of Basic Findings, *Psychological Review* 88, 375-407, 1981.

[5] Miller T.W., Sutton R.S., Werbos P.J.(eds.): *Neural Networks for Control*, MIT Press, Cambridge, 1991.

[6] Rumelhart D.E., Zipser D.E.: Feature Discovery by Competitive Learning, *Cognitive Science* 9(1)75-112, 1985.

DOPAMINERGIC MODULATION AND NEURAL FATIGUE IN DISCRETE TIME SIGMOIDAL NETWORKS

Nur Arad

Eytan Ruppin

Yehezkel Yeshurun

School of Mathematical Sciences,
Tel Aviv University, Tel Aviv 69978, Israel

ABSTRACT

The behavior of a spatially connected neural net whose neurons' dynamics are governed by a sigmoidal function is investigated. Extending previous work, we examine dynamic system phenomena in a parameter space that includes the temperature T, threshold Z, and an adaptation parameter h. Our results may provide a qualitative account of the bi-modal response of pallidal neurons recorded during movement.

The sigmoid function $S(x) = 1/\left(1 + e^{-(x-Z)/T}\right)$ is widely used as a simple model of neuronal firing. Following [1, 2] we examine the behavior of a neural network composed of sigmoidal firing neurons, whose dynamics are assumed to be discrete-time and synchronous. Wang [3] has shown that such a network, even if composed of two neurons only, may have complex dynamics, including period doubling and chaos. However, when the connections in the network are given by a real symmetric matrix, Marcus and Westervelt have shown that the only possible attractors of the dynamical system are fixed-points and 2-period cycles [1]. Blum and Wang [2] have investigated the behavior of symmetric sigmoidal ring networks, where each neuron i is connected to neurons $i-1$ and $i+1$ only. As they have shown, the period cycle attractors exist only when the gain of the sigmoidal firing function is in a certain interval. Concentrating on such a circular network, the goal of this paper is to extend the investigations of Blum and Wang by examining the effect of the neuronal threshold level, and by introducing adaptation into the neuronal dynamics.

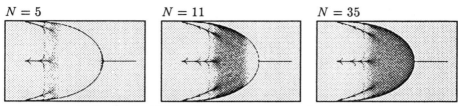

Figure 56.1   Typical bifurcation diagrams for moderate size networks. Connectivity radius is 1, $Z = 0.5$ and $h = 1$. Horizontal axis is temperature in the range $0 - 0.4$, and vertical axis is activity (in the range 0.0-1.0). Every point represents either a neuron in a steady-state or one point in a 2-cycle. The bifurcation points for all the networks is $T = 0.25$

The network is composed of a cyclic array of N neurons. Its dynamics are described by $\bar{X}_{n+1} = F(\bar{X}_n)$. Specifically: $x_i(0)$ = random transient input, $y_i(t) = S(x_i(t))$, $z_i(t) = \sum w_{ij} y_j(t)$, $x_i(t+1) = x_i(t) + h[z_i(t) - x_i(t)]$ where: $z_i(t)$ = input field of neuron i at time t, $x_i(t)$ = resulting membrane potential, $y_i(t)$ = average firing frequency of neuron i at time t, w_{ij} = weight of the connection from neuron j to i. $0 \leq i \leq N$ is the spatial variable, $t \geq 0$ is the temporal variable and $S(x) = 1/(1 + e^{-(x-Z)/T})$, where Z is the neuronal threshold, and T is the neuronal gain ('temperature').

In previous works on discrete systems cited above, the threshold was kept fixed, and h was taken as 1. We investigate the behavior of the system when h varies in the interval $[1, 2)$. As may be seen from the dynamical equations described above, h may be then viewed as a parameter representing neuronal adaptation, a pronounced characteristic of regular-spiking cortical neurons (see [4] for a review). Thus, h represents the influence of the average firing frequency of the neuron in the previous time 'bin' upon its present average firing rate. We note that $h < 1$ represents 'anti-adaptation' which seems to have no physiological counterpart; indeed, in these conditions the behavior of the network is very 'uninteresting', as the neurons firing rate always converges to a fixed point which is invariant to the initial conditions. Our work is also motivated by the recent suggestions that the activity of biological networks may be modified by the modulation of the neuron's activation function [5].

The analytic description of a single self-excitatory neuron in a steady state (ss) (with $w_{11} = 1$) is well known. By elementary examination of the logistic curve S, it is evident that for $T \geq 1/4$ a single ss exists, while for $0 < T < 1/4$ three ss exist, two of which are stable. The ss of a 2-neuron network with symmetric weights are identical to the single neuron case, since in a ss they are synchronized [6]. Moreover, in both cases, the level of neural adaptation h has

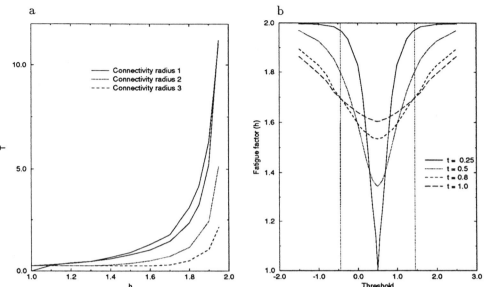

Figure 56.2 (a) Values of the bifurcation temperature as a function of h plotted for 4 different connection profiles, 3 of which are Gaussian, and one is random. In the region *under* each plot bifurcations occur. (b) Values of the adaptation factor h needed to induce oscillations as a function of Z. Note that h depends in fact on $|Z - E(X)|$. For Z In the interval between the vertical lines raising T suppresses oscillations. Outside this region the converse is true.

no effect on the long-term behavior.

In passing to larger networks, a richer situation is encountered. We concentrate on a circular array with symmetric weights $w_{ij} = w_{ji}$. If $h > 2$ or $h < 0$ linear analysis near a fixed point shows that no fixed points are possible. A major factor in the analysis of fixed points and bifurcations is the determination of the eigenvalues of the transformation F. In our case the matrix F is circulant, thus its eigenvalues can be computed as $\lambda_k = 1 - h + S(x^*) \cdot (1 - S(x^*)) \cdot \cos(2\pi k/N)/t$ for $k = 0, 1, \ldots, N - 1$. Here x^* designates a fixed point of the function S. A bifurcation point occurs when the value of $\max_k\{|\lambda_k|\}$ crosses the level surface of value 1. It can be shown that at this point the network transforms from a state where all neurons fire at the same rate to a system where each neuron oscillates in a possibly distinct 2-cycle. We show simulations of the behavior of such a network, which show close correspondence with our analytic results. Figure 56.1 depicts the long-term behavior of networks of size $N = 5, 11, 35,$

with no adaptation ($h = 1$), $Z = 0.5$ and $w_{ij} = 0.5$ for $|i - j| = 1$. It can be shown analytically that, in essence, the results of the case $N = 35$ are valid for any $N \geq 35$. Three regions of network activity are evident: for $T > 0.25$, only one ss exists and is shared by all neurons, and therefore the network is unresponsive to any transient input. For $0.1 < T < 0.25$ a multitude of 2-cycles and ss are obtained. For $T < 0.1$ the number of ss and 2-cycles is again relatively small, and a number of bifurcations emerge from each ss. For smaller networks (e.g., $N = 11$) a fourth region is present: For $0.19 < T < 0.25$, 2 stable ss are present, and all neurons converge in unison to one of them depending on the initial state.

Figure 56.2 shows the bifurcation point for various temperatures, adaptation factors and thresholds of the sigmoidal function. As can be seen from (a), the larger the connectivity radius, the higher adaptation factor is needed for a given temperature to induce oscillations. Network size has little or no effect on the bifurcation point. If random symmetric connections in the interval $[0, 1]$ are used, a similar plot is derived. (b) shows the bifurcation point as a function of h and Z for several T's. As can be seen, h and $|Z - E(X)|$ ($E(X)$ is the expectation of the neuron's membrane potential) have an opposite effect on the location of the bifurcation point: As $E(X)$ moves away from the threshold Z, a higher adaptation factor is required in order to obtain oscillatory behavior. Note that for temperatures in a given range, two distinct regions can be observed: When $|Z - E(x)|$ is small, the temperature level has a suppressing effect on the oscillatory behavior, while if $|Z - E(x)|$ increases over 1.5, raising the temperature induces oscillations.

Figure 56.3 examines the effect of threshold values on the bifurcation diagram. For all but very high thresholds the general form of the bifurcation diagram is preserved. The shifting of the single fixed point seen at high T's is due to the difference between Z and $E(X)$. For high thresholds, the fatigue factor cannot sustain oscillatory behavior. Ultimately, a single steady state is encountered. Similar behavior is seen when adaptation is varied, and the threshold is held fixed.

Assuming that the dopaminergic modulatory role of nigral cells projecting on basal ganglia structures may be modeled as threshold and gain variations, our results may account for the bi-modal response of pallidal neurons recorded following their activation; part of the neurons increase their firing rate while others decrease it below their basal rate [7]. In some basal-ganglia related movement syndromes, a rapidly changing symptomatology arises in the absence of any gross anatomical or histological damage [8], which suggests that a pathological dynamical factor is present.

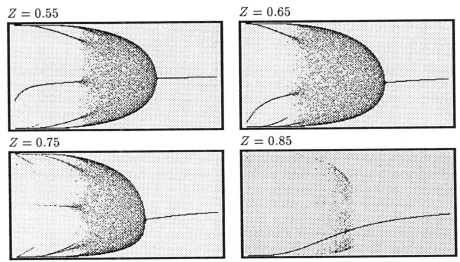

$Z = 0.55$ $Z = 0.65$ $Z = 0.75$ $Z = 0.85$

Figure 56.3   The bifurcation diagrams for a network of 11 neurons with connectivity radius 1, fatigue factor $h = 1.5$ and different Z values. The plots display the activity of the neurons (fixed points and 2-cycles) for a temperature range of $[0, 1]$.

REFERENCES

[1] C.M. Marcus and R.M. Westervelt. Associative memory in analog iterated map neural network. *Phys. Rev. A*, 41(6):3355–3364, 1990.

[2] E.K. Blum and X. Wang. Stability of fixed points and periodic orbits and bifurcations in analog neural networks. *Neural Networks*, to appear, 1992.

[3] X. Wang. Period-doublings to chaos in a simple neural network: An analytical proof. *Complex Systems*, 5:425–441, 1991.

[4] B.W. Connors and M.J. Gutnick. Intrinsic firing patterns of diverse neocortical neurons. *Trends in Neural Science*, 13(3):99–104, 1990.

[5] D. Servan-Schrieber, H. Printz, and J.D. Cohen. A network model of catecholamine effects: gain, signal-to-noise ratio, and behavior. *Science*, 249:892–895, 1990.

[6] S. Shamma. Spatial and temporal processing in central auditory networks. In C. Koch and I. Segev, editors, *Methods in Neuronal Modeling*. MIT press, 1989.

[7] J.W. Mink and W.T. Thach. Basal ganglia motor control. ii. late pallidal timing relative to movement onset and inconsistent pallidal coding of movement parameters. *Journal of Neurophysiology*, 65(2):301–329, 1991.

[8] J.F. Kelleher and J. Mandell. Dystonia musculorm deformans: A 'critical phenomenon' model involving nigral dopaminergic and caudate pathways. *Medical Hypotheses*, 31:55–58, 1990.

SECTION 8

DEVELOPMENT AND MAP FORMATION

Development of ocular dominance and orientation columns continues to be of interest to investigators of self-organization in the nervous system. Nitric oxide, feedback mechanisms, and LTP are now being postulated to modulate the development of maps. A paper on auditory brainstem maps is in section 4 (chapter 37). Related work on self-organization can be found in chapter 1 (section 1), chapter 77 (section 10), and chapter 80 (section 10).

VOLUME LEARNING: SIGNALING COVARIANCE THROUGH NEURAL TISSUE

PR Montague   P Dayan   TJ Sejnowski

CNL, The Salk Institute, PO Box 85800, San Diego, CA 92186-5800

ABSTRACT

The finding that rapidly diffusible chemical signals produced at post-synaptic terminals can cause changes in efficacy at spatially distant pre-synaptic terminals completely alters the framework into which Hebbian learning fits. These signals diffuse through the extra-synaptic space to alter synaptic efficacies under appropriate contingencies. We have analysed the resulting model of plasticity within a volume of neural tissue. As a special case, we show that rather than having the scale of developing cortical structures such as ocular dominance columns set by intrinsic cortical connectivity, it is set by a more dynamic process tying diffusion and pre-synaptic activity.

1   INTRODUCTION

The traditional view of communication between nerve cells has focused narrowly on how pre- and post-synaptic elements at a particular terminal interact. Although horizons had to be broadened to accommodate neuromodulators such as acetylcholine and serotonin, which act at a distance from their synaptic terminal of origin, a more dramatic shift is forced by recent evidence that nitric oxide (NO) is a fast acting messenger important for learning and long-term potentiation (LTP) which can readily diffuse through extra-synaptic space.

Data (*eg* O'Dell *et al*, 1991, Schuman *et al*, 1991) suggest that NO plays the rôle traditionally accorded to a retrograde signal in the familiar Hebbian synapse: pre-synaptic activity coupled with post-synaptic depolarisation leads to the latter's emission of a retrograde signal which in turn leads to the modulation of pre-synaptic efficacy. However that NO can diffuse readily through cell membranes implies a completely different theoretical account of plasticity. Any pre-synaptic terminal can feel the effects of the local concentration of NO, whether or not its own post-synaptic terminal is depolarised - the NO itself takes on a role as a Hebbian conjunction detector. This role had been assigned previously to the N-methyl-D-aspartate receptor.

We call this kind of synaptic change *volume learning* since plasticity operates within a complete, diffusion defined, region. This shift makes a theoretical difference – it is not merely the chemical instantiation of a conventionally understood process – indeed it was originally suggested on theoretical grounds (Gally *et al*, 1990). Of course, the same would hold for other similarly diffusible substances. This paper explores some of the consequences of such a mechanism of plasticity. The next section describes the model and section 3 studies what determines cortical scale in volume learning.

2 THE MODEL OF VOLUME LEARNING

Consider a region $\mathcal{R}$ containing plastic synapses $\mathcal{S}$. Each synapse $s \in \mathcal{S}$ is idealised as having a coordinate $\mathbf{r}^s \in \mathcal{R}$ and a real valued strength $w_{i^s j^s}(t)$, where i^s and j^s identify the pre- and post-synaptic cells respectively and t is time. The local concentration of NO is $\mathcal{N}(\mathbf{r}, t)$ and it changes according to:

$$\frac{\partial \mathcal{N}(\mathbf{r}, t)}{\partial t} = \mathcal{D}\nabla^2\mathcal{N}(\mathbf{r}, t) - \kappa\mathcal{N}(\mathbf{r}, t) + p(\mathbf{r}, t) \qquad (1)$$

where where ∇^2 is the Laplacian, $\mathcal{D}$ is NO's diffusion constant, κ the destruction rate, and $p(\mathbf{r}, t)$ the production rate.

NO is a small, membrane permeable molecule, for which this simplistic isotropic model of diffusion is probably not too unrealistic. Since it is an oxygen free radical, it is destructive to tissue at a high concentrations. There are protective mechanisms that help limit this toxicity. Two feedback control mechanisms are of particular interest: NO is also known as endothelial derived relaxation factor (EDRF) because of its rôle in causing the relaxation of surrounding arterioles and consequently increasing blood flow to the region surrounding its site of production. Hæmoglobin has a high affinity for NO, and so will remove it from a region – this is modeled here by the uniform destruction rate – the more NO, the more hæmoglobin, the more chelation of NO. Another theoretically interesting mechanism which we do not model here is that NO down-regulates (Manzoni *et al*, 1992) the receptor on the post-synaptic terminal which is one of the first steps in the chemical cascade that leads to its production. The more NO, the less sensitive the receptor to pre-synaptic firing, the less NO gets produced.

According to the model, NO is produced through local activity at the synapse filtered through the current weights, *ie* $p(r, t) = \mathcal{K}\left[\sum_{s \in \mathcal{S}: r=r^s} x_{i^s}(t) w^s_{i^s j^s}(t)\right]$ where $x_{i^s}(t)$ is the firing rate of the relevant pre-synaptic cell and K is a constant. This is obviously a gross simplification – even with NO, the post-synaptic terminal retains some unmodeled capacity as a Hebbian conjunction detector.

The local concentration of NO and pre-synaptic activity interact to cause weight changes. This is modeled as:

$$\frac{dw^s_{i^s j^s}(t)}{dt} = \delta\left[x_{i^s}(t) - \theta\right]\left[\mathcal{N}(r, t) - \mathcal{F}(x_{i^s}(t), \theta)\right]$$

where θ is a fixed excitatory threshold, $\mathcal{F}(x, \theta)$ is the function that picks thresholds for potentiation and depression, and δ is the learning rate.

Judicious choice of θ and $\mathcal{F}$ can lead to arbitrary Hebbian rules – including potentiation and forms of hetero- and homo-synaptic depression. Typically a higher concentration of NO would be expected to be required to cause an inactive pre-synaptic terminal to be depressed than to cause an active one to be potentiated. It is the balance between these thresholds combined with the nature of the input firing that sets the cortical spatial scale.

One facet of this class of reaction-diffusion equation (Turing, 1952) is its ability to support travelling wave solutions (*eg* Fisher, 1937) in which information is transmitted through space much faster than would be expected from diffusion alone. This effect may be important in development since the diffusion coefficients in cells of the chemicals used for non-local signaling are tiny (Murray, 1989). For this to work, there should be a short term effect of local NO concentrations within a region of co-localised pre-synaptic activity such that low concentrations enhance its release whereas high concentrations suppress it. NO production from a small patch could then rapidly spread out throughout the volume, diffusing passively wherever presynaptic terminals are quiet.

It is a moot point whether this effect is required or even desirable. Nevertheless, waves may be important in the early stages of activity controlled development when relatively global signals are needed.

3 CORTICAL SCALE

Hebbian mechanisms have frequently been proposed as governing the activity dependent development of such structures as topologically ordered maps and ocular dominance (OD) columns from lateral geniculate nucleus to visual cortex (see *eg* Montague *et al*, 1991 for references). Pre-synaptic firing consequent either on the statistics of the visual input and/or the intrinsic connections in the retina reports the input neighbourhood relations. Traditionally, the neighbourhood relations in the post-synaptic layer have been modeled as coming from intrinsic cortical connections. However volume learning offers an alternative explanation – neighbourliness is reported directly through the diffusion process. Not only does this not require these connections to be extant and functioning at the time the maps are developing, but also it turns out that pre- and/or post-synaptic weight normalisation, the unphysiological bane of traditional accounts, can be dispensed with. What remains therefore is to understand how the thresholds set cortical scales and what determines observable phenomena such as the width of OD stripes.

For the latter, we performed simulations starting from random innervation with relatively small domains controlled mostly by one or other eye, and with positive correlations within an eye and zero or anti-correlations between the eyes (Montague *et al*, 1991). They suggested that if the development rate is reasonably low, the characteristic size of a domain seems to grow up to a certain point at which it stabilises. If the development rate is too high, then one or other of the eyes can sometimes capture large chunks of cortex, or even the whole of it. Note that there is no artificial synaptic normalisation scheme preventing this from happening, as in some competing architectures we could name. This is a strong hint that there is some minimum size of a domain which is stable to the random perturbations induced in the simulation. Domains smaller than this are not stable – they need to club together with neighbouring patches.

We tested a simple version of this hypothesis using a discrete approximation and looking for the instability of regular stripes or patches of certain widths from two 1d retinæ to perturbations from the state in which cells in both eyes fire for equal times. The figure shows the concentration of NO (assuming saturated weights) at the centre of a patch of seven cells when its 'parent' eye is on for ten times steps but the opposing eye is on for eleven.

Note from the curve that the own-eye concentration rise more steeply, which arises since in the opposite-eye case the NO has to diffuse. Also once it reaches a peak, the concentrations decrease according to the decay constant of NO in the tissue – if there were no decay then this idealised system would reach steady state. Importantly, for this patch width, the curve for the opposite-eye case ultimately lies above that for the own-eye. As the size of the patch grows, the point at which the curves cross over gets later and later, and the total difference between the two curves changes sign. Depending on the precise nature of the learning rule, when this difference is negative, there will be a tendency for the opposite eye to decrease the weight of the own-eye synapses, effectively extending its local patches into the own-eye cortical turf. Conversely, if the difference is positive, the patch will tend to be stable to this perturbation. This implies exactly the phenomenon of a minimum stable stripe width.

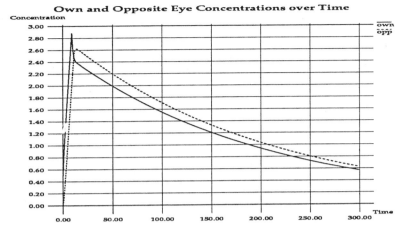

Figure 1 Concentrations for own- and opposite-eye cases over time at the middle of stripes of width 7 cells. Diffusion constant $D = 5.0$, decay rate $\lambda = 0.995$.

References

Fisher, RA (1937). The wave of advance of advantageous genes. *Ann. Eugenics.* **7**, pp 353-369.

Gally, JA, Montague, PR, Reeke, GN, Edelman, GM (1990). The NO hypothesis: Possible effects of a short-lived rapidly diffusible signal in the development and function of the nervous system. *Proceedings of the National Academy of Sciences (USA)*, **87**, pp 3547-3551.

Manzoni, O, Prezeau, L, Marin, P, Deshager, S, Bockaert, J, Fagni, L (1992). NO induced blockade of NMDA receptors. *Neuron, 8*, pp 653-662.

Montague, PR, Gally, JA, Edelman, GM (1991). Spatial signaling in the development and function of neural connections. *Cerebral Cortex* 1(3), pp 199-220.

Murray, JD (1989). *Mathematical Biology.* Berlin: Springer Verlag.

O'Dell, TJ, Hawkins, RD, Kandel, ER, Arancio, O (1991). Tests of the roles of two diffusible substances in LTP: evidence for nitric oxide as a possible early retrograde messenger. *Proceedings of the National Academy of Sciences (USA)*, **88**, pp 11285-11289.

Schuman, EM, Madison, DV (1991). A requirement for the intercellular messenger nitric oxide in long-term potentiation. *Science, 254*, pp 1503-1506.

Turing, AM (1952). The chemical basis of morphogenesis. *Philosophical Transactions of the Royal Society of London, B, 237*, pp 37-72.

58

HEBBIAN LEARNING IN FEEDBACK NETWORKS: DEVELOPMENT WITHIN VISUAL CORTEX

Dawei W. Dong

Lawrence Berkeley Laboratory
University of California, Berkeley, California 94720

ABSTRACT

There is controversy as to the development of columnar structure in primary visual cortex. Various experiments showed the important role of cortex interconnection in developing the columnar structure. We proposed a model of Hebbian development within visual cortex. In this model, the columnar structure originates from symmetry breaking in feedback pathways within an area of cortex, rather than feedforward pathways between areas.

58.1 INTRODUCTION

The basic idea described by Hebb (Hebb 1949) was that the change of a connection strength should be due to correlated activity of the pre- and post-synaptic cells.

This idea has been applied to model the development of early visual pathways (Bienenstock *et al* 1982; Linsker 1986; Miller *et al* 1989). But there are several important aspects which were not address by most prior research: 1) the focus of those models have been on feed-forward networks, while the actual neurobiology of visual cortex involves extensive feed-back circuitry (Gilbert and Wiesel, 1989); 2) the assumed Hebbian-type synaptic change of those models is of feed forward connections, while direct experimental evidence suggested Hebbian-type synaptic change of feedback connections (Bonhoeffer *et al* 1989); 3) in

those models feed forward connections developed columnar structure, while it
is observed that columnar specified feedback connections exist before function
column (Hubel and Wiesel 1963) appear in cat striate cortex (Luhmann etc
1986).

Our research (Dong and Hopfield 1992) emphasized the important role of the
feed ack connections in the development of columnar structure of visual cortex.
We developed a theoretical framework to help understand the dynamics of
Hebbian learning in feedback networks and showed how the columnar structure
originates from symmetry breaking in development of interconnections within
visual cortex.

58.2   HEBBIAN LEARNING IN FEEDBACK NETWORKS

Two kinds of dynamic processes take place in neural networks. One involves
the change with time of the activity of each neuron. The other involves the
change in strength of the connections (synapses) between neurons. When a
neural network is learning or developing, both processes simultaneously take
place, and their dynamics interact. This interaction is particularly important
in feedback networks.

The set of variables $\{V_i(t), T_{ij}(t)\}$ are used to describe the state of the network
which is undergoing the two kinds of dynamic processes at the same time. The
set of the dynamic equations is

$$
\begin{aligned}
a_i \frac{du_i}{dt} &= -u_i + g \sum_j T_{ij} V_j + I_i \\
V_i &= F(u_i) \\
B_{ij} \frac{ds_{ij}}{dt} &= -s_{ij} + H V_i V_j \\
T_{ij} &= F(s_{ij})
\end{aligned}
\tag{58.1}
$$

In above equation, a neuron i is described as an input-output device, with the
output V_i a function of the input u_i. The strength of the synaptic connection
from neuron j to neuron i is T_{ij}, which is a function of the "recent" correlation
s_{ij} of the neuron activities V_i and V_j. The input current I_i represents any additional inputs to the ith neuron besides those described by connection matrix
T_{ij}. H and g are constant.

The convergent flow to stable states is a fundamental feature of this system of joint evolution of the activity and the connections. We have developed the theory for the dynamic system described by above equations (Dong 1991a). The theory showed that there is a Lyapunov or "energy" function behind the joint dynamics, which leads to stability of the combination of synapses and activities:

$$L = -\frac{1}{2} \sum_{ij} T_{ij} V_i V_j + \sum_i \int u_i \mathrm{d}V_i - \sum_i I_i V_i + \frac{1}{2} \sum_{ij} \int s_{ij} \mathrm{d}T_{ij} \qquad (58.2)$$

These ideas and equations are applied to the development of columnar structure (Hubel and Wiesel 1963) in the first stage of cortical visual processing in mammals.

58.3 DEVELOPMENT WITHIN VISUAL CORTEX

The "center-surround" organization of the early visual system results in a inputs pattern to the visual cortex such that inputs to nearby neighbors are positively correlated. (Hubel and Wiesel 1961; Cleland *et al* 1971; Shapley and Lennie 1985). A center-surround symmetric interconnection structure is formed when the input is strong compared to the interconnections. This kind of structure has positive interconnections with circular symmetry to nearby neurons, and negative connections with circular symmetry to neurons which are further away. In this case the neuron activities V_i are largely determined by inputs I_i. Thus the correlation function s_{ij} of neuron i and neuron j will resembles the correlation of the inputs to these two neurons.

After the formation of the center-surround structure, we tested the *neuronal activity* dynamics starting from a small random activity. The stable states are patterns of bars at different orientations and at any translational position. The dynamics starting from random states exhibited equal probabilities of ending up in a pattern of bars of different orientations and at all positions. It has been shown analytically why these are the stable patterns (Dong and Hopfield 1992).

When the amplitude of the input is low, the network develops further. In the early stage of the learning process, it exhibits features similar to the center-surround case just described; the neuronal states resemble the input patterns

changing with time, and the connections from one neuron to the others becomes
excitatory for nearby cells and inhibitory for further away ones. But when the
connections grow stronger, the neuronal state begins to spend more and more
time in those patterns resembling oriented bars at various possible positions,
with the interconnections preserving the center-surround structure. Ultimately,
when this tendency has grown strong, because a little more time is spent by
chance in a particular one of these patterns, the connections become slightly
biased to that activity state. This state, as a result of learning, becomes an
even deeper attractor and the fluctuation grows. Finally the neuronal excitation
pattern stays in only one of these states, and the interconnections grow to reflect
that single pattern.

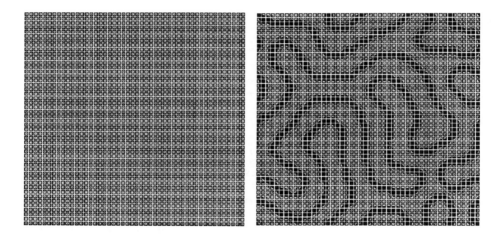

Figure Development of the interconnection structure. On the left is the
initially learned center-surround structure, on the right is the symmetry
breaking strip pattern. The simulated cortex is of 48 by 48 neurons, each
neuron connects to 5 by 5 other neurons: white represents excitatory connection, and black represents inhibitory connection. The symmetry breaking results in a stripelike pattern which has different orientation at different
locations. The characteristic period of the locally parallel bands is about
the same size as the connections region, twice the radii of anticorrelated
region.

For a larger network, the symmetry breaking results in locally parallel strips
and strips with different orientations are represented at different locations. It is
columnar structure itself. The Figure shows one simulated cortex interconnection structure before and after the development. The feedforward pathways will

form after the symmetry breaking of the interconnection development, the orientation of receptive field shifts with the same characteristic spatial frequency as the intercortical strip pattern (Dong 1991b).

58.4 CONCLUSION AND DISCUSSION

This kind of dynamical system has the unique feature of learning the correlation of input vectors under certain conditions and selecting a unique final learning result under other conditions. It is an example of "symmetry breaking" since there are several equivalent patterns possible from which it chooses one. When these dynamics are used to simulate the development of visual cortex connections, it first develops a center-surround form of connections, which is rotationally symmetric and has no chosen orientation. At a later stage, the orientation symmetry is broken, and an interconnection structure of oriented stripes develops.

In a typical neurobiological systems, the axons from one particular neuron usually form either excitatory or inhibitory connections to other neurons, but usually do not produce both kinds from a single neuron. Our model has used a single class of neuron which can form both excitatory and inhibitory connections. However, within the same layer of cortex, there are often many interneurons connecting between other neurons. Our simplified model neuron could be viewed as representing a group of neurons; the input as the total input to that group and the output as the total output from that group. Similarly, the model connections can represent the total effect of one group of neurons on the others, including effects of interneurons. In such a sense, the model described can have a mathematics in plausible correspondence to neurobiology.

Studies of the development of early vision pathways with a more biological plausible model have been completed (Dong 1991c). The basic features of the dynamic are preserved in spite of the addition of a wealth of observed electrophysiological and biophysical detail. Such a model with realistic dynamics of neuronal activity and synaptic connections learns the center-surround input correlation at the beginning of the development; keeps the correlation as the final learning result as long as the influence of the interconnections between neurons is not as strong as the inputs; and leads to the selection of an oriented pattern when the connections become strong enough at some learning stage. Thus by constructing an energy function based on symmetric connections, tractable mathematics has been constructed to understand a far more complex situation involving the development in a feedback network.

REFERENCES

Bienenstock E L, Cooper L N, Munro P W (1982) A Theory for the development of neuron selectivity... *J Neurosci* **2**: 32-48

Bonhoeffer T, Staiger V, Aertsen A D (1989) Synaptic plasticity in rat hippocampal slice cultures... *Proc Natl Acad Sci USA* **86**: 8113-8117

Cleland B G, Dubin M W, and Levick W R (1971) Sustained and transient neurons in the cat's retina and lateral geniculate nucleus *J Physiol (London)* **217**: 473-496.

Dong D W (1991a) Dynamic properties of neural network with adapting synapses *Proc IJCNN, Seattle* **2**: 255-260

Dong D W (1991b,c) Dynamic Properties of Neural Networks *Ph D thesis (Ann Arbor, MI: University Microfilms International)* Appendix C, Chapter 7

Dong D W and Hopfield J J 1992 Dynamic properties of neural networks with adapting synapses *Network: Computation in Neural Systems* **3(3)**: 267-283

Gilbert C and Wiesel T (1989) Columnar Specificity of intrinsic horizontal and corticocortical connections in cat visual cortex *J Neurosci* **9(7)**: 2432-2442

Hebb D O (1948) The Organization of Behavior: A Neuropsychological Theory *(New York: John Wiley)*

Hubel D H and Wiesel T N (1961) Integrative action in the cat's lateral geniculate body *J Physiol (London)* **155**: 385-398

Hubel D H and Wiesel T N (1963) Shape and arrangement of columns in cat's striate cortex *J Physiol (London)* **165**: 559-568

Linsker R (1986) From basic network principles to neural architecture ... *Proc Natl Acad Sci USA* **83**: p 7508, p 8390, p 8779

Luhmann H J, Martinez L, Singer W (1986) Development of horizontal intrinsic connections in cat striate cortex *Exp Brain Res* **63**: 443-448

Miller K D, Keller J B, Stryker M P (1989) Ocular dominance column development - analysis and simulation *Science* **245**: 605-615

Shapley R M and Lennie P (1985) Spatial frequency analysis in the visual system *Ann Rev Neurosci* **8**: 547-583.

THE ROLE OF SUBPLATE FEEDBACK IN THE DEVELOPMENT OF OCULAR DOMINANCE COLUMNS

Harmon S. Nine[*][**]
K.P. Unnikrishnan[*][**]

*AI Laboratory, University of Michigan, Ann Arbor, MI - 48109
**Computer Science Dept., GM Research Laboratories, Warren, MI - 48090

59.1 INTRODUCTION

Feedback is a ubiquitous feature of mammalian sensory systems. For example, many of the visual cortical areas have reciprocating connections [23] and the lateral geniculate nucleus (LGN) has been estimated to receive about ten times more input from cortical feedback projections than from retinal fibers [16]. Many studies and models have focused on the possible role of feedback in the developed visual system [6,7,9,10,15,21].

Some feedback pathways are functional even during the critical period of visual development [18]. In spite of this, all the models of cortical development to this date have been feed-forward models [14]. Here we present a model to investigate the role of feedback pathways during the development of sensory systems. In particular, we present simulations of the model to study the role of subplate feedback during ocular dominance column (ODC) development in V1.

Table I gives the time-table for development of the cat visual system. The subplate is a transient neuron population that forms below the developing cortical plate. It receives input from layer VI neurons and sends axonal projections to layer IV and the LGN. (this projection pattern is similar to that of layer VI in the developed visual system). Hence we decided to ask the question of how critical these subplate inputs can be to the development of ocular dominance columns and investigate it using computer simulations.

59.2 MODEL ARCHITECTURE

The basic architecture of our model is shown in figure 1. The LGN is represented by one ipsilaterally driven and one contralaterally driven layer of neurons. Activities of neurons within a layer are locally correlated. Activities of neurons between layers are not correlated. Cortical layer IV is represented by a layer of interconnected neurons. The subplate is represented by a fourth layer of neurons, and these receive locally summed cortical activities. The summed subplate activities

This research was supported in part by a grant and summer fellowship from General Motors Research Laboratories.

are fed back to layer IV and are used to modify the geniculocortical (GC) synaptic strengths. The subplate projections to the LGN are not considered in the model.

Table I: Developmental Calendar for Cat Visual System

	GC axons	SP	SP processes	Layer VI	Layer IV
E20					
E25		Cells born, migrate to cortex [11]			
E30					
E35			Axons grow to thalamus [12]	Cells born, migrate to cortex [11,20]	
E40	Axons grow to subplate [4,19]				Cells born [11]
E45	*(Wait in subplate [4])*			Axons arrive at LGN, wait in vicinity [13]	
E50					Cells migrate to cortex [20]
E55	Axons invade cortical plate [4]	Cells are functional neurons [3]	Axons project to layer IV, anatomical evidence for input from layer VI [3]		
E60				Axons invade LGN [13]	
E65-P00	Extensive branching in layer IV [4]				
P05					
P10					
P15					
P20		Cells die, gradual programmed death [19]			
P25					Ocular Dominance Column Development [19] *(Crit. Per. [8])*
P30					
P35					
P40					
P45					

59.3 ALGORITHM FOR SYNAPTIC STRENGTH MODIFICATION

The GC synaptic strengths are modified using the ALOPEX algorithm. The algorithm, in its many forms, has been used to investigate the role of feedback in the developed visual system [7,9,21], to learn connection strengths in artificial neural networks [17], and recent biophysical simulations [21] have established the biological plausibility of the algorithm. We have used a discrete-time version of the algorithm. At time t, the change in strength of the ith synapse is given by the equation

$$\Delta w_i^t = \mu \Delta s^{t-1} \Delta l_i^{t-1} \tag{1}$$

where μ is a constant and Δs and Δl are changes in integrated subplate feedback and local LGN activity respectively. In other words, the synaptic strength modification depends on the *temporal* correlation between changes in the subplate feedback and local LGN activity.

The GC synapses in the developing visual cortex have been shown to be of the NMDA type [17]. If we assume that the post-synaptic depolarization of these dendritic sites are provided by the subplate inputs (through non-NMDA synapses), then the local circuitry shown in figure 2 can carry out the ALOPEX algorithm. It should be noted that unlike the Hebb rule, ALOPEX uses correlation between *changes* in pre- and post-synaptic depolarizations. The local temporal buffer for this purpose could be any of the local circuit loops or time constants at the synapses.

Detailed biophysical simulations of the underlying circuitry using NEURON has shown the biological plausibility of the algorithm [2].

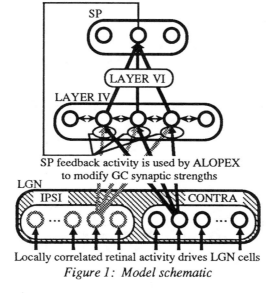

Figure 1: Model schematic

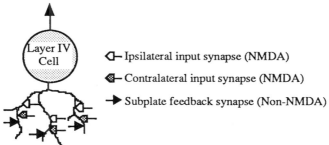

Figure 2: Neural circuitry capable of carrying out ALOPEX

59.4 RESULTS OF COMPUTER SIMULATIONS

The model was implemented in C and run on a Silicon Graphics Predator. The network is characterized by the following parameters:

(i)   The correlation coefficient of LGN neuronal activity.

(ii)   The GC arbor radius.
(iii)   The intracortical interaction function.
(iv)   Arbor radius of subplate neurons.
(v)   The transfer functions of cortical and subplate neurons (currently linear).

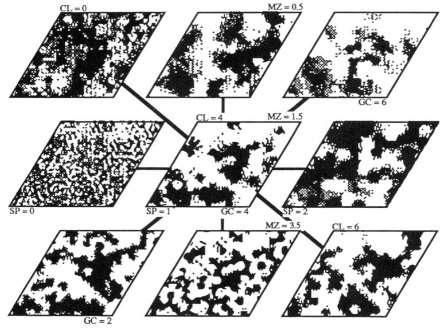

Figure 3: Results of computer simulations. Extent of local correlations (CL),
x-axis zero crossing of the Mexican hat function (MZ), subplate arbor radius (SP),
and GC arbor radius (GC) vary respectively along the downward diagonal,
vertical, horizontal, and upward diagonal lines.

We used a 35x35 hexagonal array to simulate each one of the LGN layers. This enabled us to generate locally correlated LGN activities., The intracortical interaction was implemented using a "Mexican-hat" type function normalized so that the ratio of net excitation to net inhibition is 2:1. The zero-crossing point of this function gives a parametric description of it.

Figure 3 shows the ocular dominance columns developed at the end of 20,000 iterations. From the figures, it can be seen that

(i)   Some correlation in activity of LGN neurons is necessary for forming the columns, but it is not a critical parameter that determines column width.
(ii)   Some amount of intracortical inhibition is necessary for the development of ordered columns.
(iii)   The subplate arbor radius is the most important parameter that determines the column width.

59.5 DISCUSSION

Simulations of our model show that the subplate feedback is the most critical aspect for the robust development of ocular dominance columns. This would predict that local inactivation of the subplate would lead to disruption of ODC development. Recent results of Ghosh and Shatz [5] has verified our prediction. A further prediction from our model is that limiting the subplate arbor radius would critically limit the width of ODC's. Our simulations also show that, due to the feedback in the system, the development is *not* critically dependent on the correlation in LGN activity. Recent results of Wong, Meister, and Shatz [24] show that in ferrets, the waves of activity usually seen in the developing retina disappears before the development of ODC's in the ferret.

It should be pointed out the all the feed-forward models [14] are critically dependent on input correlations for the development of ODC's. None of these models could also account for the disruption in ODC development when the subplate is ablated. Many of the feed-forward models also misuse the Hebb rule. In these models, the presynaptic activation is usually correlated with the post-soma activation (see [14] for example). In our model, this would mean that the dendritic arbors of the layer IV neurons have to be isopotential. The only detailed study that has attempted to answer this question [1] shows that cells, especially those with long dendrites, are never at isopotential. In our model, since the post-synaptic depolarizations are provided by the feedback signals, the computations are completely local, independent of dendritic time constants. The role of transient neuron populations during development has remained a mystery [18]. From our model, we would like to propose that the role of these populations may be to provide locally integrated neuronal activity from "higher" centers to the appropriate synaptic sites so that the computations could be completely local, independent of dendritic time constants. This may also be one of the roles of feedback pathways in developed sensory systems (see Janakiraman & Unnikrishnan, in this volume).

REFERENCES

[1] Brown, T.H., et. al., in *Single Neuron Computation*, T. McKenna et. al. ed. (Academic , 1992), p. 81.

[2] Fox, K., et. al., *Nature* 350, 342 (1991).

[3] Friaf, E., et. al., *J. Neurosci.* 10, 2601 (1990).

[4] Ghosh, A., and C.J. Shatz, *J. Neurosci.* 12, 39 (1992).

[5] Ghosh, A., and C.J. Shatz, *Science* 255, 1441 (1992).

[6] Harth, E., et. al., *Conc Neurosci.* 1, 53 (1990).

[7] Harth, E., et. al., *Science* 237, 187 (1987).

[8] Hubel, D.H., and T.H. Wiesel, *J. Neurophysiol.* 26, 994 (1970).

[9] Janakiraman, J. and K.P. Unnikrishnan, in *Proc. IJCNN*, III-541 (1992).

[10] Koch C., *Neurosci.* 23, 399 (1987).

[11] Luskin, M.B., and C.J. Shatz, *J. Comp. Neurol.* 242, 611 (1985).

[12] McConnell, S.K., et. al., *Science* 245, 978 (1989).

[13] McConnell, S.K., and C.J. Shatz, *Soc. Neurosci. Abs.* 14, 743 (1988).

[14] Miller, K., et. al., *Science* 245, 605 (1989).

[15] Mumford, D., *Biol. Cybern.* 65, 135 (1991).

[16] Robson, J.A., *J. Comp Neurol.* 216, 89 (1983).

[17] Sekar, N.S., and K.P. Unnikrishnan, *CSH Lrn. & Mem. Abs.*, (1992).

[18] Shatz, C.J., et. al., *CSH Symp. Quan. Biol.* 40, 269 (1990).

[19] Shatz, C.J., in *Cerebral Cortex*, A. Peters & E.G. Jones ed. (Plenum, 1988), Vol. 7, p. 35.

[20] Shatz, Carla J. and Marla B. Luskin, *J. Neurosci.* 6(12), 3655 (1986).

[21] Unnikrishnan, K.P., and K.P. Venugopal, in *Proc. IJCNN*, I-926 (1992).

[22] Unnikrishnan, K.P., and J. Janakiraman, *Soc. Neurosci. Abs.* 18, 741 (1992).

[23] Van Essen, D.C., in *Cerebral Cortex*, A. Peters & E.G. Jones Ed. (Plenum, 1988), Vol. 3, p. 259.

[24] Wong, R.O.L., et. al., *Soc. Neurosci. Abs.* 17, 186 (1991).

A COMPARISON OF MODELS OF VISUAL CORTICAL MAP FORMATION

Edgar Erwin[1], Klaus Obermayer[2] and Klaus Schulten

Beckman Institute, University of Illinois
405 N. Mathews Ave., Urbana, IL 61801

ABSTRACT

Several classes of models of visual cortical map characterization and development are compared. Characteristics and predictions of the models are compared to one another and to cortical maps observed in animals. Several models are found to predict incorrect map structure. Certain observed patterns of visual maps imply constraints on the processes which could be involved in their morphogenesis.

1 FEATURE MAPS IN VISUAL CORTEX

Individual cells in the mammalian primary visual cortex, or *striate cortex*, respond differently to features in visual input, with feature preferences determined in part by the pattern of connection between retinal light receptors and cortex. Striate cortical *receptive fields*, descriptions of features to which each cell responds, are often localized in the visual field, may be dominated by input from either eye, and usually show a preference for stimuli with a particular orientation. Several receptive field properties of neurons are arranged in the cortex in a complicated two-dimensional map such that nearby columns of neurons tend to have similar receptive fields. Due to the ordered projections from the eyes, a roughly topographic map of visual space is formed on the cortical surface. An optical imaging technique [1, 2, 4] reveals the embedded maps of ocular dominance and orientation preference over small patches of the cortex.

The details of the cortical maps vary greatly between individual animals, but certain organizing principles appear invariant. Depending on the species, ocular

---

[1] To whom correspondence should be addressed.
[2] Current address: The Rockefeller University, 1230 York Ave., New York, NY 10021

Figure 1   Tangent curves illustrating local orientation preferences near (a),(b) vortex and (c),(d) fracture singularities. (Adapted from [13].)

dominance patterns consist of bands or patches of left- and right-eye dominance. Orientation preference changes smoothly over most of the map surface, including linear regions where *slabs* of cells with a common preferred orientation are aligned alongside one another. Regions of rapid change are primarily confined to one-dimensional *fractures* (Fig. 1c,d), across which orientation preference changes by up to 90° and *vortices* where orientation preference rotates through a complete cycle of 180° around a point [1, 2]. Loop vortices (Fig. 1a) where orientation preference rotates in the same direction as motion around a central point and tri-radius vortices (Fig. 1b) where orientation preference rotates in contrary motion exist in equal numbers. These structures correspond to the simplest singularities which are possible in a ridge-type pattern (Fig. 1) [13]. Fourier power spectra of the ocular dominance and orientation preference maps have a ring, or bi-lobed distribution, for species with isotropic or anisotropic map patterns. Map patterns are correlated such that vortices tend to lie at the centers of ocular dominance bands where receptive fields are primarily monocular, and linear regions tend to lie in binocular regions with the slab borders orthogonal to the ocular dominance column borders. A successful model of striate cortical map formation must be capable of explaining the development of these patterns and the correlations between them. Further details of the spatial patterns of orientation preference and ocular dominance may be found in [1, 12].

2   VISUAL CORTICAL MAP MODELS

Although often based on different developmental principles and degree of biological detail, many models are successful at predicting or describing the structure of striate cortical maps. We attempt to find the common features of successful models, as well as unique insights from individual models. Rather than discussing the details of a series of models, which are more fully described in the original references, we have organized a discussion of the most common modelling approaches around three classes of models based on similar principles: (1) *feature map models*, (2) *correlation-based learning models*, and (3) *pattern models*. Feature map models and correlation-based learning algorithms both explain the development of feature selectivities based on the Hebbian learning hypothesis for synaptic change, but differ in the manner in which this principle is expressed. Pattern models, may also suggest developmental principles and processes, but the emphasis is on finding a concise description of the observed map patterns. A careful study of the structure of cortical maps reveals organizing principles which constrain the types of processes which could form the maps.

3   FEATURE MAP MODELS

The feature map models, which include the self-organizing feature map algorithm [6, 7, 12, 11], the elastic-net algorithm [3], and related models [16], comprise one class of developmental models. Although the details of the algorithms differ, these models share many features. They each assume a set of cortical units — either single cells, or local groups of cells — arranged in a two-dimensional lattice. The receptive field of each cortical unit is described as a feature vector, which encodes the strengths of connections to retinal receptors, either directly, or in an abstract representation. For example, a five-dimensional feature vector $\vec{w} = \{x, y, q\sin(2\theta), q\cos(2\theta), z\}$ can describe the location of a receptive field in visual space x and y, the orientation preference θ and orientation specificity q, and the ocularity z of a cortical unit. Feature map models have been used to provide a rationale for cortical map patterns, by suggesting that the maps are organized to map a high-dimensional visual feature space as completely as possible while maximizing continuity of feature preferences [3].

Feature map algorithms consider the development of cortical maps to be governed by neural activity driven either by images viewed by a young animal, or by the spontaneous firing of retinal cells, which can occur even before birth. Map formation proceeds through repetitive presentations of stimuli and modification of the receptive fields of the cells in response. The rule for receptive field modification requires that a distance measure, which must be computable by a single cortical unit, be defined between the input pattern vectors and the synaptic weight vectors. This distance determines the amount each neural unit is stimulated by an input pattern, and its form crucially affects the organization of the feature maps. Changes in the feature vectors are proportional to this distance. However, in the self-organizing feature map algorithm, competition for activity allows only units in a neighborhood around the unit with the smallest distance to modify their receptive fields, which generates the smooth map structure. Smoothness is maintained in the elastic-net model maps by a type of averaging of the receptive field properties of nearby neurons. Since the receptive fields are only adapted to more resemble the stimuli, if the stimuli show pronounced features, the specificity of the receptive fields will be high with regard to these features.

Maps produced by the feature map models closely resemble observed cortical maps in structure [3, 12], even though only the most basic properties of the maps are specified in advance. For example, ocularity varies across the cortical surface in blob- or band-like patterns. Orientation preferences vary smoothly and gradually across most areas of the cortex, except near the fractures and the loop and tri-radius vortices, and these are correlated with regions of gradual change in receptive field location. Feature specificity is high in most map areas, with the exact distribution of specificities being tunable through algorithm parameters. Power spectra of orientation and ocular dominance patterns can have a ring-like or bi-lobed distribution, and the orientation and ocularity map patterns may repeat on different length scales.

4   CORRELATION-BASED LEARNING

Correlation-based learning algorithms [8, 9, 10] comprise a second class of developmental models, which share many features with the feature map algorithms, but differ in the way the effect of retinal activity is implemented. Whereas feature-map models assume a competitive network where receptive fields change through a non-linear function of retinal input, correlation-based learning models assume a linear relationship. Thus patterns of correlation in the input will be represented in the receptive field properties of cortical cells.

In one such model of ocular dominance development [10], ocularity of each cortical cell is determined by the difference in total connection strength to receptors in each eye. Assuming that the activity in nearby cells in one retina is correlated, and that activity at greater distances or in opposite eyes is non- or anti-correlated, and taking a Gaussian or "Mexican-hat" function of intra-cortical influence, this model produces cortical patterns of ocular dominance resembling those seen experimentally. Miller [9] adapted this algorithm to study the development of orientation-selective cells and orientation preference maps. In his model a spatial "Mexican-hat" correlation function in the firing of retinal cells in the same population of ON- or OFF-center cells, with a weaker, inverted "Mexican-hat" correlation function for cells in different classes, leads to banded patterns of connection to ON- and OFF-center retinal cells, giving cortical cells an orientation preference.

When "Mexican-hat" intra-cortical interactions are included, receptive field properties become arranged in a repetitive map across the cortical surface. Maps of orientation preference show high orientation specificity and smooth variation in preferred orientation at most cortical locations, and the maps contain loop and tri-radius vortices in equal numbers. However, the organization of the maps seems to differ from the organization of cortical maps in some finer details. The difference in organization between the model and cortical maps is not immediately apparent in the maps developed with the best set of model

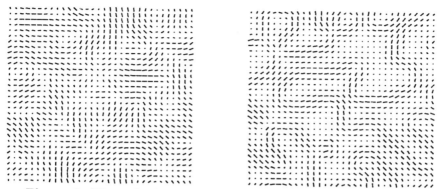

Figure 2 Maps of orientation preference from the correlation-based learning algorithm with (a) standard parameters, (b) one parameter altered. (See text)

parameters, Fig. 2a. However, varying just one parameter in the model makes the organizing principles behind the model maps more clear. The model map in Fig. 2b was developed with the parameter determining the length-scale for correlations in the firing pattern of retinal cells set 27 percent longer than its preferred value. This map contains many cells which are unspecific to orientation, and the cells with high orientation specificity are mainly located in one of several "streams" which run through the map with their receptive field orientations aligned with the "streamlines". As the parameters of the model are slowly returned to normal, the streams widen and grow together producing loop and tri-radius vortices. Nearby parallel streams grow together seamlessly, seldom producing one-dimensional fractures along which orientation preference changes abruptly. The "streamlines" in Fig. 2b suggest contour lines of a surface, whereas the curves in a general ridge system [13] and the experimental orientation preference maps cannot be associated with contour lines. If the model maps for normal parameter values are governed by the same principles, then the model may be rejected by this topological reasoning.

5 PATTERN MODELS

Pattern models attempt to find the most concise description of feature map structure, using an algorithm or formula with as few parameters as possible. The hypothesis is that studying the structure of cortical maps can help characterize types of processes which either are or are not capable of generating the observed patterns.

Several distinct pattern models have been successful at reproducing the patterns of ocular dominance [5, 14]. In one of the simplest, Rojer and Schwartz [14] demonstrated that the global patterns of ocular dominance in several species can be reproduced by convolving a two-dimensional array of random real numbers with an appropriate filter and associating the positive and negative values in the resulting array with left- and right-eye dominances. A circularly-symmetric, decaying sinusoidal filter, or a "Mexican-hat" filter gives an isotropic ocular dominance pattern with a ring-like power spectrum similar to the pattern seen in cat striate cortex, whereas a similar anisotropic filter gives a pattern of mostly parallel, branching columns with a bi-lobed power spectrum as seen in macaque cortex. A few parameters in the shape of the filter can be tuned to give ocular dominance columns with any orientation, width and degree of branching. Although convolution with a filter is sufficient to give the global patterns, tuning the distribution of ocular dominance values requires applying some local operations, such as a threshold function associating each cortical unit wholly with either the right or left eye, or a more realistic sigmoidal function.

Since the power spectrum of typical orientation column systems has a ring or bi-lobed distribution similar to the ocular dominance column spectra, a similar algorithm should be able to generate orientation column patterns. The simplest such algorithm was proposed by Rojer and Schwartz [14]. A two-dimensional array of random real numbers is convolved with a circular sinusoidal filter and the local gradient vector of the resulting pattern is computed for each array unit. Dividing the angle of each gradient vector by two gives angles $0° \leq \theta < 180°$ which may be taken as preferred orientations for cortical cells. The length of the gradient vectors gives the orientation specificity of the model cells, either directly or after being subjected to a local function, such as a sigmoid, to give the desired distribution of orientation specificities.

The pattern of orientation preferences given by this simple formula resembles in many ways the observed patterns of orientation columns. Cells with similar orientation preference are clustered together, and there are both loop and tri-radius vortices, and one-dimensional fractures. However, closer observation reveals a deficiency in the patterns. The orientation column pattern was derived from a vector field under the rather restrictive assumption that this vector field must be conservative. The properties of conservative vector fields then dictate that for any closed circuit through one of the model maps, the line integral $\oint q \cos(2\theta)dx + q \sin(2\theta)dy$ always vanishes, where $q(x, y)$ is the orientation specificity, and where the orientation preference angles $\theta(x, y)$ are measured from the x-axis. Experimental maps, and other model maps are not similarly restricted[3]. The failure of this model illustrates the value of pattern models since it reveals that *any* model which relates the lines in an orientation column map to a conservative gradient field must fail for purely topographic reasons [13], regardless of the form of the physiological implementation.

---

[3] The integral can only be approximated in maps defined only at discrete points.

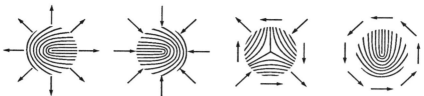

Figure 3 (a)-(d) Vortices in orientation maps and the associated vector fields. The vector field in **(d)** is non-conservative.

Although maps generated by this model differ from experimental maps in all regions, the difference is most easily demonstrated at the vortices. The inserts in Fig. 3 show four types of vortices, all of which occur in orientation column maps. Also shown are the vector fields associated with each vortex with directions given by multiplying the orientation preference angles, measured from the x-axis, by two. The loop vortices in Fig. 3a,b are associated with gradient vector fields around a local maximum or minimum in a surface, and the tri-radius vortex of Fig. 3c is associated with the gradient vector field of a saddle point. Although the loop vortex in Fig. 3d is simply a rotated version of the vortices in Fig. 3a,b, such a vortex could not appear in a map from Rojer and Schwartz' model, since the associated vector field could not be derived from the gradient of any surface. All rotated version of the tri-radius vortex, Fig. 3c, are allowed.

It is possible to predict orientation map patterns based on their power spectra, without imposing the unnecessary constraint in Rojer and Schwartz' model. Taking a ring spectrum in a two-dimensional Fourier space, with some noise in the amplitudes and with random phases of the modes, and transforming to a real space generates a two-dimensional array of complex numbers which may be taken to represent orientation preference (phase) and specificity (magnitude) [17]. An optional local function can also be applied to give the desired distribution of orientation specificities. This method produces arrays of orientation preferences that resemble experimentally observed maps, and which aren't subject to the constraint of Rojer and Schwartz' model maps. Yet the algorithm remains simple enough to be implemented by many different biological processes.

Swindale [15] has presented a model for the combined development of orientation and ocular dominance columns, which follows an algorithm similar to those used in pattern models. A vector representing local feature preference is associated with each location in a cortical lattice. Feature preferences grow from initially small, random values through an iterative process of separately convolving the orientation and ocular dominance components of the feature vectors with "Mexican-hat" filters on different length scales. The column systems are coupled by making growth of orientation preference more rapid in monocular regions of cortex. Maps developed from this simple model have linear regions, vortices and fractures in the orientation column map, and the ocular dominance pattern consists of blobs or branching parallel columns. The patterns are correlated such that vortices occur more often near the centers of the ocular dominance bands, and as a natural consequence the linear regions tend to intersect the ocular dominance bands at right angles. Like the pattern models, this algorithm is general enough to allow several physiological implementations.

6 CONCLUSIONS

Careful characterization of the structure of cortical map patterns is the first step toward understanding the rationale behind the maps and the processes responsible for creating them. The lines in cortical maps of orientation preference are organized in such a way which reveals that they could not, for example, be given by the lines in a force field or by surface contour lines. Some proposed models of cortical map formation may be rejected through such topological arguments, even before consideration of the validity of the suggested physiological mechanism.[4]

REFERENCES

[1] G. G. Blasdel. *J. Neurosci.*, 12(8):3115–3138,3139–3161 (series), Aug. 1992.

[2] G. G. Blasdel and G. Salama. *Nature*, 321:579–585, 1986.

[3] R. Durbin and G. Mitchison. *Nature*, 343:644–647, 1990.

[4] A. Grinvald, *et al. Nature*, 324:361, 1986.

[5] D. Jones, R. van Sluyters, and K. Murphy. *J. Neurosci.*, 11:3794–3808, 1991.

[6] T. Kohonen. *Biol. Cybern.*, 43:59–69, 1982.

[7] T. Kohonen. *Biol. Cybern.*, 44:135–140, 1982.

[8] R. Linsker. *PNAS*, 83:7508–7512,8390–8394,8779–8783 (series), 1986.

[9] K. D. Miller. *NeuroReport*, 3:73–76, 1992.

[10] K. D. Miller, J. B. Keller, and M. P. Stryker. *Science*, 245:605–615, 1989.

[11] K. Obermayer, H. Ritter, and K. Schulten. *PNAS*, 87:8345–8349, 1990.

[12] K. Obermayer, G. Blasdel, and K. Schulten. *Phys. Rev. A*, 45:7568–7589, 1992.

[13] R. Penrose. *Ann. Hum. Genet., Lond.*, 42:435–444, 1979.

[14] A. S. Rojer and E. L. Schwartz. *Biol. Cybern.*, 62:381–391, 1990.

[15] N. V. Swindale. *Biol. Cybern.*, 66:217–230, 1992.

[16] C. von der Malsburg. *Kybernetik*, 14:85–100, 1973.

[17] F. Wörgötter and E. Niebur. This volume. 1992.

[4] This research was supported by the National Science Foundation (grant 91-22522) and the National Institute of Health (grant P41RRO5969). Computing time on a CM-2 was provided by the National Center for Supercomputing Applications, funded by the National Science Foundation. Financial support to E. E. by the Beckman Institute is gratefully acknowledged.

FIELD DISCONTINUITIES AND ISLANDS IN A MODEL OF CORTICAL MAP FORMATION

F. Wolf, H.-U. Bauer*, T. Geisel

Institut für Theoretische Physik and SFB Nichtlineare Dynamik,
Universität Frankfurt, Robert-Mayer-Str.8-10,
W-6000 Frankfurt/Main 11, Fed. Rep. of Germany.
* present adress: CNS-Program, Caltech 216-76,
Pasadena, CA91125, USA

ABSTRACT

The existence of field discontinuities, i.e. discontinuities in cortical maps of the retina is well established by neurophysiological investigations in a variety of species [1-4]. In cat, additional geometric anomalies known as islands or type-II organization have been observed [2, 3]. The ontogenesis of these visual representations is studied in a model of activity-dependent self-organizing maps. The field discontinuities and islands are explained by the occurrence of dynamical instabilities depending on the geometry of the cortical area. A comparison with neurophysiological data from a variety of species shows that field discontinuities indeed occur in areas of elongated shape only. The approach also explains the large degree of interindividual variability observed in cats by the coexistence of several solutions of the model. These results demonstrate that field discontinuities and islands can be explained by the same developmental mechanism.

1 INTRODUCTION

Visual hemifields are mapped onto cortical areas in a retinotopic fashion, i.e. by continuous maps which preserve neighborhood relationships. Allman and Kaas [1] established that in certain cortical areas this retinotopy only holds up to a line across which the map is not continuous. Such a line is called a field discontinuity(FD). Maps exhibiting FDs are called second order transformations. Subsequent studies of retinotopy in cats [2, 3] identified additional geometric anomalies in areas 18 and 19. These were dubbed type-II organization or islands due to the island-like representation of the peripheral parts of the visual field. The number(0-3) and size of islands varied considerably between different individua. The observed FDs have been described from a phenomenological point of view in terms of coordinate transformations [5] but an explanation of their origin is still lacking.

2   VARIABLE AREAL GEOMETRY

In the following we study a model for the ontogenesis of topographic maps by activity-dependent self-organization. We investigate the mapping of a half retina onto target areas of elliptic shape with variable aspect ratio γ (i.e. the ratio of the half axes, for definition see Fig. 1b) and constant area. The maps are computed following the Kohonen algorithm for self-organizing feature maps (SFMs)[6, 7]. The retinal activity distribution $\rho(r)$ is modelled in accordance with known ganglion cell densities [8] by a power law: $\rho(r) \propto 1/(r + r_0)^2$. The maps are evaluated in the limit that the length scales of cortical cooperation σ are small compared to the size of the target area.

The resulting maps are visualized in two different ways. First we show the receptive field centers (RFCs) of the cortical units in retinal space (Fig.1a,c). This inverse projection is the usual representation of a SFM. Secondly we show the projection of a spherical polar coordinate system of the visual field (Fig.2a) onto the cortical area (Fig.1b,d,2c-d). The latter representation is frequently used in physiological mapping studies. The range of the polar coordinate system matches the extent of the visual field represented in cat A18, a typical visual area containing a FD. The following results are a consequence of our approach which treats the output geometry as a variable parameter.

3   GEOMETRY INDUCED INSTABILITIES

Mapping the model retina first onto a circular target area ($\gamma = 1.0$) leads to a continuous representation of the sensory field. Increasing γ to moderate values $1 < \gamma \leq 2.0$ leaves the continuous mapping of the sensory field unchanged (Fig.1a,b). As the aspect ratio of the target area is further increased, a first FD of growing depth is formed (Fig.1c,d). This effect is due to an instability of the RFC-distribution at the areal border [9]. It corresponds to a transition from a continuous map to a second order transformation. The signature of the field discontinuity is a notch in the RFC-distribution along the horizontal meridian (HM) , which corresponds to a split representation of the HM in the target area.

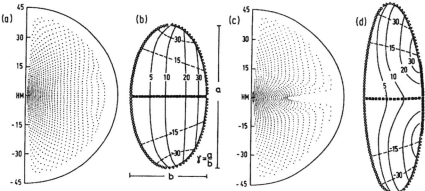

Figure 1 Geometry dependent transition from a continuous to a discontinuous representation of the visual field. (a) Distribution of RFCs of a map onto a target area with aspect ratio $\gamma = 2.0$ (b) Same map as (a), but depicted by the projection of a visual hemifield onto the target area. (For the meaning of the different coordinate lines, see Fig. 2a.) (c) same as (a), but for an area with $\gamma = 3.5$. The distribution of the RFCs exhibits a deep notch along the horizontal meridian. (d) same as (b), but for the map of c). The notch of (c) results in a double representation of the peripheral part of the horizontal meridian along the boundary of the target area, a field discontinuity.

Comparing these results with neurophysiological experiments we indeed find second order transformations in areas of elongated shape only. In primates area V2 forms a narrow belt and always contains a field discontinuity [1, 10]. In rodents second order transformations show up in the elongated V2 of the mouse [4], but not in rat, whose extrastriate cortex is subdivided in a set of rather compact areas [11]. Area 17 of the cat provides an example of an elongated area which exhibits a continuous representation of the retina [2]. Area 17 is approximately twice as long as wide, a shape which is to be considered subcritical in our model.

In the regime of larger elongations $\gamma > 4.0$, the configuration with one FD looses stability and we find a coexistence of stationary mappings with a variable number of FDs (Fig.2). One class of solutions contains two FDs in the periphery of the visual field. Since isoazimuth lines crossing these FDs form loops in the cortical map (Fig.2c) this configuration corresponds to an insular representation of the neigbourhood of the peripheral HM. Possible solutions

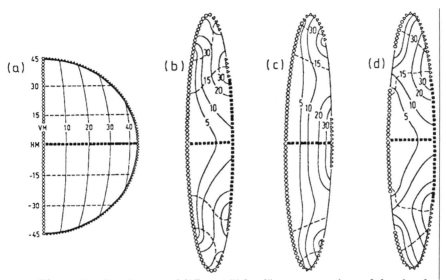

Figure 2 Coexistence of different "island" representations of the visual field for a aspect ratio of $\gamma = 6.0$. (a) Spherical coordinate system of the visual hemifield, used to visualize the projection of the visual field onto the cortex, (b)-(d) As Fig. 1a) for three different solutions coexisting at $\gamma = 6.0$. (b) Map showing three FDs that is two "islands" in the periphery of the visual field, (c) solution exhibiting a pair of FDs leading to an "island" representation of the peripheral part of the HM, (d) "island" representation of the area centralis similar to that observed in cat area 19 by Albus and Beckmann [3].

differ in the final position of the FDs depending on initial conditions and details of the formation process [9]. Coexisting with these maps we find a second class of solutions containing three or more FDs. This situation corresponds to a cortical representation exhibiting two or more islands (Fig.2b). As the described solutions show a continuous variability of FD depth and location as well as a variability in the number of islands the coexistence of different solutions to the map formation problem provides an explanation for the variability of FDs and islands observed in cat areas 18 and 19 [2].

Another class of solutions arises in the presence of the primary FD by the formation of two minor FDs at the vertical meridian (Fig.2d). This leads to an insular representation of the area centralis. Albus and Beckmann have observed such a configuration in A19 of the cat [3]. They report that in some animals the lateral border of the area does not form a continuous representation of the vertical meridian, but one in which central and peripheral parts of the vertical meridian are interrupted by representations of interior parts of the visual field.

Models of activity-dependent map formation have previously been shown to account for a variety of properties of visual cortex, including local topographic phenomena like orientation columns and ocular dominance stripes [12]. The present study demonstrates that within this framework one can also understand the differences in global topographic organization of visual areas. In our model the occurence of second order transformations is explained by a simple and biologically meaningful parameter, areal elongation. The variability observed in cat visual cortex arises as an intrinsic property of our model. Our results provide further support for the presence of an activity-dependent self-organizing process of map formation in the visual cortex.

REFERENCES

[1] Allman, J. M. & Kaas, J. H., Brain Resarch **76**,247-265 (1974).

[2] Tusa, R. J., Rosenquist, A. C. & Palmer L. A., J. Comp. Neur. **185**,657-678 (1979).

[3] Albus, K. & Beckmann, R., J. Physiol. **299**,247-256 (1980).

[4] Wagor, E., Mangini, N. J. & Perlman, A. L., J. Comp. Neur. **193**,187-202 (1980).

[5] Mallot, H. A., Biol. Cyb. **52**,45-51 (1985). Mallot, H. A., v.Seelen, W. & Giannakopoulos, Neural Networks **3**,245-263 (1990).

[6] Kohonen, T., *Self-Organization and Associative Memory*, (Springer, Berlin, 1987).

[7] Ritter, H., Martinetz, T., & Schulten, K., *Neuronale Netze*, (Addison Wesley, 1990).

[8] Orban, G., *Neuronal Operations in the Visual Cortex*, (Springer, Berlin, 1984).

[9] Wolf, F., diploma thesis, Universität Frankfurt 1992, (unpublished).

[10] Van Essen, D.C. & Zeki, S.M., J. Physiol. **277**,193-226 (1980). Kaas, J.H., *Comparative Neurology of the Telencephalon*, ed. Ebbesson, S.O.E., 483-501, (Plenum Press, New York, 1982).

[11] Espinoza, S. G. & Thomas, H. C., Brain Research **272**,137-144 (1983).

[12] Von der Malsburg, Ch., Biol. Cyb. **14**,85-100 (1973). Miller, K.D., Keller, J.B. & Stryker, M.P., Science **245**,605-615 (1989). Durbin, R. & Mitchison, G., Nature **343**,644-647 (1990). Obermayer, K., Blasdel, G.G., & Schulten, K., Phys. Rev. A **45**,7568-7589 (1992).

62

ORIENTATION COLUMNS FROM FIRST PRINCIPLES

Ernst Niebur[†]
Florentin Wörgötter[‡]

[†] *Computation and Neural Systems*
California Institute of Technology 216-76
Pasadena, CA 91125, USA

[‡] *Institut für Physiologie*
Ruhr-Universität Bochum
D-4630 Bochum, Germany

ABSTRACT

The input to primary visual cortex is embedded in a high dimensional feature space whose dimensions include position in visual space, preferred orientation, ocular dominance *etc.* Since all cells in a column with axis perpendicular to the cortical surface have approximately the same properties [4], this space is effectively mapped on a two-dimensional space, the cortical plane. This dimension reduction leads to complex maps which so far have evaded intuitive understanding. We show that their most salient features can be understood from a few basic design principles, in particular, local correlation, isotropy and homogeneity. These principles can be defined most easily in the Fourier domain where they correspond to a power spectrum which has a rather simple annulus-like structure. After inverse Fourier transformation, we obtain maps of orientation column structures which are very similar to the experimentally observed orientation column maps of the cat. We show this by comparison with maps which were obtained by optical imaging methods. We expect that many of the models which have been developed to explain the mapping of the preferred orientations (e.g., [5, 7, 6, 8, 10, 11, 13]) can be subsumed under our approach.

Recently it became possible to extract features of cortical cell behavior using optical imaging techniques [1, 2, 3, 12]. A map of the preferred orientations in the visual cortex of monkey obtained this way is shown in Fig. 1a. One of the obvious features of the map shown is its *periodicity* [4]: for most points, the preferred orientation is repeated in a certain distance, which we call λ. Generally, cells with all preferred orientations are found in an area of linear dimension λ around any point and visual space appears to be mapped in a repeating pattern onto the cortex. A module of cells in which "all" features

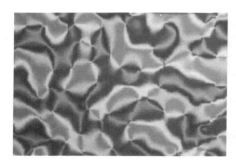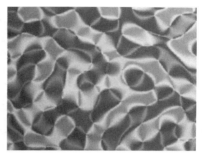

Figure 1 (a), left. Map of preferred orientations in area 18 of the macaque monkey measured with optical imaging by Blasdel (pers. comm.). Different gray levels correspond to different preferred orientations. (b), right. Map of preferred orientations, as computed from an annulus spectrum. Gray-level coding as in (a).

(e.g. preferred orientation, ocular dominance, color, velocity, etc.) of visual space are represented at least once is called a *hypercolumn* [4] whose width is λ. Hypercolumns seem to be arranged in "modules" in which adjacent x,y-locations in the visual field are projected onto adjacent hypercolumns in the cortex. This leads to a complete representation of all features of one location in the visual space in a locally confined cortical module while representing adjacent locations in adjacent modules.

There are many ways to achieve a modular organization. It is neither required to arrange the features periodically with one predominant frequency nor is it necessary to have an orderly arrangement within the individual modules, by only making sure that each module has at least one feature detector of each characteristic. Therefore, periodicity does not seem to be an *a priori* concept of cortical design but rather a derived quantity. What are then the basic cortical design principles of which periodicity may possibly be a consequence?

We propose that these principles are (positive) local correlation, homogeneity and isotropy, all with respect to a length scale λ, which is the only parameter in our framework. We define a map as being *locally correlated* in this sense if, on average, variations on a length scale much smaller than λ are significantly smaller than those on a length scale λ. *Homogeneity* means, that no systematic differences prevail between map locations over distances much larger than λ. *Isotropy* implies the same with respect to directions in the cortical plane or, for this matter, the visual field.

The spectrum of a system with the aforementioned characteristics is an annulus with radius λ^{-1}. Local correlation leads to the absence of spectral components for frequencies much larger than the radius of the annulus (i.e. for spatial frequencies $>> \lambda^{-1}$). Any non-zero components for high spatial frequencies would lead to short-range variations in the map, which are generically (i.e., except at isolated points, the so-called singularities) not observed. Homogeneity makes the spectrum zero inside of the annulus because any non-zero components for low spatial frequencies would lead to systematic differences (inhomogeneities) between adjacent hypercolumns which are also not observed. Isotropy is reflected in the spectrum by the fact that the statistical distribution of the non-zero components is the same in all directions around the origin. Local correlation and homogeneity (i.e. missing high and low frequency components) will lead to a band-pass characteristic and consequently to a periodicity with only one predominate frequency.

In order to test if an annulus spectrum is sufficient to produce realistic column structures or if more information is hidden in the details of the amplitudes or phases of the spectra of orientation column structures, we performed inverse Fourier transform of simple annulus spectra, which have zero amplitudes everywhere except on an annulus of radius $\approx 1/\lambda$. On this annulus, amplitude and phase have random values. In Fig. 1b, we show the cortical map which is obtained from this spectrum by a standard procedure[1]. Features of measured (Fig.1a) and computed map (Fig.1b) are very similar. Local correlation, homogeneity and isotropy can be tested statistically for the different maps. In measured as well as in maps computed from our model we found that the three features are represented at a highly significant level.

We found this scheme to be robust: we obtain realistically looking maps for a wide range of variations in amplitudes, phases and the width of the annulus which can be varied by about a factor of 10 without significant disturbance of the maps. Similar robustness was found when using different probability distributions for the amplitudes (uniform or Gaussian) and phases, or by using a different method for obtaining the (vectorial) angles of preferred orientation from the (scalar) result of the Fourier transformation than we did [9]. This robustness might explain why so many different developmental models are ca-

---

[1] The preferred orientations $\phi(x, y)$ are computed from the two-dimenstional Fourier spectrum $F(k_x, k_y)$ as [11]:

$$\phi(x,y) = 1/2 \arctan \frac{Im[IFT(F(k_x, k_y))]}{Re[IFT(F(k_x, k_y))]} \tag{1}$$

where $Re[...]$ and $Im[...]$ describe the real and imaginary parts of their arguments, and IFT denotes the inverse Fourier transform.

pable of producing "good-looking" maps. From our results, this is expected as long a they are consistent with the basic properties of local correlation, homogeneity and isotropy.

Homogeneity, local correlation, and isotropy are properties of natural images, i.e. properties of the input of the visual system, and it might be advantageous for an information processing system if its structure reflects the properties of the input signals. Two remarks are in order here: (1) Neither the visual input nor the human visual system are *completely* isotropic, and similar statements are probably true for homogeneity and local correlation. Our results should rather be taken as a general framework for leading-order effects than as a detailed model for particular features. (2) We neglect distortions by the "complex-logarithm transformation" of visual images that emphasizes the foveal region with respect to the periphery.

Independent of the properties of visual input, one expects homogeneity to be a useful feature in any parallel system, since it allows to replicate one module many times for parallel information treatment. Local correlation is found in all cortical areas reflecting the tendency of neurons to work in an environment in which they are surrounded by other neurons whose properties vary in a smooth, orderly manner. There seem to be less compelling reasons for strict isotropy except from conceptual and developmental simplicity, and, indeed, this property is not always found in perfect form. Ocular dominance columns in cat and orientation columns in monkey are better described by an anisotropic spectrum [9].

Thus, the complicatedly looking maps of orientation preferences might be based on few design principles which in turn rely on very general developmental mechanisms utilizing the input structure of the system. If this conclusion is true, such a more intuitive insight into the cortical design also explains the robustness of the cortical network during development and while suffering from damage. Very little structural information appears to be the basis of a highly complex performance which still evades theoretical analysis.

Acknowledgement. We thank Gary Blasdel for sharing his data with us prior to publication and Christof Koch, Klaus Obermayer, Ken Miller and Ulf Eysel for helpful discussions. E.N. is supported by the Air Force Office of Scientific Research and F.W. by the Deutsche Forschungsgemeinschaft. We also acknowledge support by the Air Force Office of Scientific Research, the James S. McDonnell Foundation and by a NSF Presidential Young Investigator Award to Christof Koch.

REFERENCES

[1] G.G. Blasdel. Personal communication.

[2] G.G. Blasdel and G. Salama. Voltage-sensitive dyes reveal a modular organization in monkey striate cortex. *Nature*, 321:579–585, 1986.

[3] T. Bonhoeffer and A. Grinvald. Iso-orientation domains in cat visual cortex are arranged in pinwheel-like patterns. *Nature*, 353:429–431, 1991.

[4] D.H. Hubel and T.N. Wiesel. Receptive fields and functional architecture of monkey striate cortex. *J. Physiol.*, 195:215–243, 1968.

[5] R. Linsker. From basic network principles to neural architecture: Emergence of orientation columns. *Proc. Natl. Acad. Sci. USA*, 83:8779–8783, 1986.

[6] K.D. Miller. Development of orientation columns via competition between ON- and OFF-center inputs. *Neuroreport*, 3:73-76, 1992.

[7] M.M. Nass and L.N. Cooper. A theory for the development of feature detecting cells in visual cortex. *Biological Cybernetics*, 19:1–18, 1975.

[8] K. Obermayer, H. Ritter, and K. Schulten. A principle for the formation of the spatial structure of cortical feature maps. *Proc. Natl. Acad. Sci. USA*, 87:8345–8349, 1990.

[9] A.S. Rojer and E.L. Schwartz. Cat and monkey cortical columnar patterns modeled by bandpass-filtered 2d white noise. *Biol. Cybern.*, 62:381–391, 1990.

[10] N.V. Swindale. A model for the formation of orientation columns. *Proc. R. Soc. Lond. B.*, 215:211–230, 1982.

[11] N.V. Swindale. Iso-orientation domains and their relationship with cytochrome oxidase patches. In *Models of the Visual Cortex*, pages 452–461. Rose, D.; Dobson, V.G., John Wiley and Sons Ltd., 1985.

[12] D.Y. Ts'o, R.D. Frostig, E. Lieke, and A. Grinvald. Functional organization of primate visual cortex revealed by high resolution optical imaging of intrinsic signals. *Science*, 249:417–423, 1990.

[13] C. von der Malsburg and J.D. Cowan. Outline of a theory for the ontogenesis of iso-orientation domains in visual cortex. *Biological Cybernetics*, 45:49–56, 1982.

A MODEL OF THE COMBINED EFFECTS OF CHEMICAL AND ACTIVITY-DEPENDENT MECHANISMS IN TOPOGRAPHIC MAP FORMATION

Martha J. Hiller

MIT Artificial Intelligence Laboratory, Room NE43-825
545 Technology Square, Cambridge, Massachusetts 02139

ABSTRACT

Chemical and activity-dependent mechanisms work together in the model presented in this paper to allow rapid development of detailed topographic maps. The model resolves difficulties of previous models in specifying map orientation, attaining rapid convergence without loss of map continuity, and describing a biologically plausible mechanism to produce the coactivation pattern necessary for Hebbian learning. It matches experimental results on the ability to form rough, but not refined, maps in the absence of electrical activity and expansion of heavily stimulated areas. Although a high-level model is used, the goal is to explore organizational principles of biological development.

1 INTRODUCTION

The presence of chemical markers guiding topographic map formation has been demonstrated in many experimental preparations [1, 2]. These mechanisms are sufficient to produce a rough map topography in the correct orientation for the final map. However, electrical activity is necessary for map refinement and for segregation into ocular dominance columns [3]. Firing correlations in neighboring retinal ganglion cells during development [4] are a likely source of topographic information for map refinement.

The experimental evidence thus points to cooperation between chemical and activity-dependent mechanisms for the formation of topographic maps. However, previous computational models, with one exception [9], have attempted to find a single unified map formation process, and fail to exploit the advantages

of combining multiple mechanisms. There are several problems to be solved by
a computational model of biological map formation: achieving rapid convergence to a continuous final map, specifying map orientation, and describing a
plausible biological implementation. No existing model [5, 6, 7, 8, 9] solves all
these problems. Biological implementations, particularly of the assumed pattern of target activity,[1] are often not considered, and models that do include
them [5, 6] are out of date.

2   THE MODEL

Although some form of chemoaffinity has been found in all biological systems
[1], details vary from one system to another. Instead of modelling a specific
chemical mechanism, therefore, an attempt is made to determine the constraints
they must satisfy to ensure an accurate final map when combined with activity-
based refinement. Chemoaffinity is assumed to produce an initial rough topography, but does not play a role in the dynamics of map refinement, which are
mediated by activity-dependent mechanisms. Connections are strengthened via
a Hebbian mechanism patterned after long-term potentiation (LTP) [10] and
limited by competition. The biological underpinnings of competition are not
well understood, but it is necessary to explain experimental results such as map
expansion [1].

One-dimensional source and target cell layers are used to reduce computational
complexity so longer-range interactions can be studied. Two types of synapses
are included: modifiable excitatory projections from source to target layers,
and unmodifiable excitatory lateral projections within the target. Initial connections from each source cell cover a broad but finite region in the target,
close to (and including) the desired final location, with initially random connection strengths. The accuracy of the initial map is defined in terms of the
maximum position error (distance between the arbor's center and the desired
final map location). The actual position error of each initial arbor is chosen
randomly in a uniform distribution over the range of possible position errors.
Lateral connections are localized to a neighborhood around each neuron, and
their strength decreases with distance. Input activity occurs in spatially and
temporally localized clusters of neurons. Each input activity cluster is simulated as a single "trial" of the learning algorithm. Active neurons within a
cluster are given unit activity.

Other computational models ignore the transient response to an input stimulus,

---

[1] That is, the strength of activation in response to a given input stimulus, as a function of
location in the target cell layer.

and instead use a hypothetical steady-state frequency response as the target activity. LTP operates at a very fast time scale,[2] however, so the dynamics of the transient response are important. This simulation gives a first-order approximation of the target response to spreading activation through lateral connections. The dynamics of this response are simulated as follows:

1. Individual target neurons are modelled as simple linear thresholded elements.

2. Each "trial" is divided into time steps. In each time step neurons whose input is above threshold fire a single action potential. Input from the source layer arrives at the target in the first time step of each trial.

3. In successive time steps, inputs arrive through lateral connections from target neurons that fired in the previous time step. Lateral synaptic strengths are set in such a way that the active region in the target shrinks over time.[3]

4. Activation is summed over the duration of the target response to get the target activity a_j for the trial.[4]

Connection strengths are modified after each trial with a learning rule that combines Hebbian learning (in the numerator) and competition in the form of input normalization (in the denominator):

$$w_{ij}(t+1) = \frac{w_{ij}(t) + \alpha a_i(t) a_j(t) A_{ij}(t)}{1 + \dfrac{1}{K} \displaystyle\sum_{k=0}^{m-1} \alpha a_i(t) a_k(t) A_{ik}(t)} \tag{1}$$

where $w_{ij}(t)$ is the connection strength between source neuron i and target neuron j at trial t, α is the learning rate, a_i and a_j are the source and target activity, $A_{ij} = 1$ if there is a synapse between source i and target j, and 0 otherwise, and K is the "competition limit", that is, the total synaptic strength over all inputs to a given target neuron.

---

[2] For example, in the model of LTP described in [11], the strength of the LTP response is proportional to the peak Ca^{++} concentration; their simulation shows this concentration peaking and subsiding within 100-200 msec. after presentation of the input stimulus.

[3] If most of the cells within some region in the target are activated, the activity dies out in a wave moving inward from the edges of that region. To accomplish this, the strength of lateral connections must meet the following constraint: $\frac{\Lambda}{2} < \theta < \Lambda$, where Λ is the summed synaptic strength from all lateral inputs to a given target neuron, and θ is the cell's firing threshold.

[4] Parameters in this simulation were set to give maximum burst lengths of 10-20 spikes. At 100 Hz., this response would last up to 200 msec., which is of the same order of magnitude as the LTP time window.

a.

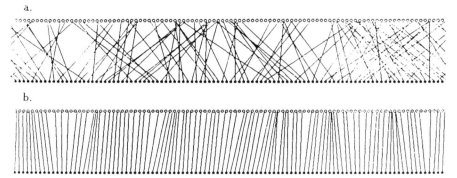

b.

Figure 1   Examples of initial and final projections. Filled circles are
source neurons, open circles are target neurons, lines are drawn from each
source neuron to the centroid of its axonal projection. a. initial state
with position error of ±20. b. typical result after 10.000 input stimulus
presentations.

3   RESULTS

Tests were run on maps of length 100, with ends wrapped to reduce edge effects.
The location of the input activity cluster for each trial was chosen at random
in a uniform distribution over the input map. Each test ran for 10,000 trials.
In all cases, the final map orientation is correct, and the map is continuous
at a global level. Convergence to the final location is extremely fast, being
essentially complete within 1000 trials (an average of ten presentations of each
input cluster).

The centroid of the axonal projection from each source neuron, weighted by
connection strength, is used as a measure of its location in the target layer.
Examples of initial and final maps are shown in figure 1. The average "spacing
error". the variance in distance between centroids of neighboring source cells,
is used to evaluate the local ordering of the map. The effect of the initial
position error and the input cluster size was tested, and results are graphed in
Figure 2. The accuracy of the final map was independent of the initial position
error, for values up to ±20 neurons. Follow-on tests showed fast and accurate
convergence with initial errors up to ±40 (results not shown). Larger input
cluster sizes converge faster, and with the smallest cluster size, the final map
was somewhat disordered.

4   ANALYSIS

The lateral connections control the target response to a given input stimulus.
In the adult, these follow a characteristic "mexican hat" pattern, with short-
range excitatory and mid-range inhibitory connections. However, their pattern

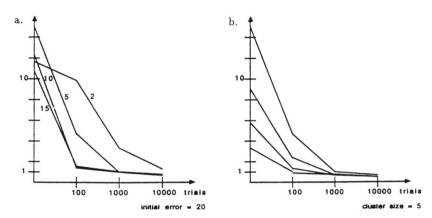

Figure 2 Average spacing error. a. fixed cluster size of 5, with several initial maximum position errors. b. fixed initial maximum position error of ±20, with several cluster sizes.

during development is unknown. An interesting question, then, is what kind of lateral connections are most useful for map formation.

Map refinement can be thought of as a process of noise reduction. The location of the initial arbors provide a rough estimate of the desired target location (i.e. they are localized to a neighborhood around it, with some error); neighbors of a source cell provide similarly rough estimates of nearby target locations. When several cells fire in a cluster, their projections can be averaged to obtain a more accurate estimate of the correct target location. However, as the centroid of the arbor converges to the correct location, the standard deviation will increase, until in the limit the synaptic strength is constant over the entire arbor [12]. To compensate for this effect, a mechanism is needed to selectively enhance the target response in the center of the region activated by an input cluster.

A purely excitatory pattern of lateral connections is used because it is ideal for producing this type of activity. The only constraints are that the input activity be sufficient to initiate firing in the target, and that there be no large gaps in connectivity within an arbor. With a "mexican hat" pattern, the desired activity is difficult to obtain. A prediction of this model, therefore, is that map topography is determined before the development of an inhibitory surround.

The noise reduction model also explains why we have problems with small cluster sizes and large initial errors. With a small cluster size, we have fewer "samples" to average over, so the process converges more slowly. However, as the arbors narrow, their position becomes fixed. If this occurs before we obtain an accurate estimate of the desired location, the map will remain somewhat disordered. This problem can be solved either by increasing the size of the activity clusters, or by reducing the learning rate.

Another effect of target activity is on the convergence properties of the learning algorithm. Erwin et al. [13] have, for Kohonen nets on one-dimensional inputs, determined the optimal target activity pattern to ensure rapid convergence and map continuity. Although their results have not been proven to generalize to this model, they still provide useful criteria for its analysis.

To ensure a continuous map, the activity pattern must be broader than the position error, and it must be convex. For an optimal convergence rate, the activity pattern must be as narrow as possible, and sloped toward the center, so that neighboring neurons have maximally differing activity levels. The breadth of the target response in this model is determined by the arbors, which are broader than their position error. As the arbors narrow and approach their final location, the breadth of the response also narrows, thereby tailoring itself to the current optimum. The shape of the response is a sum of direct activation from the source, which is (probabilistically) gaussian, and spreading activation through lateral connections, which is triangular except for some rounding at the peak. Both functions are convex by Erwin et al.'s definition, and their sum is also (probabilistically) convex. The triangular component of the activity pattern gives it a slope that meets the second criterion for rapid convergence. The activity pattern which arises naturally from intercellular interactions in this model thus meets all the requirements given in [13] for obtaining a continuous final map with an optimal convergence rate.

5   EXTENSIONS TO THE MODEL

The form of competition used above does not match any known biological mechanism. However, in the limit where incremental changes are small, the competition term (from the denominator of equation 1) can be simplified as follows:

$$
\frac{1}{1 + \dfrac{1}{K} \displaystyle\sum_{k=0}^{m-1} \alpha a_i(t) a_k(t) A_{ik}(t)} \approx \frac{1}{1 + \dfrac{\alpha N}{K} a_i(t)} \tag{2}
$$

where N is the expected activity in the inputs to an active target neuron.[5] This is a generalization of the activity-dependent decay used by Oja [14], and

---

[5] In the case where all neurons within a source activity cluster have unit activity, this is equal to the expected number of active inputs.

gives the dependence on the size of the input activity cluster.[6] This suggests a mechanism whereby all synapses presynaptic to a given target neuron decay in strength at a rate controlled by target depolarization.[7] This simplification was tested to demonstrate that it does, in fact, have a similar effect on the model. One difference is that, if the size of the input activity clusters changes, the total synaptic input to the target neurons will move to a new equilibrium point reflecting the change.

Tests were also run to replicate biological experiments showing map expansion and compression, following changes in input stimulation and map lesions (see [12] for details).

6   SUMMARY AND CONCLUSIONS

Map formation in this simulation is accurate and robust, and provides a good match to experimental data, despite the use of an extremely simple neuron model. It may be that developmental phenomena such as map formation are relatively insensitive to detailed properties of the cell, either because their success is so important to the organism, or because these properties are in such flux during development.

Chemoaffinity, in this model, is assumed to create a rough initial map that is refined by an activity-based mechanism. The accuracy of the initial map is surprisingly unimportant in achieving an accurate final map, as long as the axonal projections of neighboring source neurons overlap. Neighbor correlations in the source map, combined with an LTP-like learning process and a competitive mechanism, rapidly refine the map into a detailed topography.

The result of map refinement depends on the target activity pattern. The model described in this paper results in an activity pattern that is ideal for producing rapid convergence to a detailed map.

REFERENCES

[1] S. B. Udin and J. W. Fawcett. Formation of topographic maps. *Ann. Rev. Neurosci.*, 11, 1988.

[2] C. J. Shatz and D. W. Sretavan. Interactions between retinal ganglion cells during the development of the mammalian visual system. *Ann. Rev. Neurosci.*, 9, 1986.

---

[6] In Oja's model, the size of an activity cluster is not represented.

[7] For example, a voltage-dependent process that affected the turnover rate of postsynaptic receptors might have this effect.

[3] C. J. Shatz. Impulse activity and the patterning of connections during cns development. *Neuron*, 5, 1990.

[4] L. Maffei and L. Galli-Resta. Correlation in the discharges of neighboring rat retinal ganglion cells during prenatal life. *Proc. Natl. Acad. Sci. USA*, 87, 1990.

[5] D. J. Willshaw and C. von der Malsburg. How patterned neural connections can be set up by self-organization. *Proc. R. Soc. Lond. B.*, 194, 1976.

[6] C. von der Malsburg and D. J. Willshaw. How to label nerve cells so that they can interconnect in an ordered fashion. *Proc. Natl. Acad. Sci. USA*, 74(11), 1977.

[7] T. Kohonen. *Self-Organization and Associative Memory*. Springer-Verlag, 1988.

[8] K. Obermayer, H. Ritter, and K. Schulten. A principle for the formation of the spatial structure of cortical feature maps. *Proc. Natl. Acad. Sci. USA*, 87, 1990.

[9] K. Miller. Ocular dominance column development: Analysis and simulation. *Science*, 245, 1989.

[10] M. B. Kennedy. Regulation of synaptic transmission in the central nervous system: Long-term potentiation. *Cell*, 59, 1989.

[11] T. H. Brown, A. M. Zador, Z. F. Mainen, and B. J. Claiborne. Hebbian modifications in hippocampal neurons. In M. Baudry and J. L. Davis, editors, *Long-Term Potentiation*. MIT Press, 1982.

[12] M. J. Hiller, in preparation.

[13] E. Erwin, K. Obermayer, and K. Schulten. Self-organizing maps: Stationary states, metastability and convergence rate. *Biol. Cybern.*, 67, 1992.

[14] E. Oja. A simplified neuron model as a principal component analyzer. *J. Math. Biol.*, 15, 1982.

SECTION 9

ASSOCIATIVE MEMORY AND LEARNING

Associative memories and learning rules feature prominently in the field of artificial neural networks. Some of the work presented here bridges the gap between artificial neural systems and biological "wetware". The section starts with a chapter on NMDA receptors and ends with two models for Alzheimer's disease, a disease process that affects cognition and memory in humans. Readers should check chapters 40 and 42 in section 5, chapter 55 in section 7, and chapter 57 in section 8 for related topics.

NMDA-ACTIVATED CONDUCTANCES PROVIDE SHORT-TERM MEMORY FOR DENDRITIC SPINE LOGIC COMPUTATIONS

Roderick V. Jensen* and Gordon M. Shepherd

*Department of Neurobiology, Yale University School of Medicine
New Haven, CT 06510*

** Department of Physics, Wesleyan University,
Middletown, CT 06457*

ABSTRACT

Numerical models of dendritic trees indicate that the nonlinear interaction of postsynaptic potentials in active spines can generate simple logic operations such as AND, OR and NAND gates. However, because the spine head EPSP's closely follow the underlying, short-duration (1-3 ms), synaptic conductances, previous studies concluded that the precise timing of synaptic inputs is critical for these logic operations. We show that this severe temporal limitation on dendritic computation can be overcome by the inclusion of slow (100-200 ms), voltage-dependent, NMDA-receptor mediated conductances in the spine heads. Our numerical simulations show that this simple mechanism provides a short term memory ($\sim$ 100 ms) for logical AND gates with time-delayed inputs on one or more spines.

1   NUMERICAL SIMULATIONS

Active conductances in dendrites may permit elaborate computational processing of multiple synaptic inputs long before these signals reach the soma. Numerical, multi-compartmental models of dendritic trees indicate that voltage-dependent conductances localized in spine heads or on dendritic branches can amplify individual synaptic inputs[1,2]. Moreover, the nonlinear summation of the postsynaptic potentials has been shown to be capable of performing simple logic operations such as AND, OR, and NAND gates.[3,4]

For example, an AND gate can be realized by two simultaneous EPSP's in nearby active spines.[3,4] Even though the individual EPSP's may be sub-threshold for the generation of an action potential in the active spine heads, the combined EPSP's may mutually reinforce to depolarize the active membranes in each spine to firing. (Similar results were obtained when the active membrane was localized on the dendritic branches.[4])

However, because the spine head EPSP's closely follow the underlying, short-duration (1-3 ms), synaptic conductances, previous studies concluded that the "precise timing of inputs is critical for the summation of individual responses to occur"[3]. In other words, the "memory" of each synaptic event is restricted to the few millisecond duration of the event itself. Here we show that this severe temporal limitation on dendritic computation can be overcome by the inclusion of slow (100-200 ms), voltage-dependent, NMDA-receptor mediated conductances[5] in the spine heads.

We used a 16 compartment model of a $700\mu m$ segment of tapered dendrite that was structurally identical to the models studied by Shepherd and Brayton[3] and Shepherd, Woolf, and Carnevale[4]. Four $1\mu m \times 1\mu m$ spine heads were connected by $1\mu m \times 0.1\mu m$ spine necks at different points along a $1\mu m$ diameter dendrite. Two spines were attached to the same dendritic segment and the other two were $50\mu m$ and $100\mu m$ away. The membrane resistance and capacitance of all compartments was choosen to be $R_m = 2000\ Ohm - cm^2$ and $C_m = 1\ \mu F/cm^2$ and the internal resistance was $R_i = 80\ Ohm-cm$ giving a membrane time constant of $\tau = 2\ ms$ and an electrotonic characteristic length of $250\ \mu m$ in the $1\ \mu m$ diameter segments of dendrite.

The numerical simulations were performed using both the SABER general purpose simulation program[6] and the GENESIS neural network simulator[7] with identical results. We first reproduced the earlier simulations [3,4] of an AND gate using active sodium and potassium conductances in the spine heads and an AMPA synaptic input of the form of a modified α-function

$$g_{AMPA} = \bar{g}_{AMPA}(1 - e^{-t/0.3})e^{-t/3} \tag{1}$$

with a peak conductance of $\bar{g}_{AMPA} = 2\ nS$ on a $3\ ms$ time scale with a $70\ mV$ reversal potential (relative to rest). We then added a simple model of the slow, voltage dependent NMDA conductance to the spine heads of the form[8]

$$g_{NMDA}(t) = \bar{g}_{NMDA}(1 - e^{-t/3})e^{-t/200}(6.05e^{0.06(V-70)}) \tag{2}$$

with a $75\ mV$ reversal potential.

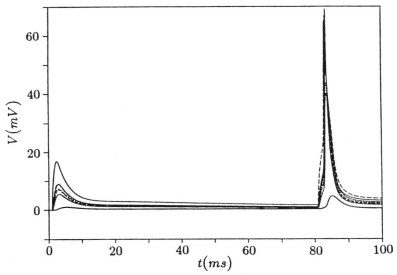

Figure 1 The membrane voltage is plotted as a function of time at 7 different locations along the dendrite. One synaptic input with $\bar{g}_{NMDA} = 2.0 \ nS$ is applied on spine head 1 (top solid curve) at $t = 1 \ ms$ and a second synaptic input is applied on the adjacent spine head 2 (top dashed curve) at $t = 81 \ ms$. The bottom solid curve shows the membrane potential at the soma at the end of the 700 μm dendrite.

Many different simulations were performed to examined the effects of temporal delays in the synaptic inputs to the spine heads. For example, Fig. 1 shows results for the membrane potentials at different locations along the dendritic tree for two synaptic inputs on adjacent spines with an 80 ms delay. In this case the peak NMDA conductance was as large as the AMPA conductance, $\bar{g}_{NMDA} = 2 \ nS$. The first synaptic input on spine head 1 at $t = 1 \ ms$ is insufficient to cause the active membrane to fire. The NMDA current is activated during the course of the AMPA current but is inactivated by the strong voltage dependence in Eq. 2 when the voltage returns to rest. However, because of the long 200 ms time constant, the first NMDA conductance is reactivated when the spine-head voltage is raised again after 80 ms by the synaptic input to the second spine head. In this case the combinations of the new AMPA and NMDA conductances in the second spine head and the reactivation of the NMDA conductance in the first are sufficient to raise the membrane above threshold in both spine heads causing the active membranes to fire there, as well as in the electrotonically close spine heads 50 and 100 μm away. The net result is a nonlinear summation of the first synaptic input AND the second input 80 ms

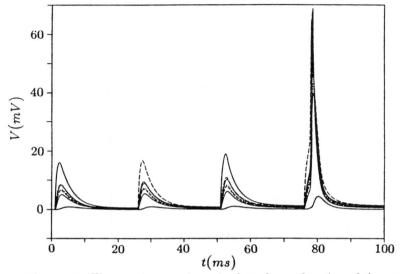

Figure 2 The membrane voltage is plotted as a function of time at 7
different locations along the dendrite. In this case one synaptic input
with $\bar{g}_{NMDA} = 0.5$ nS is applied on spine head 1 (top solid curve) at
$t = 1$ ms, a second synaptic input is applied on spine head 2 (top dashed
curve) at $t = 26$ ms, a third input is applied again on spine head 1 at
$t = 51$ ms, and a final input arrives at spine head 2 at $t = 76$ ms causing
the active membrane in all the electrotonically close spine heads to fire.

later leading to a greatly enhanced voltage pulse at the soma located at the
end of the dendritic tree.

Many variants of this simple mechanism for time-delayed logic computations
are possible. For example, Fig. 2 shows the spine head and dendritic potentials
for a series of four synaptic inputs to two adjacent spines spaced 25 ms apart.
Since our model of the NMDA conductance assumes that new channels are
opened for $\sim 100 - 200$ ms following each synaptic event, the $\bar{g}_{NMDA}$ can be
reduced to 0.5 nS in this case and the nonlinear summation of the four synaptic
inputs can still lead to firing after 75 ms.

Moreover, since the individual synapses are temporally separated, it is no longer
necessary for the inputs to the AND operation to be applied on separate spines.
A temporal AND gate can be realized using two or more delayed synapses on a
single spine head analogous to paired pulse facilitation. Successive subthreshold
synaptic pulses separated by as much as 100 ms can still lead to firing due

to the build-up of the latent, voltage-dependent NMDA conductances. The numerical result for two synaptic impulses on the same spine head with a 80 *ms* time-delay is virtual identical to the case shown in Fig. 1. Even though the membrane voltage may return to values close to the resting potential between pulses, the "memory" of the previous pulse is preserved by the activated, but Mg^{2+} blocked, NMDA receptors.

NMDA conductances have long been implicated as a mechanism for the induction of long-term memory (LTP) in neuronal networks.[9] Our numerical simulations show that the unique voltage and time dependence of the NMDA receptors may also provide an important short-term memory ($\sim$ 100 ms) for spine logic operations with time-delayed inputs on multiple spines. In particular a sequence of excitatory synaptic inputs in nearby spines, separated spatially by distances shorter than the characteristic length for electrotonic decay in the passive dendrite and temporally by times shorter than the lifetime of the NMDA receptor, can be as effective as simultaneous synaptic inputs. Very recently, Mel[10] has shown how the spatial "clustering" of NMDA mediated synapses can allow a single neuron to perform complex pattern discrimination tasks. Our results show that the additional temporal "clustering" of NMDA mediated synapses (in a $\sim$ 100 *ms* time window) can further expand the computational capabilities of a single neuron.

This work was supported by NIDCD, ONR, and NSF.

[1] D.H. Perkel and D.J. Perkel, *Brain Research* 325, 331 (1985).

[2] J.P. Miller, W. Rall, and J. Rinzel, *Brain Research* 325, 325 (1985).

[3] G.M. Shepherd and R.K. Brayton, *Neuroscience* 21, 151 (1987).

[4] G.M. Shepherd, T.B. Woolf, and N.T. Carnevale, *Journal of Cognitive Neuroscience* 1, 284 (1989).

[5] M.L. Mayer and G.L. Westbrook, *Progress in Neurobiology* 28, 197 (1987).

[6] *SABER User's Guide*, (Analogy, Inc.,Beaverton, Oregon, 1988).

[7] M. Wilson and J.M. Bower, *Genesis: The Caltech Neural Network Simulator*, 1991.

[8] F. Pongracz, N.P. Poolos, J.D. Kocsis, and G.M. Shepherd, Journal of Neurophysiology (in press).

[9] P. Churchland and T. Sejnowski, *The Computational Brain*, (MIT Press, Cambridge, MA, 1992).

[10] B.W. Mel, *Neural Computation* 4, 502 (1992).

65

A MODEL OF CORTICAL ASSOCIATIVE MEMORY BASED ON HEBBIAN CELL ASSEMBLIES

Erik Fransén, Anders Lansner and Hans Liljenström

SANS – Studies of Artificial Neural Systems
Dept. of Numerical Analysis and Computing Science
Royal Institute of Technology, S-100 44 Stockholm, Sweden

ABSTRACT

A model of cortical associative memory, based on Hebb's theory of cell assemblies, has been developed and simulated. The network is comprised of realistically modelled pyramidal–type cells and inhibitory fast spiking interneurons and its connectivity is adopted from a trained recurrent artificial neural network. After–activity, pattern completion and competition between cell assemblies is readily demonstrated. If, instead of pyramidal cells, motor neurons are used as excitatory neurons in the network, spike synchronization can be observed but after–activity is hard to produce. Geometry dependent time delays below 10 ms have little effect. After–activity is facilitated by increased levels of serotonin and disrupted by low levels. Our results support the biological plausibility of Hebb's cell assembly theory.

65.1 BACKGROUND

In his classical book Hebb described a functional unit which he called a cell assembly [1]. This was a group of cells strongly connected through excitatory synapses. Such an assembly could emerge as a result of activated Hebbian synapses. Hebb proposed that the assembly so formed thereafter could serve as an internal representation of the corresponding object in the outside world. Hebb's cell assembly theory has been further elaborated [2, 3] and is still, in its general aspects, compatible with experimental findings. We also see a connection to abstract networks. A prototypical recurrent artificial neural network used as an auto–associative content–addressable memory can be regarded as a mathematical realization of Hebb's basic idea. A "memory trace" in such a network corresponds closely to a cell assembly. The dynamic recall process converging to a low energy, stable state is analogous to the triggering of activity in a cell assembly. Earlier investigators [4, 5] have found, for example, spike synchronization but no signs of after–activity of the kind hypothesized by Hebb.

However, in some of our own investigations, after–activity could indeed be produced provided that the model motor neurons were replaced by pyramidal–cell type neurons [6, 7].

65.2   CELL MODEL AND NETWORK CONNECTIVITY

A general purpose simulator, SWIM, intended for numerical simulation of networks of model neurons with a Hodgkin–Huxley type formalism, has been used [8]. In the present study, two different types of excitatory neurons were simulated, *i.e.* the "P–cell" modelled after a typical cortical pyramidal cell and the "MN–cell" with properties derived from a motor neuron. "FS–cells" modelled after cortical fast spiking cells were used as inhibitory interneurons.

The simulated network was composed of fifty pairs of one excitatory cell and one inhibitory interneuron. The interneuron receives input from excitatory cells in other pairs belonging to other assemblies and inhibits its companion excitatory cell. Excitatory cells in one and the same assembly are connected. Values for synaptic strengths were adopted from a recurrent Bayesian ANN [9] trained with 8 random patterns with 8 active units in each. The Bayesian learning rule produced excitatory synapses within the patterns and inhibitory ones between them.

65.3   ASSEMBLY OPERATION RESULTS

When stimulating all excitatory cells in an assembly comprised of P–cells as the excitatory cells, with 0.4 nA for about 40 ms, the effect was *after–activity* for about 400 ms. For an isolated cell such a stimulation resulted in two or three spikes. Now, due to the mutual synaptic excitation, the cells continue to fire. The firing frequency gradually decreases due to accumulated calcium entering through Ca^{2+}– and NMDA–channels. Calcium opens Ca^{2+} dependent K^+–channels which counteracts synaptic excitation.

Mutual excitation between neurons in the cell assembly also provides the network with a capability for *pattern completion*. With two P–cells out of eight stimulated just transient activity occurred, while with one more cell receiving stimulation activity in the entire assembly was triggered (Fig. 1a).

There is a quite potent lateral inhibition between assemblies in our network. This gives rise to *competition* between assemblies. When five cells in one assembly and three cells in another were stimulated, the first assembly prohibited the activity in the second assembly and completed its own missing cells (Fig. 1b). In another simulation, when the stimulus used activated a part of a cell assembly together with some randomly activated cells, the spurious cells were quickly silenced and the missing ones activated. This demonstrates the *noise tolerance* of the network operation.

Worth noting in this context is the rather short time required for the activation of a complete assembly. Almost without exception, even in cases of

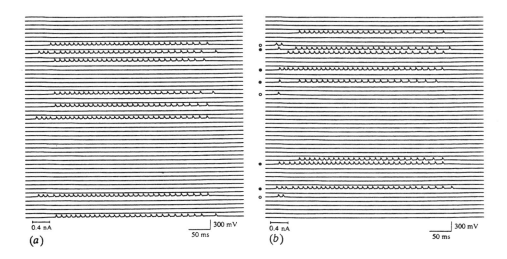

Figure 1 (a) Pattern completion and after–activity in an assembly with three cells stimulated. Each trace shows the activity of one P–cells. (b) Competition between two assemblies. The stimulated cells in the two different patterns are marked with open and closed circles.

conflicting input, a clean "interpretation" was stable 50–70 ms after stimulus onset in our simulations (see *e.g.* Fig. 1a, b). These *reaction times* are also short enough to support sustained processing times below the 100 ms observed in some experiments on perceptual tasks *e.g.* visual object identification [10].

The capability to produce after–activity of an assembly consisting of motor neurons was also investigated. After–activity was observed only in exceptional cases. One explanation seems to be that the prominent AHP of the MN–cell to a large extent masks the EPSP produced by the mutual excitatory synapses. This effect is enhanced by the high firing threshold of the MN–cells and by their tendency to *spike synchronize*. To investigate if the assembly size had been too small assemblies with 100 cells were tested. Qualitatively the same behavior was seen. At least under the conditions studied here, the P–cell assembly did not seem to give very significant spike synchronization. If continuously stimulated, an assembly consisting of MN–cells also displayed some pattern completion tendencies.

In an extended model, *time delays* have been introduced to represent separation of the cells in space. For a network with P–cells as the principal cells, an average delay of up to 10 ms was possible without any significant changes in assembly operation. With longer time delays performance began to degrade. The spike synchronization tendencies of the MN–cell network decreased already at delays of 3–4 ms. Thus, the important assembly operations do not seem very sensitive to short time delays, which means that a cell assembly may be distributed over a relatively large area and still be operating reliably.

We model the *neuromodulatory effects* of serotonin by decreasing the conductance through the calcium dependent potassium channels thus decreasing the AHP amplitude [11]. This mechanism provides a means for modulating cell

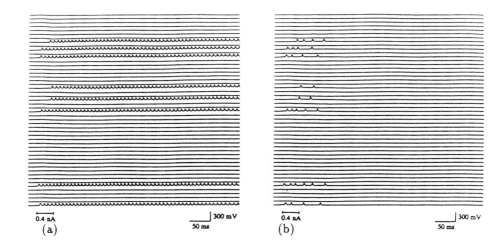

Figure 2 (a) An assembly of P–cells with increased AHP giving no after–activity.
Stimulation on 5 out of 8 cells. (b) An assembly of P–cells with reduced AHP showing
enhanced assembly operation.

properties in a way that dramatically influences the assembly–related network
operations [12]. When the AHP is reduced triggering is facilitated and the duration of after–activity is prolonged (Fig. 2a). When the AHP is increased these
features weaken and disappear (Fig. 2b). The behavior in this case resembles
more that of the motor neuron network.

65.4 CONCLUSIONS

The simulation results presented here show that Hebbian cell assembly related
activity could readily be produced by a network with cortical pyramidal cells as
principal excitatory cells. Such a network would also display pattern completion, noise tolerance and competitive phenomena in cases of conflicting inputs
in much the same way as a recurrent ANN. One important prerequisite is that
the model, of the excitatory neurons used, is provided with properties of cortical pyramidal cells rather than spinal motor neurons. When effects of serotonin
were modeled the operation was strongly affected. With high levels assembly
related activity was enhanced, with low levels it was supressed. If the time
delays in the network were below 10 ms, performance was little affected. Our
simulation results support the biological feasibility of Hebb's cell assembly theory and could eventually provide a bridge between recurrent artificial neural
network models and biological models of associative memory.

65.5 ACKNOWLEDGEMENTS

Financial support from NFR (grants no. F-FU 6445-300-302 and F-TV-9421-307) and NUTEK/STU (grant no 87-00321P) is gratefully acknowledged.

REFERENCES

[1] Hebb, D. O., *The Organization of Behavior*, New York: John Wiley, 1949.

[2] Palm, G., *Neural Assemblies. An Alternative Approach to Artificial Intelligence*, Berlin: Springer Verlag, 1982.

[3] Amit, D. J., Evans, M. R., and Abeles, M., "Attractor neural networks with biological probe records," *Network*, vol. 1, 1990, pp. 381–405.

[4] Rochester, N., Holland, J., Haibt, L., and Duda, W., "Tests on a cell assembly theory of the action of the brain, using a large digital computer," *IRE Trans. Information Theory*, vol. IT-2, 1956, pp. 80–93.

[5] MacGregor, R. and McMullen, T., "Computer simulation of diffusely connected neuronal populations," *Biol. Cybernetics*, vol. 28, 1978, pp. 121–127.

[6] Lansner, A., "Information processing in a network of model neurons. A computer simulation study," Tech. Rep. TRITA-NA-8211, Dept. of Numerical Analysis and Computing Science, Royal Institute of Technology, Stockholm, Sweden, 1982.

[7] Lansner, A. and Fransén, E., "Modelling Hebbian cell assemblies comprised of cortical neurons," *Network*, vol. 3, 1992, pp. 105–119.

[8] Ekeberg, Ö., Wallén, P., Lansner, A., Tråvén, H., Brodin, L., and Grillner, S., "A computer based model for realistic simulations of neural networks. I: The single neuron and synaptic interaction," *Biol. Cybernetics*, vol. 65, no. 2, 1991, pp. 81–90.

[9] Lansner, A. and Ekeberg, Ö., "A one-layer feedback, artificial neural network with a Bayesian learning rule," *Int. J. Neural Systems*, vol. 1, no. 1, 1989, pp. 77–87.

[10] Thorpe, S. and Imbert, M., "Biological constraints on connectionist modelling," in *Connectionism in Perspective* (Pfeifer, R., ed.), Berlin: Springer Verlag, 1989.

[11] Wallén, P., Buchanan, J., Grillner, S., Hill, R., Christenson, J., and Hökfelt, T., "Effects of 5-hydroxytryptamine on the afterhyperpolarization, spike frequency regulation, and oscillatory membrane properties in lamprey spinal cord neurons," *J. Neurophysiol.*, vol. 61, 1989, pp. 759–768.

[12] Fransén, E. and Lansner, A., "Modulatory effects on the operation of Hebbian cell assemblies," (in preparation), 1992.

66

SELF-TEACHING THROUGH CORRELATED INPUT

Virginia R. de Sa
Dana H. Ballard

Dept. of Computer Science, University of Rochester, Rochester, NY 14627

ABSTRACT

Previous work has shown that competitive learning coupled with a top-down teaching signal can produce compact invariant representations. In this paper we show that such a teaching signal can be derived internally from correlations between input patterns to two or more converging processing streams with feedback. Such correlations arise naturally from the structure present in natural environments. We demonstrate this process on two small but computationally difficult problems. We hypothesize that the correlations between and within sensory systems enable the learning of invariant properties.

66.1 INTRODUCTION

Unsupervised neural network learning algorithms, which are limited to classifying patterns based only on the similarity of their input representations, are unable to form position invariant and other image invariant representations. Previous work [3] has shown, using the architecture in Figure 66.2a), that competitive learning coupled with a top-down teaching signal can learn these task-relevant representations of the input patterns as shown for the XOR problem in Figure 66.1.

For this problem the network must learn to separate the two sets of input patterns, $\{(-1,+1),(+1,-1)\}$ and $\{(-1,-1),(+1,+1)\}$. Figure 66.1a) shows graphically the weights of neurons in the first layer of an unsupervised competitive network. One weight is shared between two of the closest patterns, but as these patterns are from different classes, future layers cannot separate the classes. Figure 66.1b) shows the effect of adding another weight from a teaching signal. The feedback signal adds an extra dimension to the weight space and influences the relative distances between patterns in the new augmented space allowing an appropriate representation (see Figure 66.1c).

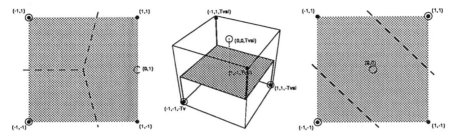

Figure 66.1 Input patterns are represented by the small dark dots, weights are represented by the larger open circles, and the dashed lines represent the partition of the input space among the neurons.
a) The patterns and weights for the XOR problem with 3 hidden-layer neurons and no teach input. The solution has allocated one weight to be shared between two of the closest patterns. The other two weights are allocated to the other patterns.
b) The patterns and weights for the XOR problem with 3 hidden-layer neurons and a teach input. The third dimension comes from the feedback from the output layer. The solution has allocated one weight to be shared between two neurons with the same feedback. The other two weights are allocated to the other patterns.
c) The result of removing the Teach input after training as in b).

The algorithm's dependence on an externally derived teaching signal, however, is unsatisfactory from a biological perspective. In this paper we demonstrate that an internal teaching signal can be derived from correlations between input patterns to several networks. That is, a collection of semantically correlated input patterns can collectively teach themselves [1].

Consider an infant learning to recognize his parents. The infant receives many visually dissimilar views of his mother and father as well as many voice samples of different words in different tones from both parents, yet he must learn to recognize his parents' faces and voices. We assume that he cannot simply memorize every visual and auditory instance (corresponding for example to one hidden layer neuron per pattern) and that he is not born with appropriate "mother's face" type feature detectors but instead must develop appropriate invariant representations. In addition he has no external teaching signal classifying each instance of visual and auditory data.

What he can use is a benevolent environment in which the sensations of the two (and more) modalities are correlated. An image of his mother's face usually appears with an instance of his mother's voice and likewise for his father's face and voice [2]. We hypothesize that these correlations between and within sensory systems can drive the development of appropriate representations in each modality[3].

---

[1]This is similar to [5] and [2] except that we do not have one modality train the others—all modalities cooperate to train themselves.

[2]There are also temporal correlations within modalities, but our algorithm currently does not make use of this information.

[3]This idea is put forth in [6], however as in [3] the simultaneous development of the sensory representation and association is not demonstrated.

66.2   ARCHITECTURE AND ALGORITHM

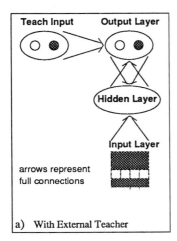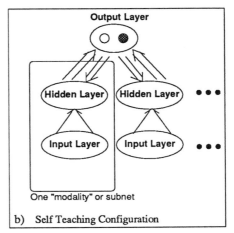

Figure 66.2   a) The architecture used in [3]. The hidden and output layers are competitive networks. The arrows represent full connections in the directions shown
b) The modified self-teaching architecture. The teaching input has been replaced by another sub-network. Together the networks train each other according to the correlations in their patterns

We demonstrate these ideas using the architecture shown in Figure 66.2b). Incoming sensations to each sensory modality are represented as activation in an input layer. They project (through separate hidden layers) to a common output layer. This common output layer projects back to all hidden layers and serves as the shared extra dimensions just as the externally taught output layer does in Figure 66.2a).

The network operates exactly as that in [3] [4] except that the output activation is determined by the activations of the hidden layers. Weight updates are similar except that the weights of the hidden layer neurons active on the upward pass (with no active feedback) are moved away from the input patterns[5].

These updates allow neurons responding to temporally correlated patterns to develop similar connections to and from the output neurons while at the same time encouraging neurons to respond to patterns receiving the same feedback. The key to the algorithm is that all patterns within one class in in each modality occur with a unique distribution of patterns in the other modalities, and these distributions serve as a type of stochastic teaching signal for each other. This problem is non-trivial though, as we are not simply

---

[4]Each layer within each modality net performs a competitive winner-take-all calculation, meaning that only one neuron is active in each layer at each time. In the output layer the winning neuron is given activation Tval (all other winning neurons are given activation 1).

[5]This is similar to LVQ[4] and can be seen as an abstracted form of the BCM [1] weight update rule in that during the upward pass (where activations are not as strong due to lack of input on the descending connections) weight updates are anti-Hebbian and during the downward pass they are Hebbian.

learning a mapping from static hidden layer representations to the output neurons, but this mapping is changing the hidden layer representations themselves. The correct representations must develop in concert with the correct connections from the hidden to the output layers.

66.3   RESULTS

This approach solves two small but computationally difficult problems. The XOR problem was tested for networks of two, three, four and five "sensory modalities". At random one of the four input patterns from the two sets of patterns $\{(-1,+1),(+1,-1)\}$ and $\{(-1,-1),(+1,+1)\}$ was presented to the first network. For the original tests, the other networks received, with equal probability, either the same pattern or the other pattern from the same set (Note that in this case all modalities receive inputs from the same pattern space).

The networks were able to learn the correct representation strictly through correlations between the two inputs. Figure 66.3a) shows the performance for different combinations of output activation strength (Tval) and number of modalities. For these graphs correct performance was defined as the development of a correct mapping for each input modality, thus higher modality nets start at a disadvantage in that there are more input nets to develop correctly. The figure shows that as the value of Tval becomes more useful [6] the nets with more modalities perform better. That is, more networks provide a more reliable output and when this output is allowed to be effective more modalities can offset their more stringent requirements [7].

Subsequent trials tested the effect of relaxing the correlations on the two-modality network. For these trials the second network received a pattern from the correct class with various probabilities [8]. Figure 66.3b) shows the graceful degradation in performance for correct correlation probabilities between 100 and 50%.

We also tested the algorithm on the task of distinguishing horizontal and vertical lines. Again both "modalities" were given inputs from the same pattern class (There were 8 possible lines in a 4×4 pixel array.) and the goal was to learn to activate a different output neuron for horizontal and vertical lines. A network of two modalities was consistently able to learn the lines problem with 6 hidden layer neurons in both sub-nets. A more efficient, with only 4 neurons per sub-net hidden layer, encoding was achieved by adding a third sub-net.

66.4   SUMMARY

In summary, experiments show that association between different input streams can

---

[6]higher values result in hidden neurons ignoring the input and lower values are not strong enough to counteract the input similarities

[7]These figures were made for simulations of 400 time steps. For longer simulations all the plots migrate up — given sufficient time all the tested combinations gave 100% correct performance over 400 trials.

[8]This corresponds to occasionally hearing father's voice while seeing mother's face.

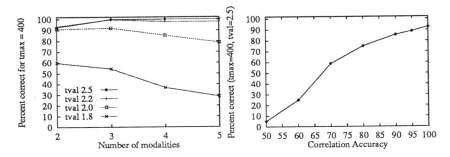

Figure 66.3 a) The effect of output activation strength Tval on convergence accuracy (percent correct performance as calculated from 400 random trials) for different numbers of modalities in the XOR problem.
b) The effect of correlation accuracy on convergence accuracy for a 2-modality net. X% of the time the second modality received a pattern from the same class and 100-X% a pattern from the other class.

influence the representations of all the streams. This problem is hard in that the network must detect correlations and simultaneously change its encoding which looks for the correlations. Thus, the whole system must bootstrap itself to achieve both the right representations and the right association.

Acknowledgements

We would like to thank Randal Nelson and Robert Jacobs for helpful conversations. This work was supported by a grant from the Human Frontier Science Program and a Canadian NSERC 1967 Science and Engineering Scholarship.

REFERENCES

[1] E. L. Bienenstock, L. N. Cooper, and P. W. Munro. Theory for the development of neuron selectivity: orientation specificity and binocular interaction in visual cortex. *J. Neurosci.*, 2:32–48, 1982.

[2] G. A. Carpenter, S. Grossberg, and J. H. Reynolds. Artmap: Supervised real-time learning and classification of nonstationary data by a self-organizing neural network. *Neural Networks*, 4:565–588, 1991.

[3] V. R. de Sa and D. H. Ballard. Top-down teaching enables task-relevant classification with competitive learning. In *IJCNN International Joint Conference on Neural Networks*, volume 3, pages III–364—III–371, 1992.

[4] T. Kohonen, G. Barna, and R. Chrisley. Statistical pattern recognition with neural networks: Benchmarking studies. In *Proc. IEEE Int. Conf. on Neural Networks, ICNN-88*, volume 1, pages I–61—I–68, 1988.

[5] P. Munro. Self-supervised learning of concepts by single units and "weakly local" representations. Technical report, School of Library and Information Science, University of Pittsburgh.

[6] E. Rolls. The representation and storage of information in neuronal networks in the primate cerebral cortex and hippocampus. In Durbin, Miall, and Mitchison, editors, *The Computing Neuron*, chapter 8, pages 125–159. Addison-Wesley, 1989.

67

MULTI-LAYER BIDIRECTIONAL AUTO-ASSOCIATORS

Dimitrios Bairaktaris

Human Communication Research Centre,
University of Edinburgh, 2 Buccleuch Place, Edinburgh EH8 9LW, Scotland

ABSTRACT

This paper presents a Content-Addressable memory network, which is based on the Bidirectional Memory network architecture and the employment of randomly generated hidden representations. The capacity and recall performance of the proposed system were investigated under different connectivity patterns. For certain conditions, the network's recall performance is invariant to connectivity levels. A multi-layer version of the proposed network allows storage and recall of correlated patterns. This is achieved by means of an expansion process which relates very closely to what it is believed to be one of the primary hippocampal functions.

67.1 INTRODUCTION

Bidirectional Associative Memories (BAMs) were firstly introduced by Kosko [1] as an efficient way of improving the performance of the basic associative memory system presented in [2]. The function of a BAM system is to associate a set of input vectors (f) to a set of output vectors (g) using the covariance encoding rule [3]. The covariance encoding rule used for BAMs cannot be directly applied to the auto-associator paradigm, because there is only a single set of memory vectors (f) available. To overcome this problem each memory pattern matched to a randomly generated (g) pattern. The architecture of a BAM network is shown in Figure 1. A BAM network comprises a number (K) of external nodes which matched the size of the memory pattern vectors, fully connected to a number (L) of internal nodes.

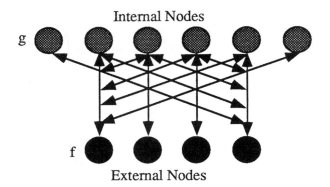

Figure 67.1   Two Layer Auto-Associator BAM network.

Association between a pair of patterns is established by changing the weight values on the connections between the external and internal nodes. The final weight matrix W is constructed incrementally by associating all M pairs of patterns:

$$\mathbf{W} = \sum_{a=1}^{M} \mathbf{W}^a = \sum_{a=1}^{M} \mathbf{f}^a \mathbf{g}^a \qquad (67.1)$$

During synchronous two-stage recall memory pattern $\mathbf{f}^a$ is presented to the external units and recall of $\mathbf{F}^a$ is as follows:

$$\mathbf{F}^a = \mathbf{W}^T \mathbf{W} \mathbf{f}^a \qquad (67.2)$$

Provided that: $\mathbf{f}^a \, \mathbf{f}^i = 0$, $\forall$ i$\neq$a, $\mathbf{F}^a = \mathbf{f}^a$ (see [2] [5]). The theoretical storage capacity (C) of the network [4] is:

$$C = \frac{\sqrt{N}}{4 \log \sqrt{N}} \qquad (67.3)$$

where N (=K L) is the total number of connections in the network. The storage capacity of the network was tested for maximum number of memories with completely error free recall (see Figure 2).

67.2   CONNECTIVITY PATTERNS

Unlike a Hopfield network, where the total number of connections is directly equivalent to the square of the number of nodes, the two layer BAM auto-

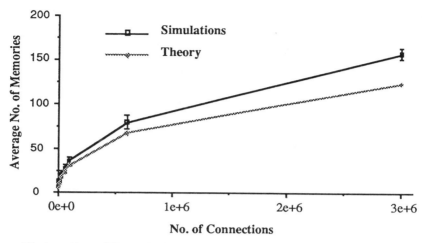

Figure 67.2 Theoretical and empirical storage capacity of a two layer BAM network.

associator network has an extra degree of freedom. A fixed number of connections can be the result of different product combinations of the number of the external and internal nodes. We define (R) to be the ratio between the number of external and internal nodes as follows: $R = \frac{K}{L}$. In our simulations we examined what effect the ratio (R) has on the storage and recall capacity of the network. The total number of connections in the networks we simulated was kept constant at N = 100000 and the network's recall performance was tested for a load of M = 100. These simulations have shown that recall performance decreases as the value of R increases. Noise in recall, due to the correlation of the memory patterns, is offset by xpanding the representation of the memory pattern in the internal layer. A similar expansion of patterns may occur between the Entorhinal Cortex (EC) and Dentata Gyrus (DG) in the human hippocampus. Unlike our fully connected network model, EC and DG are not fully connected. An approximation of a reduced connectivity pattern is achieved, by randomly removing connections between the external and internal nodes according to a varying % value. At each connectivity level the network was simulated for a range of ratio (R) $(1.1 \geq R \geq 0.001)$ values. Simulation results, shown in Exhibit 3, demonstrate that for $0.2 \geq R \geq 0.1$ the system's recall performance is invariant to the level of connectivity. This invariance ratio is very close to the one observed between areas EC and DG [6] and it may be essential during periods of changing connectivity (eg. synaptic growth).

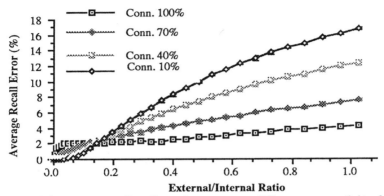

Figure 67.3 Recall performance at different connectivity levels remains constant when $0.2 \geq R \geq 0.1$.

67.3   MULTI-LAYER BAM NETWORKS

Similarly to the two layer BAM network a multi-layer BAM network with two or more levels of internal representation can be constructed. In this case associations are established only between two consecutive layers of the network as it is described in Section 2. A similar anatomy, although far more complicated, is observed in the hippocampal formation (areas EC, DG, CA3 and CA1). It was shown in [7] that multi-layer associative memory networks have increased storage capacity. Our simulations demonstrate that their extra capacity can be used to store and recall correlated memory patterns. Recall in any associative memory system is subject to noise if the memory patterns are not random. A two-layer BAM network and a three-layer BAM network were used to store a set of correlated (approx. 50% correlation) patterns. The simulation results shown in Figure 4, clearly demonstrate that the two-layer network fails to improve recall performance as the number of connections increases, whereas the three-layer network clearly improves its performance. This result suggests that the multi-layer networks, can be used for effective storage and recall of individual memories which have a significant number of features in common.

REFERENCES

[1] Kosko, B., "Bidirecional Associative Memories,"IEEE Transactions on Systems, Man, and Cybernetics, Jan/Feb 1988, Vol 18, No 1.

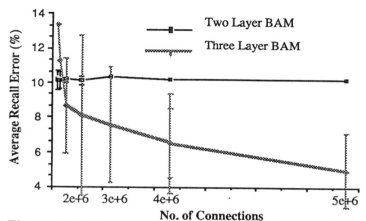

Figure 67.4 The two-layer BAM network fails to recall correlated memories. whereas a three-layer BAM network succeeds.

[2] Anderson, J.A., and Hinton, G.E., "Models of Information Processing in the Brain," in Hinton and Anderson, J.A., Parallel Models of Associative Memory, Lawrence Erlbaum Associates Inc., New Jersey 1981.

[3] Hopfield, J.J., "Neural Networks and Physical Systems with Emergent Collective Computational Abilities," Proc. National Academy of Sciences, Vol. 79, pp. 2554-2558, April 1982.

[4] McEliece, J.R., Posner, C.E., Rodemich, R.E., and Venkatesh, S.S., "The Capacity of the Hopfield Associative Memory," IEEE Transactions on Information Theory, Vol. 33, No. 4, July 1987.

[5] Bairaktaris, D., "A Model of Auto-Associative Memory that Stores and Retrieves, Successfully, Data Regardless of their Orthogonality , Randomness or Size," Proc. of the Hawaii International Conference on Systems Sciences, January 1990.

[6] McNaughton, B.L, and Nadel, L., "Hebb-Marr Networks and the Neurobiological Representation of Action in Space," in Gluck, M.A., and Rumelhart, D.E., Neuroscience and Connectionist Theory, Lawrence Erlbaum Associates Inc., New Jersey 1990.

[7] Baum, E.B., Moody, J., Wilczek, F., "Internal Representations for Associative Memory," Biological Cybernetics, Vol. 59, pp.217-228, 1988.

A NEURAL MODEL FOR A RANDOMIZED FREQUENCY-SPATIAL TRANSFORMATION

Yossi Matias

Eytan Ruppin

School of Mathematical Sciences,
Tel Aviv University, Tel Aviv 69978, Israel

ABSTRACT

We examine a random neural network model of the cortex, composed of neurons having short membrane time constants and stochastic dynamics. We show that such limited memory resources suffice for a *frequency-spatial transformation (FST)*: Depending on the frequency of the input signal, different neural assemblies generate sustained cortical activity that persists after the input stimuli is removed. These assemblies may be only indirectly connected to the input region via the cortical mesh of connections. The FST scheme proposed demonstrates that random neural networks may respond specifically to different input stimuli.

The existence of both spatial (place) and frequency (temporal) coding of information in the central nervous system has been conventionally claimed [1]. Consider a scenario where input fibers from some cortical region project upon a specific group of neurons in another cortical region. A natural question is whether information encoded at different rates of neuronal firing can cause the activation of (and thus be 'mapped' into) distinct neuronal populations, although the latter are only *indirectly* connected to the input group via the cortical mesh of connections.

This question could be answered in a straight forward manner by temporal summation, if one would assume neurons with a variety of large membrane time constants (see [1]). Yet, cortical neurons may be characterized by short membrane time constants [2], of the order of very few inter-spike intervals [3].

An alternative solution that has been proposed involves a network with a precisely wired architecture of connections incorporating delay-lines [4]. However, it is claimed [5] that evidence in support of such a broad spectrum of delays is lacking. Moreover, when viewing the architecture of some cortical regions on a local microscopic scale, a description of their geometry in statistical terms seems the only appropriate one [6].

We present a frequency-spatial transformation (FST) scheme which is compatible with cortical regions that are composed of randomly connected neurons, with essentially short membrane time constants (MTCs). It is shown that by incorporating a stochastic component into the neurons' dynamic behavior, a network with memory in a more limited sense than those previously described suffices to perform an FST. The FST scheme is presented and analyzed in a feed forward layered network, yet, simulations show that (with an appropriate setting of the neurons' threshold value) our analysis pertains also to the general case of a randomly connected network.

Consider a feed-forward network, where the input signal is applied to the first layer, denoted as the *input* group (see Fig. 68.1). The input signal consists of a train of spikes arriving at a given frequency f, separated by time interval $\Delta = 1/f$. The network is composed of memoryless neurons having short MTCs whose firing activity is a function of only the last arriving input spike. The main idea underlining the FST is that the input signal is 'spread' into signals of different frequencies, each obtained in a distinct layer; neurons at layer i of the FST network have a certain probability P_i to fire in response to an input spike arriving at the input group. Thus, the input frequency f is reduced at layer i to the average frequency $f \cdot P_i$. It is assumed that only a certain fixed frequency band, denoted as f_s, is *significant*. The way in which f_s is actually expressed in the network is discussed further on.

Each neuron in layer i is connected to a randomly selected subset of size s_i in the preceding layer $i-1$ (where 'layer 0' is the input group), and has a threshold T_i. Every neuron of the input group transmits each input spike with probability $P_0 \leq 1$. Let Y_i be the number of spikes received by a neuron in layer i, as a response to a new input spike. We assume that s_i is much smaller than the number of nodes and therefore there is very small dependency between nodes in the same layer. Therefore, Y_i is approximately a Binomial with parameters (s_i, P_{i-1}). Using the Normal approximation to the Binomial distribution, we have $P_i \approx \text{Prob}\,(Y_i > T_i) \approx \frac{1}{\sqrt{2\pi}} \int_{t_i}^{\infty} e^{-y^2/2} dy$, where $t_i = \frac{T_i - s_i P_{i-1}}{\sqrt{s_i P_{i-1}(1 - P_{i-1})}}$. Assuming that the neurons have similar parameters, a monotonically conforming FST is obtained, ensuring the uniqueness of the frequency-spatial mapping.

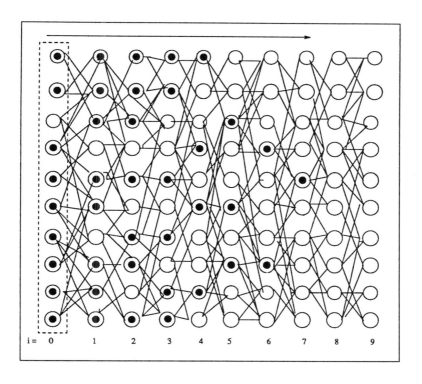

Figure 68.1 A schematic drawing of an FST in a feed-forward layered network. The input layer ($i = 0$) is marked on the left. Every neuron has a randomly selected set of s synaptic connections, projecting from neurons of the previous layer. The spike propagation is illustrated by marking in black the neurons that were activated by a given spike of the signal train. With every input spike, a similar scenario of neuronal activity, gradually decaying with its propagation along the network is evident, leading to a continuing decrease of the firing frequency measured along the network.

We propose that sustained activity is generated in a layer if its neurons' expected firing frequency is within a *significant frequency band* $f_s(1 \pm \epsilon)$. We denote such layer the *activated layer*. Our proposal is based on the assumption that cortical neurons have a superexcitability period of length $\Delta_s = \frac{1}{f_s}$. Such superexcitability may originate from the neuron's intrinsic membrane conductance properties [7] or from self-excitatory cycles (involving a few neurons) with a period of length Δ_s. Such cycles may be composed of vertical connections existing between neighboring connections existing between neighboring neurons belonging to different cortical layers, i.e., orthogonal to the propagation of activity along the cortical sheet. (It should be noted that if a random network is sparse enough, cycles of a short length are indeed predominant.) The Δ_s periodic enhancement of the neurons' firing probability leads to a situation where the number of neurons in the activated layer firing *synchronously* at frequency f_s increases until some critical mass is achieved, and the firing becomes stimulus independent. Recalling that the frequencies obtained are average rather than exact, an activated layer indeed has the highest chance to become active: Suppose that k evenly spaced spikes, separated by intervals of Δ_s, are required for sustained activity to emerge. It is easy to see that the probability P_s of such sustained activity to occur is highest at the activated layer, and that the ratio $\frac{P_s(l)}{P_s(j)}$ (l is the activated layer and $j \neq l$) increases exponentially in k.

To examine the performance of the FST scheme in a cortical-like random network, several simulations have been carried out, performed on the random network sketched in figure 68.2. Each neuron has a set of s incoming connections, determined randomly with probability $\phi(z) = \sqrt{1/(2\pi)} \exp(-z^2/2)$, where z is the distance between the neurons. All neurons have an identical threshold. The input signal is applied to some spatially continuous group of neurons, and the resulting firing frequency is measured at consecutive, equally spaced distances from the input group. Every input spike activates several neural assemblies as it propagates along the network. These dynamically generated sets of synchronously firing neurons may be considered as 'layers' in a virtual layered network. The simulation results show that a monotonic FST is obtained. However, the firing frequency 'spread' achieved does not have a uniform rate, as most of the frequency decrease occurs along a fairly narrow region of the network. As shown in figure 68.2, a more homogeneous spreading of the input frequency along the network is achieved by replacing the neurons' threshold function by the sigmoidal non-linearity $f(x) = 1/(1 + \exp[-(x - T/g)])$. As the sigmoidal gain g is increased, the rate of frequency decrease along the network is attenuated. When the gain is increased further, the FST mapping does not remain monotonic any more. A fairly homogeneous frequency decrease is maintained even when each neuron has a distinct threshold and gain

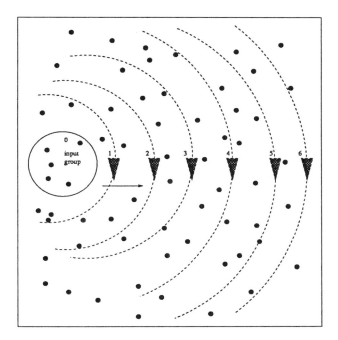

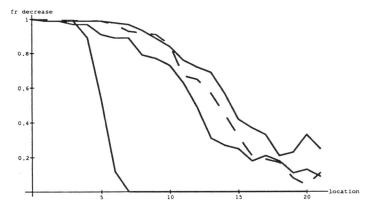

Figure 68.2 Top: A schematic drawing of part of a random network. The input group is circled on the left. The wedge-shaped icons denote the location of the simulated 'electrode', measuring the average firing frequency at increasing distances $1d, 2d, 3d \ldots$. The dotted arcs stand for the dynamically generated virtual 'feed-forward layers'. Bottom: A plot of the average frequency obtained (y-axis) as a function of the spatial distance nd between the measured region and the input group (x-axis). The three full curves display (from left to right) the frequency reduction obtained with a sigmoidal function, with gain values $g = 1.1, 1.5, 1.9$ correspondingly. The dashed curve presents the FST obtained with $\sim$ 10% perturbations of the mean threshold and gain values.

value determined randomly within some bounded interval. The proposed FST scheme demonstrates the potential role of randomness, both in structure and in dynamics, in cortical information processing.

REFERENCES

[1] E.R. Kandel and J.H. Schwartz. *Principles of Neural Science*. Elsevier, 1985.

[2] M. Abeles. *Local Cortical Circuits, An Electrophysiological Study*. Springer-Verlag, 1982.

[3] T.J. Sejnowski. Open questions about computation in cerebral cortex. In J. McLelland and D. Rummelhart, editors, *Parallel Distributed Processing*, volume 2, pages 372–389. MIT Press, 1986.

[4] D.C. Tam. Temporal-spatial coding transformation: Conversion of frequency-code to place-code via a time-delayed neural network. In *International Joint Conference on Neural Networks*, pages I–130–133, January 1990.

[5] S. Shamma. Spatial and temporal processing in central auditory networks. In C. Koch and I. Segev, editors, *Methods in Neuronal Modeling*. MIT press, 1989.

[6] V. Braitenberg and A. Schuz. *Anatomy of the Cortex: Statistics and Geometry*. Springer-Verlag, 1991.

[7] S.A. Raymond and J.Y. Levittin. *Aftereffects of activity in peripheral axons as a clue to nervous coding*. Raven Press, 1978.

69

OUTLINE OF A THEORY OF ISOCORTEX

Mark R. James
Doan B. Hoang

University of Sydney, NSW 2006, Australia

ABSTRACT

A complete understanding of brain function requires a description of three levels of its anatomy: The input/output and adaptive behaviour of its cells, the local neural networks of subcortical nuclei and cortical areas, and the ways in which these specialised brain regions interact. Here, we focus on the second of these, and outline an anatomically plausible model of six-layered, homotypic cortex (isocortex) that includes mechanisms for stable pattern selectivity and sequence learning. Important information processing concepts elicited include automatic gain control, the decoupling of discrimination and association, the role of top-down processing, and buffering for unidirectional temporal chaining.

1 NEURONAL CONNECTIVITY

Cortex is viewed as a three-dimensional array of interacting modules, each the diameter of a minicolumn (Mountcastle, 1978), and the depth of a cortical layer. The pattern of intrinsic and extrinsic connections between modules in a cortical area (e.g. V4, MT) is outlined in Fig. 1a. This has been distilled from the work of neuroanatomists such as Lund, Burkhalter, Wiesel, and Gilbert. The synaptic strength of all excitatory synapses are considered modifiable while the strength of inhibitory synapses are considered fixed, although both are susceptible to activity-dependent pruning during development.

2   The L4-L6 Neural Field

Afferents entering a cortical area form excitatory synapses with layer 4 (L4) and layer 6 (L6) modules ($\mathcal{E}$ and $\mathcal{F}$ on figure respectively) over a significant portion of the area. The final shape and size of individual afferent projections is determined by activity dependent connection pruning during development. L4 modules send an excitatory connection to the L6 module in the same minicolumn ($\mathcal{A}$), and send short-range (up to 250 μm) excitatory projections to other L4 modules ($\mathcal{H}$). L6 modules send a bowl-shaped inhibitory projection to L4 that tapers off approximately 1 mm from its home minicolumn ($\mathcal{B}$). This projection includes a strong inhibitory projection to the L4 module in the same minicolumn. L6 modules also form long range excitatory synapses with each other ($\mathcal{C}$), effectively sending the inhibition further afield. L6 also sends excitatory connections to wide regions of layer 1 (L1) of the cortical areas from which the afferents origninated ($\mathcal{D}$), contacting in layer 1 the apical dendrites of neurons in layer 2/3 (L3), layer 5 (L5).

The combination of the cooperative ($\mathcal{H}$) and competitive ($\mathcal{B}$ – via $\mathcal{A}$, $\mathcal{F}$, and $\mathcal{C}$) interactions in L4 produce a level of competition between L4 modules that is small between adjacent modules, reaches a peak at a range of about 500 μm, and tapers off beyond that. This short and long range competition ensures different L4 modules become tuned to different patterns of afferents. The strong competition at 500 μm leads to the patchy pattern of afferent connectivity that is often seen.

In competitive neural network models, synaptic weight saturation due to avalanche competitive selection (as synapses strengthen) must be avoided. One biologically plausible model that avoids weight saturation is the BCM model (Bienenstock et al., 1982). In this model, synaptic strengths of recently active synapses are either potentiated or depressed as a function of current postsynaptic membrane voltage. As well, the threshold voltage for synaptic potentiation is a increasing function of recent postsynaptic activity. The existence of a threshold for the direction of synaptic weight change has been observed in the visual cortex (Artola et al., 1990), however clear evidence for a molecular mechanism implementing the sliding threshold (Bear et al., 1987) has not been seen. Also, as pointed out by Grossberg (1982), the activity dependent sliding threshold can lead to unstable pattern coding.

In the model described here, the problems with the activity dependent threshold are solved by using adaptive positive ($\mathcal{D}$) and negative (L6 to L4 in the same minicolumn) feedback loops to set the proper operating point for the L4 neurons.

The positive feedback to the areas from which the afferents originated ensures that significant activity in those areas leads to a significant, but capped, level of superthreshold activation in L4. The wide extent of the feedback ensures that input level relativities are maintained. These feedback projections can be adaptive to allow learning of top-town influences. The L1 synapses are far from the cell body, resulting in a weak and slow action that mimics a threshold shift.

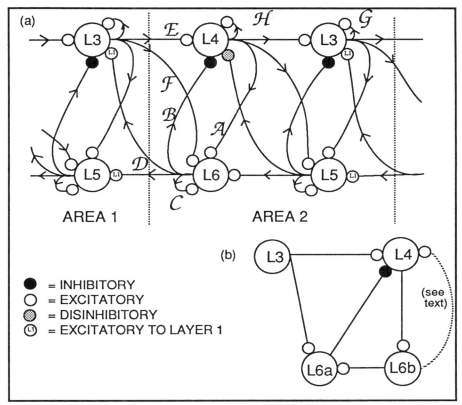

Figure 1 Interactions between cortical layers

The negative feedback from L6 to L4 acts with the feedforward connections to L6 ($\mathcal{F}$) to hold constant the output of the L4 cell against variations in the mean strength of its afferent input by normalising the total synaptic weight of the afferent synapses.

Recent work by Barrow and Budd (in press) has shown that a neuron under adaptive recurrent inhibition displays output normalization when a simple homosynaptic potentiation plus decay learning rule is used. Such a learning rule is not suitable for long term memory storage in the neocortex. Using the mathematical method of Barrow and Budd, we found that for normalization to be observed under the learning rule used here (postsynaptically controlled homosynaptic potentiation or depression), not only must the feedforward excitation $\mathcal{F}$ be present, but details of the layer 6 connectivity must be considered to ensure stability. The circuit shown in Fig. 1b learns to normalise the output of the L4 cell by balancing the levels of excitation and inhibition reaching the cell.

While there is evidence for positive feedback from L6 to L4 (Ferster and Lindström, 1985), (see Fig. 1b), which may play a role in amplifying afferent input, there is also evidence for the adaptive inhibition described here. Cortical excitatory neurons fire at slow rates, suggesting a fine balance of excitation and inhibition; receptive fields of visual neurons are initially large after an afferent burst, and then quickly become more selective (Dinse *et al.*, 1991); and the adaptation of neurons to prolonged input. The inhibition is likely mediated, in part, by $GABA_B$ receptors.

As observed in cortical activation studies, the response of an area to more familiar inputs will be more restricted than that to novel inputs. As a pattern is repeatedly presented, those modules that respond weakest will get their active synapses weakened, sharpening subsequent responses. Weakened synapses can be eliminated during development. No heterosynaptic depression is employed to prevent forgetting of old patterns which have not recently occurred.

Not shown in the figure are cells implementing local feedforward and feedback inhibition, and basket cells. The respective function of theses cells could be to allow effective afferent strengths to fall to and below zero, to mediate disinhibition and gain control of weak inputs, and to bootstrap and add noise to competitive interactions.

3   The L3-L5 Neural Field

The pattern of L4-L6 connectivity re-occurs in a second processing stage in each area, in which L4 axons take on the role of the afferents, layer 2/3 (L3) as L4, and layer 5 (L5) as L6. Thus L5 acts as an automatic gain controller (AGC) of the L3 modules. There is a similar substructure to L5 as was shown for L6 in Fig. 1b.

There are two connectivity differences within this second processing field:

1. Neurons within an L3 module form excitatory synapses with each other to implement short term memory (STM) via recurrent excitation. The persistence of this excitation will be dependent on the ability of the neurons to integrate the excitation. Thus, the longest persistence will be in the "magnopyramidal" cortical areas. There are long intrinsic connections between L3 modules, which implement pattern associations. These connections are paralleled by long L3 to L5 connections (Ojima *et al.*, 1990) to balance them for AGC purposes. There are no long range connections between L5 modules. The afferent inputs from L4 are stronger than the associative inputs. The L4 to L3 projection is narrow.

2. The feedback connection from L5 to L4 is disinhibitory rather than excitatory so that L4 activity is only modulated rather than created. This input can help form topographic maps in L4.

Thus each cortical area has two processing stages. The L4-6 bottom-up feedforward system which performs a sparse recoding of afferent patterns, independent of top-down and associative influences; and the L3-5 associative memory with short term

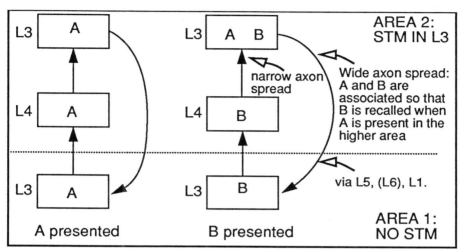

Figure 2 Sequence Recall

memory capabilities. Layer 4-6 can act as a buffer between two L3s, allowing learning and recall of sequences by limiting pair associations to those forward in time (Fig. 2). Often L5 contributes to the inter-areal back projections, eliminating the need to go through L6.

4 Orienting Attention

The large cells of Meynert in layer 5 form a separate system. They encode patterns that warrant the capturing of attention, and send their output to the cortical attention fields. Some neuromodulatory feedback, to flag interesting or boring events, facilitates weight modification on these neurons (e.g. nicotinic ACh).

References

Artola, A., *et al. Nature* **347** (1990), 69-72.

Barrow, H. G., and Budd, J. M. L. In Alexander, I., and Taylor, J. (Eds.) *Artificial Neural Networks II: Proceedings of the International Conference on Artificial Neural Networks.* Elsevier, in press.

Bear, M. F., *et al. Science* **237** (1987), 42-48.

Bienenstock, E. L. *et al. J. Neuroscience* **2** (1982), 32-48.

Dinse, H. R., *et al.* In Kruger, J. (Ed.) *Neuronal Cooperativity.* Springer-Verlag, 1991, pp. 68-104.

Ferster, D., and Lindström, S. *J. Physiol. (London),* **367** (1985), 233-252.

Grossberg, S. In Amari, S., and Arbib, M. A. (Eds.) *Competition and Cooperation in Neural Nets.* Springer-Verlag, 1982.

Mountcastle V.B. In *The Mindful Brain.* The MIT Press, Cambridge, MA, 1978, pp. 7-50.

Ojima, H., *et al. Society for Neuroscience Abstracts* **16** (1990), 721.

A NETWORK FOR SEMANTIC AND EPISODIC ASSOCIATIONS SHOWING DISTURBANCES DUE TO NEURAL LOSS

Michael Herrmann[1]

Eytan Ruppin[2]

Marius Usher[3]

[1] *Dept. of Computer Science, Leipzig University, F.R.G.*
[2] *School of Mathematical Sciences,*
Tel Aviv University, Tel Aviv 69978, Israel.
[3] *CNS 216-76, Caltech, Pasadena, CA 91125.*

ABSTRACT

We study an Attractor Neural Network that stores natural concepts, organized in semantic classes. The concepts are related by both semantic and episodic associations. When neurons characterized by large synaptic connectivity are deleted, semantic transitions among concepts decay before the episodic ones, in accordance with the findings in patients with Alzheimer's disease.

In this work we propose a model for the dynamics of concepts activation, that is biologically motivated and generates some basic characteristics of normal memory function, such as semantic and episodic associations. Our model is based on a Transient Attractor Neural Network. It exhibits successive stages of competition and *spreading of activation* among concepts' activities, accounting thus for both the serial and parallel aspects of thought processes.

We consider a network of N neurons, characterized by two–valued variables $S_i \in \{0, 1\}$ corresponding to a non–active or active state. Each neuron is subject to a dynamical threshold variable θ_i [2]. According to our model, the neurons correspond to properties or attributes, and the distributed patterns of neural activity stand for concepts. Semantic classes are represented by vectors $\xi^\mu = (\xi_1^\mu, \ldots, \xi_N^\mu)$, where $\xi_i^\mu = 1$ if at least one of the concepts in the μ–th class has the i–th property, and $\xi_i^\mu = 0$ otherwise. Similarly, we represent specific concepts by vectors $\xi^{\mu\nu}$, which are stochastically selected, but subject to the constraint that concepts in different classes have very few attributes in common.

While patterns in the same semantic class have more common attributes, there is a very low probability that a specific attribute will be shared by all the concepts in the class. The strengths of the synaptic connections are defined as in [3]:

$$J_{ij} = \frac{1}{(1-p)pN} \sum_{\mu=1}^{L_1} \sum_{\nu=1}^{L_2} (\xi_i^{\mu\nu} - p)(\xi_j^{\mu\nu} - p) \tag{70.1}$$

The dynamic equations for the variables S_i are given via the post synaptic potentials $h_i(t+1) = \sum_{j=1}^{N} J_{ij} S_j(t) - \theta^0 - \theta_i(t) - \lambda(p - M(t))$.

$$S_i(t+1) = \begin{cases} 1 & \text{with probability} 1/(1 + exp(-h_i(t+1)/T)) \\ 0 & \text{with probability } 1/(1 + exp(h_i(t+1)/T)) \end{cases}, \tag{70.2}$$

where T is the thermal noise level, θ^0 denotes a constant threshold and λ is a confinement parameter which regulates the network overall activity $M = \langle S_i \rangle$ at the level of the mean activity of the patterns, p.

The equation for threshold dynamics is [2] :

$$\theta_i(t+1) = \theta_i(t)/c + bS_i(t+1) \tag{70.3}$$

According to this equation, while a neuron is active, its dynamic threshold $\theta_i(t)$ increases asymptotically to the value $\theta_{max} = cb/(c-1)$ and deactivates the corresponding neuron, thus accounting for neuronal fatigue. The parameters b and c ($c > 1$) represent the rate of increase of the dynamic thresholds and of the threshold's decay. The role of the dynamic thresholds is to provide a mechanism of motion in the concept space; neurons that are active for a relatively long time are deactivated temporarily, and the network's state evolves into a new pattern.

We define the normalized retrieval qualities of the patterns as

$$m^{\mu\nu}(t) = \frac{1}{p(1-p)N} \sum_{i=1}^{N} (\xi_i^{\mu\nu} - p)S_i(t) \tag{70.4}$$

The latter are macroscopic thermodynamic variables that are the relevant correlates of the concepts' activation in the network. A concept will be considered *activated* when its corresponding retrieval quality exceeds some *consciousness threshold* (that is chosen to be 0.9), whereas the retrieval qualities of all other patterns are below an *unconsciousness threshold* (chosen to be 0.5).

The dynamic behavior of the model is illustrated by simulations of a network which stores 3 families of 3 concepts each, i.e. $L_1 = L_2 = 3$. A characteristic

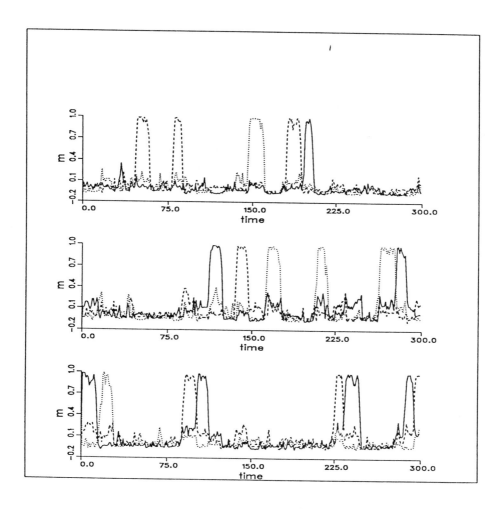

Figure 70.1 Illustration of the dynamic behavior of the model. The Y-axes represent retrival qualities and the X–axis represent time. Semantic bias is evident.

example in terms of the retrieval qualities is shown in Fig. 1. Each stripe presents the retrieval qualities of concepts in a different semantic class. Due to the competition among the patterns, most of the time the network's activity is dominated by one pattern, leading to a seriality effect. Due to threshold adaptation, the activity of the dominant pattern decays and the network's state converges to another pattern. Semantic transitions in the network are more frequent than the random transitions occurring across classes, leading to a *semantic bias* in associative transitions, which is characteristic of human memory function. We performed a numerical analysis of the network behavior in which we have varied the temperature T. The average fraction of intraclass transitions turned out to be higher for low noise (see Fig. 2.). For high T values, the fraction of semantic transitions approaches the baseline of randomness (represented by a dotted line in Fig 2.), given by $(L_2 - 1)/(L_1 L_2 - 1) = 1/4$. The deterioration of the semantic bias as the temperature increases may be of significance, in light of the claims that aminergic modulation of the responsiveness of target cortical neurons (modeled as a temperature change) may generate 'loosening of associations' in REM dreams [1].

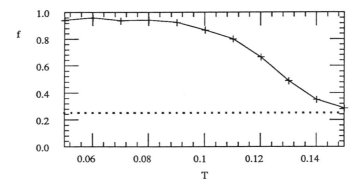

Figure 70.2 Average fraction of semantic transitions. The Y Axis represents the fraction of semantic transitions and the X axis represents T. Dotted line gives the baseline of random tansitions

To incorporate episodic associations, we have added a new term to the synaptic connectivity matrix described in equation 70.1, generating explicit projections that link the corresponding patterns. The projection strength κ is determined so that in an undamaged network 'semantic' and 'episodic' transitions occur with the same probability. As an approximation we choose κ equal to the incoming field of a neuron belonging to an nonactivated pattern in the same class of the currently activated pattern, i.e. $\kappa = p_1^2 p_2 / (1 - p_1)$. A random deletion of some fraction of the neurons leads to a concomitant decrease in both the semantic and episodic transitions to noise level, without any advantage to any specific type of transitions. Following the findings that neuronal loss in Alzheimer's disease (AD is primarily limited to a specific subpopulation of large neurons [4], we have conducted another simulation, where only neurons with a large connectivity tree (i.e., whose sum of excitatory connections was large) were deleted. We observe that as neuronal deletion proceeds, the fraction of semantic transitions out of the total number of transitions occurring in the network, actually rises (Fig3). It was also found that semantic transitions deteriorate earlier than the episodic ones as AD progresses [5]. However, although a rise in the fraction of episodic transitions is observed, it should be noted that that the absolute number of all types of transitions constantly decreases. This decrease may account for the paucity of speech and thought production observed in advanced stages of AD [6].

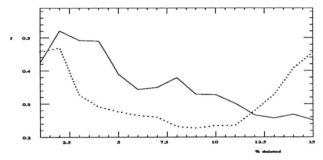

Figure 70.3 The relative fraction of semantic and episodic transitions when large neurons are specifically deleted. Full (dotted) line shows the fraction of semantic (episodic) transitions

REFERENCES

[1] Mamelak N., & Hobson J. A. (1989) Dream bizarreness as the cognitive correlate of alterated neuronal behavior in REM sleep. *Jour. of Cognitive Neuroscience* **1**, 3, 201–222.

[2] Horn D., & Usher M. (1989) Neural Networks with Dynamic Thresholds. *Phys. Rev. A*, **40**, 1036–1044.

[3] Tsodyks M. V. (1990) Hierarchical Associative Memory in Neural Networks with Low Activity Level. *Mod. Phys. Lett. B* **4**, 259–265.

[4] Hyman B. T., Van Hoesen G. W., Damasio A. R., & Barnes C. L. (1984) *Science* **225** 1168–1170.

[5] Granholm E., & Buters N. (1988) Associative encoding and retrieval in Alzheimer's and Huntington's disease. *Brain–Cogn.* **7(3)**, 335–347.

[6] Adams R. D., & Victor M. (1989) *Principles of Neurology.* McGraw–Hill, New York.

SYNAPTIC DELETION AND COMPENSATION IN ALZHEIMER'S DISEASE: A NEURAL MODEL

David Horn

Eytan Ruppin

Marius Usher

Michael Herrmann

*School of Mathematical Sciences,
Tel Aviv University, Tel Aviv 69978, Israel*

ABSTRACT

We use a neural network model to investigate how the interplay between synaptic deletion and compensation determines the pattern of memory deterioration, a clinical hallmark of AD. We show that, in parallel with the experimental data, memory deterioration can be much delayed by strengthening the remaining synaptic weights. Using different dependencies of the compensatory strengthening on the amount of synaptic deletion various compensation strategies can be defined, corresponding to the observed variation in the progression of AD.

The clinical course of Alzheimer's disease (AD) is generally characterized by a progressive deterioration of the patient's clinical status. However, both slow and rapid progressive forms of AD have been reported, exhibiting a large variation in the rate of the disease progression [1]. Recent neuroanatomical studies have found a significant decrease in the synapse to neuron ratio in AD patients, due to *synaptic deletion*. *Synaptic compensation*, manifested by an increase of the synaptic size, was found to take place concomittantly, reflecting a functional compensatory increase of synaptic efficacy [2].

Concentrating on memory degradation, a clinical hallmark of AD, we investigate these synaptic changes in a neural network model of associative memory. Our model is based on the feedback network proposed by Tsodyks & Feigel'man (TF) [3], where all N neurons have a uniform positive threshold T. Each neuron is described by a binary variable $V = \{1, 0\}$ denoting an active (firing) or passive (quiescent) state, respectively. The postsynaptic potential of neuron i is

$$h_i = \sum_{j \neq i}^{N} W_{ij} V_j. \tag{71.1}$$

$M = \alpha N$ distributed memory patterns η^{μ} are stored in the network. Every one of the neurons composing each memory pattern is chosen to be 1 (0) with probability p $(1 - p)$ respectively, with $p \ll 1$. The weights of the synaptic connections are

$$W_{ij} = \frac{1}{N} \sum_{\mu=1}^{M} (\eta^{\mu}{}_i - p)(\eta^{\mu}{}_j - p). \tag{71.2}$$

Initially, in the network's 'pre-morbid' state, the memories have maximal stability, achieved by choosing $T = p(1 - p)(\frac{1}{2} - p)$. This *optimal threshold* lies in the middle between the two possible mean values of the postsynaptic potentials.

We model synaptic deletion by randomly deleting some of the incoming synapses of every neuron, leaving each neuron with $l = (1-d)N$ input connections, where $d < 1$ is the *deletion factor*. Synaptic deletion results in deterioration of the network's performance that can be counteracted by synaptic compensation, modeled by multiplying the weights of the remaining synaptic connections by a *compensation factor* $c > 1$. The *optimal performance compensation* (OPC) factor is $c = \frac{1}{1-d} = \frac{N}{l}$ since it leads back to the original postsynaptic potential values, but with larger variance. The OPC strategy is biologically motivated by studies of biopsied cortical tissue of mild AD patients, indicating that the increase in synaptic size suffices to maintain the total synaptic contact area per unit volume (TSA) at early stages of the disease [4].

Both the case of deletion without compensation and the OPC belong to a class of *fixed k* compensatory strategies, defined by

$$\hat{W}_{ij} = cW_{ij} \qquad c = 1 + (\frac{1}{1-d} - 1)k = 1 + \frac{dk}{1-d} \tag{71.3}$$

where $k \leq 1$ is a positive constant determining the compensation magnitude. These strategies all display a similar sharp transition from the memory-retrieval phase to a non-retrieval phase, as shown in figure 71.1. Varying the compensation magnitude k merely shifts the location of the transition region. For every finite network this transition manifests itself as a range of d in which the network's performance deteriorates, as displayed in figure 71.2.

Young AD patients are likely to have high compensation capacities, and therefore can implement an OPC strategy ($k = 1$, in figure 71.1). Indeed, they have

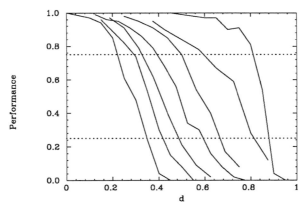

Figure 71.1 Performance of a network with fixed k compensation. The initial state is a corrupted version of a stored memory pattern, and the percentage of successful convergence to the correct memory is measured. The simulation parameters are $N = 800$ neurons, $\alpha = 0.05$ and $p = 0.1$. The curves represent (from left to right) the performance of fixed strategies with increasing k values, for $k = 0, 0.25, 0.375, 0.5, 0.625, 0.75, 1$. The horizontal dotted lines represent performance levels of 25% and 75%.

been reported to have a rapid clinical progression, accompanied by severe neuronal and synaptic loss [5]. A similar clinical pattern of rapid memory decline, already manifested with less severe neuroanatomical pathology, was found in very old patients. We therefore propose that in these old patients, the rapid clinical decline results from a severe lack of compensatory resources ($k = 0$, in figure 71.1).

Rapid cognitive decline characterizes a minority of AD patients. Most patients show a continuous, gradual pattern of cognitive decline [6], taking place along *a broad span* of synaptic deletion [4]. Hence, this performance decline cannot be accounted for by a standard neural network model with no synaptic compensation, or by a network employing fixed compensation. We propose that most AD patients employ a *variable* compensation strategy, where k is varied with d so that the system stays at the upper boundary of the critical region of figure 71.2, thus avoiding, or postponing, the transition in the network's performance. Such an *optimal resource compensation* strategy (ORC) may be necessary in face of restricted synaptic regeneration resources. Another form of variable compensation may lead to the 'plateau' phenomenon that has been reported clinically [7]. Examples of the performance with gradually decreasing and 'plateau' strategies are presented in figure 71.3.

The variable compensation strategies which we have discussed rely on the fact

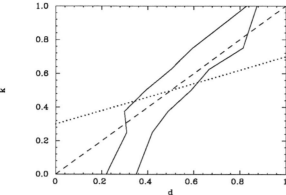

Figure 71.2 The critical transition range in the (k, d) plane. The full curves represent performance levels of 75% and 25%. The straight lines describe the variations employed in two variable compensations presented in figure 4.

that there is some span in the (k, d) plane over which deterioration takes place, as shown in figure 3. It can be shown that even in the infinite size limit, the corresponding performance curve will always retain a finite width, if the initial condition is defined as a spectrum of noisy inputs. Variable compensatory strategies may hence play an important role in maintaining partial performance capabilities in the brain.

Synaptic deletion and compensatory mechanisms play a major role also in the pathogenesis of Parkinson disease [8]. The significant incidence of AD patients having accompanying extra-pyramidal parkinsonian signs [9] naturally raises the possibility that such patients may have a decreased synaptic compensatory potential in general. Moreover, the cognitive deterioration of these AD patients is faster than that of AD patients without extra-pyramidal signs. This fits well with our proposal that a severely deteriorated synaptic compensation capacity leads to an accelerated rate of cognitive decline in AD patients.

We have also incorporated neuronal loss (which is known to occur at a level below 10% [6]) in our simulations in addition to synaptic deletions. This did not change the qualitative results. We conclude therefore that the important factors are indeed the number of synapses retained and the compensation strategy employed, whose interplay may lead to various patterns of performance decline. Although being defined 'globally' as 'strategies', it should be noted that synaptic compensation may take place via local feedback mechanisms: the decreased firing rate of a neuron being gradually deprived of its synaptic inputs due to the progressing deletion, may enhance the activity of cellular processes strengthening its remaining synapses. Indeed, in our model, the decline in the network's

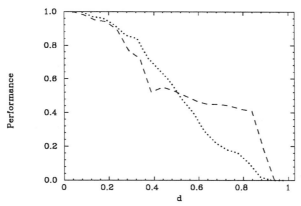

Figure 71.3   Performance of a network with *variable* compensation. The dotted, gradually decreasing curve, is obtained with the variation $k = 0.3 + 0.4d$. The dashed curve, manifesting a 'plateau', follows from the rule $k = d$.

performance resulting from synaptic deletion is coupled with the decrease in the network's overall activity. Finally, we wish to emphasize that neuroanatomical data are gathered via cortical biopsies, performed in the treatment of some AD patients. Hence, our model can be examined experimentally by obtaining a detailed record of the relation between synaptic changes and memory decline, as AD progresses.

REFERENCES

[1] D. A. Drachman, B. F. O'Donnell, R. A. Lew, and J.M. Swearer. The prognosis in alzheimer's disease. *Arch. Neurol.*, 47:851–856, 1990.

[2] C. Bertoni-Freddari, P. Fattoretti, T. Casoli, W. Meier-Ruge, and J. Ulrich. Morphological adaptive response of the synaptic junctional zones in the human dentate gyrus during aging and alzheimer's disease. *Brain Research*, 517:69–75, 1990.

[3] M.V. Tsodyks and M.V. Feigel'man. The enhanced storage capacity in neural networks with low activity level. *Europhys. Lett.*, 6:101 – 105, 1988.

[4] S. T. DeKosky and S.W. Scheff. Synapse loss in frontal cortex biopsies in alzheimer's disease: Correlation with cognitive severity. *Ann. Neurology*, 27(5):457–464, 1990.

[5] L. A. Hansen, R. DeTeresa, P. Davies, and R.D. Terry. Neocortical morphometry, lesion counts, and choline acetyltransferase levels in the age spectrum of alzheimer's disease. *Neurology*, 38:48–54, 1988.

[6] R. Katzman. Alzheimer's disease. *New England Journal of Medicine*, 314(15):964–973, 1986.

[7] R. Katzman. Clinical presentation of the course of alzheimer's disease: the atypical patient. *Interdiscipl. Topics. Gerontol.*, 20:12–18, 1985.

[8] D.B. Calne and M.J. Zigmond. Compensatory mechanisms in degenerative neurologic diseases. *Arch. Neurol.*, 48:361–363, 1991.

[9] Y. Stern, D. Hesdorffer, M. Sano, and R. Mayeux. Measurement and prediction of functional capacity in alzhemer's disease. *Neurology*, 40:8–14, 1990.

SECTION 10

CORTICAL DYNAMICS

Recently, oscillations and synchrony in cortical field potentials have resurfaced as interesting phenomena related to perception, learning and information processing. Oscillations were first described in the olfactory system by Adrian in the 1950's, and the literature on olfactory processing continues to include numerous papers on oscillations (see chapter 40 in section 5). The section concludes with a chapter on dissipative structure formation, a class of behavior between simple attractor dynamics and chaos, as a basis for information processing in the brain.

EFFECTS OF INPUT SYNCHRONY ON THE RESPONSE OF A MODEL NEURON

Venkatesh N. Murthy and Eberhard E. Fetz

Physiology and Biophysics, SJ-40
University of Washington, Seattle, WA 98195

ABSTRACT

For a model neuron with 3 conductances, we studied the dependence of the average spike rate on the degree of synchrony in its synaptic inputs. The effect of synchrony was determined as a function of three parameters: number of inputs, average input frequency and the size of unitary EPSPs.

1 INTRODUCTION

It is generally believed that the firing rates of neurons carry information, but additional information could also be coded in the relative timing of spikes among groups of neurons. In the latter mechanism, synchronous activity, i.e., the near-simultaneous occurrence of spikes in multiple neurons, is of particular relevance. Indeed, cross-correlation histograms between spike trains from mammalian neocortical neurons often exhibit a central peak [4]. Recently, widespread synchronous cortical activity has been observed in the visual cortex of cats [3, 5] and in sensorimotor cortex of awake monkeys [8]. Afferent inputs to a neuron clearly cause more effective depolarization if they arrive synchronously. Postsynaptic potentials have finite time constants and their amplitudes are often a small fraction of the threshold depolarization necessary to evoke a spike. Therefore, near-simultaneous occurrence of excitatory postsynaptic potentials (EPSPs) would be more effective in depolarizing cells to threshold than asynchronous arrival. Relatively few studies have investigated this mechanism parametrically, using simulations [1, 2, 6, 9].

A primary aim of our simulations was to determine quantitatively how the output of a biophysically realistic model neuron depends on the degree of synchrony in its inputs. We systematically explored the effect of four basic parameters on the average output frequency (f_o): (1) **N**, the number of active inputs converging on the model neuron, (2) f_i, the average frequency of the inputs, (3) **w**, the synaptic strength, or equivalently, the size of the unitary EPSP, (4) **s**, the degree of synchrony among the afferents (varied from 0 to 100%).

2 METHODS

The simulated neuron modeled a cortical pyramidal neuron with five compartments, a soma and four apical dendritic compartments. The membrane time constant was 15 msec. Three Hodgkin-Huxley type membrane currents were included in the soma: fast sodium, delayed rectifier potassium and an A-like potassium current to allow low-frequency firing in response to injected current. The dendritic compartments had no active currents. The parameters for the kinetics of the currents were similar to those of the three-channel model of Lytton and Sejnowski [7]. The maximal conductances of the three types of channels were adjusted to reproduce action potential waveforms and frequency-current relation seen in cortical pyramidal neurons. The intrinsic parameters of the model were then fixed at these values for all further simulations. Excitatory synaptic inputs occurred on the second dendritic compartment and produced a conductance change with a timecourse similar to the fast glutamate synapses in the cortex. The strength of the input connections was varied by changing the maximal synaptic conductance. Inhibitory inputs were not included in these simulations. Each input was modeled as a Poisson spike train; all inputs had the same mean rate in a given simulation. Synchronization was simulated by lumping the synchronized inputs into one source, whose strength was increased proportionately. The synchrony among the inputs was measured by a synchrony factor **s**, which was varied from 0 to 1 in steps of 0.1. For example, at "0.7 synchrony", 70% of the afferents had identical spike trains. The average number of spikes per second (f_o) computed over 5 s of simulation was used as the index of the output. All simulations were performed using the GENESIS program on a Sun 4 workstation [10].

3 RESULTS

For each combination of the four parameters, **N**, f_i, **w** and **s**, we determined the average frequency of the output over 5 seconds of simulation. The results are discussed in two sections, each describing the effect of increasing synchrony on f_o when one variable was systematically varied, with the other two fixed. In general **N** and f_i need not be independently varied since each input is a Poisson process and so is the sum of Poisson processes. Therefore, only the variation in f_i is discussed, although simulations were also performed for variation in N.

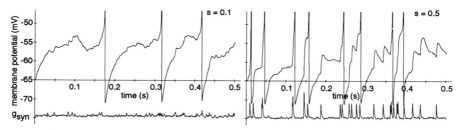

Figure 1 Membrane potential of the soma compartment and the total synaptic conductance (**g_{syn}** in arbitrary units). For both panels $N = 100$, $f_i = 50$ Hz and $w = 100\mu$V. The degree of synchrony in the inputs were 10% (left) and 50% (right). Action potentials are clipped.

(1) *s and w varied, N and f_i held constant:* Fig. 1 shows the membrane potential trajectories and the synaptic conductance changes for two levels of input synchrony with the other parameters fixed ($N = 100$, $w = 100$ μV and $f_i = 50$ Hz). At low level of synchrony ($s = 0.1$), the membrane trajectories were was relatively smooth and the firing was regular. At a synchrony level of 0.5, the membrane potential was strongly affected by the synchronized input, which produced large EPSPs. The average firing rate was higher than in the low-synchrony condition. Fig. 2 shows a typical family of curves depicting the dependence of f_O on w and s, for $N = 100$ and $f_i = 50$ Hz. For low values of

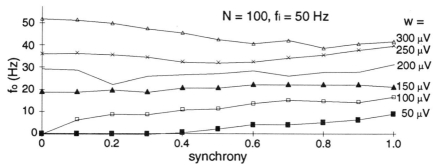

Figure 2 Variation of output frequency (f_O) with synchrony and size of EPSP.

w (< 125 μV), the output increased monotonically with increasing synchrony. For instance, with 100 inputs firing at 50 Hz, each input producing an EPSP of 100 μV, no spike was evoked in the model neuron with $s = 0$, and at 100%

synchrony ($s = 1$), the average output was 17 Hz. For larger EPSPs (> 300 μV) the output frequency actually decreased with increasing synchrony. This occurred when the EPSP size was such that, at low synchrony f_O was greater than f_i. At very high synchrony the inputs are effectively lumped into one source, and since each EPSP can lead to no more than one spike, f_O cannot be greater than f_i. In fact, f_O is less than f_i since the synchronized input sometimes occurs during the afterhyperpolarization following a previous spike and fails to evoke another spike. At low levels of synchrony the individual synaptic currents add up more smoothly and lead to a relatively regular firing rate, which can be greater than f_i. For the same reason the variance of f_O increased with increasing synchrony. When the inputs were highly synchronized, the output spikes were more tightly linked with the inputs and therefore tended to have the same Poisson statistics as the inputs. Simulations for other values of N and f_i produced results qualitatively similar to those in fig. 2.

(2) *s and f_i varied, N and w held constant:* At low values of f_i, increasing synchrony led to an increase in f_O. For sufficiently large values of f_i, f_O was greater than f_i at $s = 0$. In this regime, increasing s resulted in a decrease in f_O for the same reason discussed above (i.e., at high synchrony f_O cannot exceed f_i).

It is of interest to determine the values of N, f_i and w for which synchrony in the inputs can enhance the firing rate of the cell. Synchrony was considered to have a significant effect on f_O if f_O increased by at least 10% when synchrony was changed from 0 to 1. With this criterion, synchrony had a significant effect on f_O if the product (P) of the three factors N, f_i and w, was less than 0.5 Vsec^{-1}. That is: $P = Nwf_i \leq 0.5$ Vsec^{-1}. This particular cut-off for P depends on the parameters of the model cell. For instance, if the time constant of the membrane is reduced, synchrony is effective for a greater range of P.

4   DISCUSSION

Our primary objective was to determine parametrically the conditions under which synchrony in synaptic inputs can lead to an increase in the average output firing rate of a biologically realistic model of a cortical neuron. Interestingly, the simulations revealed that greater synchrony does not always increase the average firing rate. When input firing rates, number of inputs and unitary EPSPs are small, synchrony can significantly increase the output firing rate; for these parameters anynchronous inputs generate relatively smooth membrane potential trajectories, whereas synchronized inputs can cause large deviations in the membrane potential that trigger spikes (Fig. 1). However, when N,

f_i and **w** are large, asynchronous inputs generate a relatively smooth steady depolarization that can produce output rates higher than the firing rates of the individual inputs. In contrast, highly synchronized inputs generate large compound EPSPs that cause one spike at most, so f_o cannot be greater than f_i. Consequently, in this parameter regime, greater synchrony can actually decrease f_o. This situation will change if some intrinsic membrane properties (such as calcium currents) can generate multiple spikes per EPSP.

Our simulations have relevance to biological neurons *in vivo*. Unitary EPSPs in cortical pyramidal neurons are in the range of a few hundred μV [4]. However, *in vivo* intracellular recordings in awake monkeys show large membrane potentials fluctuations up to 10 mV, suggesting synchronous occurrence of PSPs (Matsumura, Chen and Fetz, unpublished observations). Pyramidal cells typically fire one spike and repolarize much like our model neuron. Assuming an average input firing rate of 25 Hz and unitary EPSPs of 100 μV, synchrony would increase the frequency of target neurons, if the number of active inputs is less than 200, as seems likely. Although the results presented were obtained under steady state conditions, preliminary simulations suggest similar results for time-varying inputs: transient changes in f_i cause bigger changes in f_o if synchrony is increased in parallel, at least when $\mathbf{P} < 0.5 \text{ Vsec}^{-1}$.

REFERENCES

[1] Abeles, M. *Isr. J. of Med. Sci.*, 1982. 18:83-92.

[2] Bernander, O., Douglas, R.J., Martin, K.A.C. and Koch, C. (1991) *Proc. Natl. Acad. Sci. USA.*, 88:11569-11573.

[3] Eckhorn, R., Bauer, R, Jordan, W., Brosch, M., Kruse, W., Munk, M. and Reitboeck, H.J. (1988) *Biol. Cybern.*, 60:121-130.

[4] Fetz, E.E., Toyama,K and Smith, W.S. (1991) *Cerebral Cortex.* vol 9. ed. Alan Peters and E.G. Jones (Plenum Press), pp.1-47.

[5] Gray, C.M., König, P., Engel, A.K. and Singer, W. (1989) *Nature*, 338:334-337.

[6] Kenyon, G.T., Fetz, E.E. and Puff, R.D. (1990) In *Advances in Neural Information Processing Systems 2*, Ed: Touretzky, D. (Morgan Kaufmann, San Mateo, CA), pp 141-148.

[7] Lytton, W.W. and Sejnowski, T.J. (1991) *J. Neurophysiol.*, 66:1059-1079.

[8] Murthy, V.N. and Fetz, E.E. (1992) *Proc. Natl. Acad. Sci. USA.*, 89:5670-5674.

[9] Reyes, A.D. (1990) *Ph.D. Thesis*, University of Washington, Seattle.

[10] Wilson, M.A. and Bower, J. (1989) In: *Methods in Neuronal Modeling*, Ed: C. Koch and I. Segev. MIT Press.

ANALYSIS AND SIMULATION OF SYNCHRONIZATION IN OSCILLATORY NEURAL NETWORKS

Xin Wang*, E.K. Blum and P.K. Leung

*Center for Applied Mathematical Sciences and
Neural, Informational and Behavioral Program
University of Southern California, Los Angeles, CA 90089.*
*Corresponding author's address: Mathematics Department, USC

ABSTRACT

Electrophysiological recordings in visual cortex give evidence of oscillatory dynamics which undergoes a synchronization in response to certain correlated stimuli. We present two types of models of neural networks and show by analysis and computer simulation that these model networks have oscillatory dynamics which exhibits synchronization when subjected to inputs similar to the experimental stimuli. Effects of noisy inputs on synchronization are also simulated as well as chaotic oscillations.

73.1 INTRODUCTION

Motivated by recent experimental observations suggesting synchronization of oscillations in cat visual cortex [GEKS90, EKGS90, EBJ+88, GEKS92] and earlier observations in olfactory cortex [FS85] we propose two types of models of oscillatory neural networks that can account for the observed phenomena. Our two models differ from those in [KS91, F88, FS85, vdMS86] and various others. One is based on a more abstract discrete-time neuron model and the other on a more realistic (biologically) spiking compartmental neuron model. The former has the advantage of being rigorously mathematically analyzable. The latter has, so far, defied our rigorous analysis and was therefore studied by detailed computer simulation. Results of our analysis and simulations are consistent with experimental observations and indicate that such models can suggest possible neuronal coupling mechanisms and network architectures underlying cortical functions that produce the synchronization observed. The parameters of the model networks can be adjusted to obtain phase-locking and frequency-locking under a variety of input conditions.

73.2   THE MODELS OF INDIVIDUAL OSCILLATORS

The discrete-time network model employs sigmoidal neurons and is described in detail in [BW92, Wan91, Wan92ab]. It is based on a 2-neuron oscillator defined by the equations,

$$
\begin{aligned}
x(t+1) &= f_\mu(w_{xx}x(t) - w_{xy}y(t) - \theta_x + J_x(t)),\\
y(t+1) &= f_\mu(w_{yx}x(t) - w_{yy}y(t) - \theta_y + J_y(t)),
\end{aligned}
$$

where $f_\mu(z) = \tanh(\mu z)$ and the synaptic weights (w's) are non-negative real numbers. x is thus an excitatory neuron and y is inhibitory. θ's are thresholds and J's external inputs. It is proved in [Wan92ab] that for many choices of w's this 2-neuron recurrent network undergoes a Hopf bifurcation as μ increases through a value μ^* when $\theta = 0$ and $J = 0$. For example, taking $w_{xx} = 2, w_{yy} = 0, w_{xy} = 2, w_{yx} = 1$, we get $\mu^* = .706$ as the value of the "gain" at which (0,0) becomes an unstable equilibrium and a stable quasi-periodic orbit is spawned. By varying the weights different quasi-periodicities (rotation numbers) are obtained. For $w_{xx} = a = w_{xy}$ and $w_{yx} = b = w_{yy}$, with $b > a > 0$ and $b/a > 2$, as μ increases there occurs period-doubling to chaos. See [Wan91, 92a].

The second model is also based on a 2-neuron oscillator but using compartment model neurons. Each neuron consists of passive dendritic compartments, a soma and a Hodgkin-Huxley axon hillock. The two somas are coupled through modified Eccles-type synapses. The dynamics is simulated using the CAJAL simulator as described in [BK91, BKT91]. The compartments have biologically realistic dimensions and electrophysiological parameter values (e.g. conductances, Nernst potentials, etc). By appropriate choices of the synaptic parameters (e.g. conductances and synaptic-propagation delays from pre- to post-synaptic neurons) the 2-neuron network can be shocked to self-oscillate, generating two repetitive spike trains that are slightly out of phase and in the 40-80Hz range. We also simulate a single-neuron forced oscillator model.

73.3 COUPLED OSCILLATORS

To model synchronization of oscillations occurring in neighboring regions of cortex we simulated various networks (chains) of coupled oscillators. Starting with a set of 2-neuron discrete-time oscillators, we considered four possible coupling modes: excitatory-excitatory (EE), connecting with nonnegative weights the outputs of the x (excitatory) neurons to the x neurons themselves as in [FS85, BW92]; inhibitory-excitatory (IE), connecting y (inhibitory) neurons with nonpositive weights to x neurons; and similarly IE and II. In [Wan92], it is proven that for weak coupling, EE and EI produce in-phase-locked oscillation, whereas IE and II produce out-of-phase-locked oscillation. Also, for EE and II, the locked frequency ω_L of the pair is less than the "natural" oscillator frequency ω_N, whereas for EI and IE $\omega_L > \omega_N$. This result is verified by simulation of a one-dimensional chain of 50 such coupled discrete-time oscillators (Fig. 1). We then carried out corresponding simulations with the spiking model of

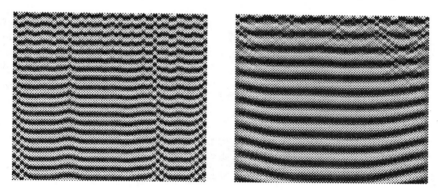

Figure 1 Synchronization effect of increasing coupling strength (left: weak; right: strong) in a 1-D chain of 50 EE coupled discrete-time oscillators, verifying Wang's theorems: in-phase-locking (with end effect) and $\omega_L < \omega_N$ (Gray scale: black state $x = -1$, white state $x = 1$; increasing time: top-down)

2-neuron self-oscillators. Using the CAJAL simulator, it was possible to adjust synaptic parameters to obtain $\omega_L < \omega_N$ by about 10% for EE coupling, but the IE coupling did not yield significant $\omega_L > \omega_N$. The CAJAL simulations were repeated for two coupled single-neuron forced oscillators driven respectively at 40 and 50 Hz, obtaining frequency-locking with strong EE coupling. The same locking was also obtained with random pulse inputs at 50 Hz $\pm$ 40% and with constant inputs with 50% noise.

Figure 2   1-D ring of EE coupled discrete-time oscillators with a single
light bar input (left) and two light bar input (right)

Finally, for both models, simulations were performed on chains of oscillators
subjected to two steady inputs spaced along the chain to represent two separated light bar stimuli as in [GEKS90, EBJ+88]. For close spacing starting in
a random phase relation, the chains eventually synchronized (phase-locked) as
shown in typical Figs. 2 and 3. With 30% noisy light bars the spiking neuron

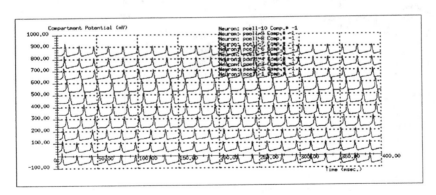

Figure 3   Phase-Locking of 10 coupled spiking oscillators with two light
bar input at 1, 2, 3, 8, 9, 10

model only partly synchronized (not shown). Likewise, for discrete-time chaotic
oscillators, the chain was only partly synchronized by the light bars (Fig. 4).

These initial results in our ongoing research suggest possible coupling modes

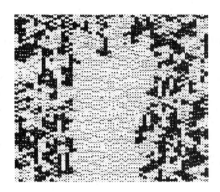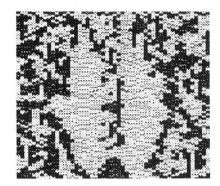

Figure 4 Synchronization effect of a single light bar (left) and two light bar (right) inputs in a 1-D ring of EE diffusively coupled chaotic oscillators

and strengths that could produce synchronization of neural oscillators under various distributions of stimuli such as those reported in [GEKS92, EBJ+88]. They also demonstrate an interesting close relationship between discrete-time models and realistic spiking models (see [WB92]) that points the way to new analytical techniques for studying spiking models.

REFERENCES

[BK91] E.K. Blum and P.K. Khademi. CAJAL-91: A biological neural network simulator. To appear in *AMNS2*. ed. F.H. Eeckman, Kluwer Academic Publisher.

[BKT91] E.K. Blum and P.K. Khademi and R.F. Thompson. Modeling and simulation of a simplified cerebellar neural network for classical conditioning of the rabbit eyeblink response. To appear in *AMNS2*. ed. F.H. Eeckman, Kluwer Academic Publisher.

[BW92] E.K. Blum and X. Wang. Stability of fixed points and periodic orbits and bifurcations in analog neural networks. *Neural Networks*, vol. 5, 577-587, 1992.

[EBJ+88] R. Eckhorn, R. Bauer, W. Jordan, M. Brosch, W. Kruse, M. Munk, and H.J. Reitboeck. Coherent oscillations: a mechanism of feature linking in the visual cortex? *Biological Cybernetics*, 60:121-130, 1988.

[EKGS90] A.K. Engel, P. König, C.M. Gray and W. Singer. Stimulus-dependent neuronal oscillations in cat visual cortex: inter-columnar interaction as determined by cross-correlation analysis. *European J. of Neuroscience*, vol. 2, 588-606.

[F88] W.J. Freeman. Nonlinear neural dynamics in olfaction as a model for cognition. In: *Dynamics of Sensory and Cognitive Processing by the Brain*, ed. E. Basar, 1929, 1988, Springer.

[FS85] W.J. Freeman and C.A. Skarda. Spatial EEG patterns, nonlinear dynamics and perception: the neo-sherringtonian view. Brain Res. Rev., 10:147-175, 1985.

[GEKS90] C.M. Gray, A.K. Engel, P.König, and W. Singer. Oscillatory responses in cat visual cortex exhibit inter-columnar synchronization which reflects global stimulus properties. *Nature*, 338 (23): 334-337, 1989.

[GEKS92] C.M. Gray, A.K. Engel, P. König, and W. Singer. Synchronization of oscillatory neuronal responses in cat striate cortex: temporal properties. *Visual Neuroscience* 8, 337-347, 1992.

[KS91] P. König and T.B. Schillen. Stimulus-dependent assembly formation of oscillatory responses: I. synchronization. *Neural Computation*, 3:155-166, 1991.

[vdMS86] C. von der Malsburg and W. Schneider. A neural cocktail-party processor. *Biological Cybernetics*, 54:29-40, 1986.

[Wan91] X. Wang. Period-doubling to chaos in a simple neural network: An analytical proof. Complex Systems, 5:425-441, 1991.

[Wan92a] X. Wang. Discrete-time dynamics of coupled quasi-periodic and chaotic neural network oscillators. *Proc. of IJCNN'92, Baltimore*, 1992.

[Wan92b] X. Wang. Discrete-time neural networks as dynamical systems. *Ph.D. Thesis, University of Southern California*, 1992.

[WB92] X. Wang and E.K. Blum. Discrete-time versus continuous-time models of neural networks. *Journal of Computer and System Sciences*, 45:1-19, 1992.

ALTERNATING PREDICTABLE AND UNPREDICTABLE STATES IN DATA FROM CAT VISUAL CORTEX

K. Pawelzik, H.-U. Bauer*, T. Geisel

Institut für Theoretische Physik and SFB Nichtlineare Dynamik,
Universität Frankfurt, Robert-Mayer-Str.8-10,
W-6000 Frankfurt/Main 11, Fed. Rep. of Germany.
* present adress: CNS-Program, Caltech 216-76,
Pasadena, CA91125, USA

ABSTRACT

Visual inspection of data from cat visual cortex reveals, that the recently discovered 50-Hz-oscillations do not consist of a permanent oscillatory signal during the whole period of stimulation, but instead display interchanging stochastic and regular phases, both of varying duration times. In order to base this observation on firm statistical grounds, we assume that the signal should be predictable during regular epochs, unpredictable during the irregular epochs. Employing the method of time-resolved mutual information, we calculate the time-resolved predictability within the LFP-signal from a single electrode. Periods of high temporal coherence can clearly be discriminated from periods of stochastic behaviour. An analysis of the spatial coherence with the same method yields, that synchronous activity between two electrodes is largely restricted to episodes of high temporal coherence.

1 INTRODUCTION

The problem of how the brain manages to associate different features belonging to the same object is of fundamental importance for the theory of neuronal representations. It has been argued on theoretical grounds, that synchronous and time structured neuronal activities provide a solution to this binding problem [1]. Experimental evidence for such neuronal responses was obtained only a few years ago, when local field potentials (LFP) as well as multi-unit activities in cat visual cortex were shown to yield synchronous oscillatory responses upon suitable stimulation [2,3].

These experiments caused an avalanche of experimental as well as theoretical work. Many simulations of systems mimicking the experimental setup were performed, most of them based on coupled oscillators [4,5,6]. Models of coupled oscillators, however, are not likely to explain the detailed spatio-temporal dynamics seen in the experiment. In Figs. 1,2,4 we show the time course of local field potentials from cat visual cortex recorded by two electrodes which were located 6mm apart in area 17. The responses were caused by stimulation of the receptive fields with a long light bar. The details of this experiment are described in [2]. The corresponding correlation functions display the decaying variations in the 50 Hz range which gave rise to the notion of 50 Hz oscillations. Upon visual inspection of the raw data, however, the local field potentials appearantly are not be very well described by permanent sinusodial oscillations. Instead one might identify brief epochs of more or less oscillatory acitivity, which are interrupted by irregular activity.

The detailed analysis of this property of the data is the aim of our contribution. We assume that the data can be described by a system which switches between a stochastic and a regular state. These states should be distinguishable by different degrees of predictability of the time series. With this hypothesis in mind we determine the time-resolved predictability of the single LFP and the time-resolved dependency of one LFP on the other. As a criterion for predictability we use the time-resolved mutual information [7,8], which is an unbiased measure of statistical dependence. It does not rest on any assumptions on the form of the time series. In particular it does not assume that the statistical dependence is mediated by oscillatory signals.

In this context one should mention an observation made by C. Gray et al., who analyzed the dynamics of common oscillations between 2 electrodes [9]. Using a sliding crosscorrelation technique and fitting Gabor functions to the crosscorrelograms these authors found the amount of synchronicity to be strongly time-dependent. A time-dependent synchronicity, however, is not incompatible with models, which contain oscillators with time-varying frequency. It is imaginable, that such oscillators produce locally a permanent rather oscillatory signal, but have strongly time-dependent correlograms.

2   METHOD

While the average mutual information **M** is a well known quantity, it was discovered only recently, that it can be analyzed in a time-resolved way [7,8].

We discuss very briefly, how it can be calculated efficiently from a given time-series.

Let x_i, y_i denote different time series (e.g. of LFPs) where $i = 1, ..., N$ denotes the time index. Then we define "states" (or "patterns") as vectors of successive values (in nonlinear dynamics known as delay coordinates)

$$\mathbf{x}_i = (x_i, x_{i+\tau}, ..., x_{i+(m-1)\tau}),$$
$$\mathbf{y}_i = (y_i, y_{i+\tau}, ..., y_{i+(n-1)\tau}).$$

Furthermore we define by $\mathbf{z}_i = (\mathbf{x}_i, \mathbf{y}_i)$ the compound states for the co-occurrence of the states $\mathbf{x}_i$ and $\mathbf{y}_i$.

In order to calculate the mutual information between $\mathbf{x}$ and $\mathbf{y}$ one has to coarse grain the state-spaces in a suitable way. An efficient coarse graining is given by (hyper-)boxes of a fixed size ε around the states. For the computation of predictability within one time series we choose $\mathbf{y}_i = \mathbf{x}_{i+\Delta t}$, for the determination of spatial coherence we construct $\mathbf{x}_i$ and $\mathbf{y}_i$ from the two time series, resp.. With these ingredients we can now estimate the probabilities p to fall into a box with the temporal index i in the state-spaces:

$$p_i^{\mathbf{x}} = \frac{1}{N} \sum_{j=1}^{N} \Theta(\varepsilon - | \mathbf{x}_i - \mathbf{x}_j |),$$

$$p_i^{\mathbf{y}} = \frac{1}{N} \sum_{j=1}^{N} \Theta(\varepsilon - | \mathbf{y}_i - \mathbf{y}_j |),$$

$$p_i^{\mathbf{z}} = \frac{1}{N} \sum_{j=1}^{N} \Theta(\varepsilon - | \mathbf{z}_i - \mathbf{z}_j |),$$

where Θ is the Heaviside step function and $| ... |$ denotes the maximum norm.

The mutual information then is given by

$$\mathbf{M} = \frac{1}{N} \sum_{i=1}^{N} \mathbf{M}_i \quad \text{with} \quad \mathbf{M}_i = \log \frac{p_i^{\mathbf{z}}}{p_i^{\mathbf{x}} p_i^{\mathbf{y}}}.$$

This formula provides an efficient algorithm for the calculation of the mutual information for a fixed partition size ε. In principle ε has to be chosen small. In

contrast to Shannon's original formula we scan the phase space by following the
time series, replacing the average over phase space by an average along the time
series. i.e. a time average. By leaving out the last averaging step, we can thus
compute the mutual information in a time resolved way and detect a possible
switching in the underlying system. The detection precision of this method with
regard to such switching events can be shown to increase exponentially with the
size of the (delay) time window necessary for the construction of the state space
[8]. This is in contrast to the algebraic increase found for a sliding correlation
technique. Therefore the time resolved mutual information M_i can result in a
superior time resolution in the localization of such events as compared to the
sliding crosscorrelation technique. The average in Shannons original formula
[10] has been replaced by a time average, a procedure which is more efficient
when probabilities have to be calculated from time series.

3   RESULTS

We now turn to the application of this method to LFPs of the above mentioned
long bar experiment. We used the data of ten repetitions of the stimulus
presentation which provided $N = 10000$ values in total. Parts of the data are
shown in Figs. 1,2,4.

The described method for estimating the mutual information is equivalent to dimension algorithms from nonlinear dynamics. Applications of these algorithms
to the LFP data reveal, that no low dimensional chaotic attractor is present.
Instead we find, that M increases most until $(m - 1)\tau > 20ms$ indicating that
the system is characterized by a stochatic process involving a memory of at least
$20ms$. While these results are discussed in Ref. 8 Figs. 1 and 2 show the results
of a detection of activities coherent in time for two different electrodes. Plotted
are 2000 ms of the data corresponding to two repetitions of the experiment.
Periods of activity which lead to a value of $M_i > 0.5$ were judged as temporally
coherent (this includes the width of the embedding time window). The methods indicates periods of increased temporal coherence, which are interupted by
uncoherent periods. The "windows of coherence" coincide with the parts of
the data which a visual inspection would regard as oscillatory. Even though
this aspect of the result seems rather trivial, we would like to remark, that
the results also rule out, that there is a hidden temporal coherence within the
irregularly looking epochs. This comes about because the mutual information
method does not rely on any specific waveform of the time series.

In Fig.3 we now further investigate the nature of the LFP-signal during the
coherent episodes. This is done by averaging the contributions M_i for the

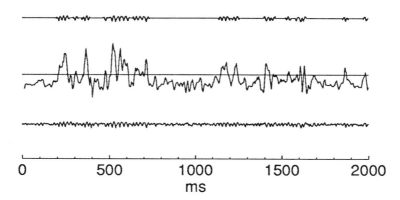

Figure 1 Analysis of temporal coherence in a time series of local field potentials. Plotted are (from below): the local field potential ξ_i, the time resolved mutual information M_i (computed with $x_i = \xi_i, m = 4, \tau = 3, y_i = \xi_{i+12}, n = 4$), and the local field potential x_j during the " windows of coherence " (i.e. the values ξ_j for which $j \in \{j = i, i+1, ..., i+\Delta t/M > 0.5 \}$, with the stochastic interplays being indicated by straight lines for ξ_i.

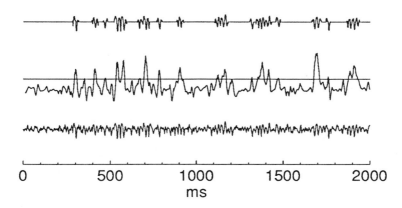

Figure 2 As Fig. 1, but for the second electrode (i.e. $x_i = \chi_i, y_i = \chi_{i+12}$).

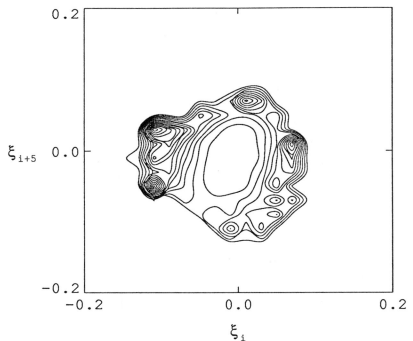

Figure 3 Contour plot of the contributions to M averaged depending on the successive values of the LFP ξ at time i and $i + 5$. Contributions from a sinousodial signal of period T would lie on circle.

activity at time i and at time $i + T/4$ where $T = 20ms$ is the period of a 50Hz oscillation. If the coherent episodes consisted of a pure sinusodial wave we would see a circle in this representation. We find however, that the average contributions are small after the amplitude was maximal (lower left corner). This indicates a low degree of predictability for this phase of activity. The limit cycle can be left most easily here.

Fig. 4 finally shows the amount of coherence of the activities of the two electrodes from Figs. 1 and 2. Plotted are again 2000 ms of both LFPs signals, M_i and the contributing windows (from below). Because of the lower average M we here chose $M_i > 0.2$. All other parameters are as in Figs. 1 and 2. We find, that episodes of spatial coherence are relatively rare, and mostly appear where both LFPs are temporally very coherent.

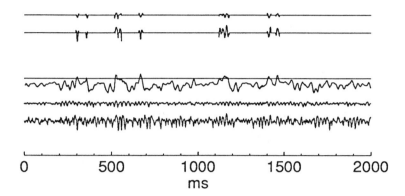

Figure 4 Results of an analysis of spatial coherence. Plotted are (from below): 2000 ms of the LFP χ at electrode 1, the corresponding LFP ξ at electrode 2, the time-resolved mutual information M_i (computed with $x_i = \chi_i, m = 4, \tau = 3, y_i = \xi_i, n = 4$), and contributing episodes (i.e. x_j and y_j for times $j \in \{j = i, i+1, ..., i+(m-1)\tau / M_i > 0.2 \}$) .

4   CONCLUSION

Our findings confirm the observation of a fast switching between irregular and coherent activity. This result stands in contrast to the notion of an ongoing oscillation which has been the basis of many models. Instead, our results suggest that a limit cycle is visited only very briefly and can be jumped off every cylcle (note the low contributions in the lower left region in Fig. 3). A possible explanation of this phenomenon is the coexistence of a limit cycle and a fixed point, with the fixed point beeing near to a certain part of the cycle such that a switching can occur there most easily [11]. Furthermore our method indicates, that synchronous activity between two electrodes is mostly restricted to periods of temporal coherence within both signals, a fact which underlines the possible importance of temporal structure for the binding problem. We expect, that future applications of our method will help to clarify this point.

5   ACKNOWLEDGEMENT

We thank the authors of Ref. 2 for most fruitful discussions and for providing the data. This work was supported by the Volkswagen Stiftung and the Deutsche Forschungsgemeinschaft through SFB 185, TP A10".

6 REFERENCES

1. C. v.d.Malsburg, Am I Thinking Assemblies, in: Brain Theory, ed. G. Plam, A. Aertsen, Springer, 161-176 (1986).

2. C.M. Gray, P. König, A.K. Engel, W. Singer, Oscillatory Responses in Cat Visual Cortex Exhibit Inter-Columnar Synchronization which Reflects Global Stimulus Properties, Nature **338**, 334-337 (1989).

3. R. Eckhorn, R. Bauer, W. Jordan, M. Brosch, W. Kruse, M. Munk, H.J. Reitböck, Coherent Oscillations: A Mechanism for Feature Linking in the Visual Cortex, Biol. Cyb. **60**, 121-130 (1988).

4. H.G. Schuster, P. Wagner, Prog. Theor. Physics **81**, 939 (1989).

5. E. Niebur, D.M. Kammen, C. Koch, in: Nonlinear Dynamics and Neuronal Networks, ed. H.G. Schuster, VCH (Weinheim), 173 (1991).

6. P. König, T. Schillen, Stimulus-Dependent Assembly Formation of Oscillatory Responses: I. Synchronization, Neur. Comp. **3**, 155-166 (1991), and T. Schillen, P. König, II: Desynchronization, Neur. Comp. **3**, 167-177 (1991).

7. K. Pawelzik, Nichtlineare Dynamik und Hirnaktivität, Verlag Harri Deutsch, Frankfurt (1990).

8. K. Pawelzik, H.G. Schuster, F. Wolf, in preparation.

9. C.M. Gray, A.K. Engel, P. König, W. Singer, Temporal Properties of Synchronous Oscillatory Neuronal Interactions in Cat Striate Cortex, in: Nonlinear Dynamics and Neuronal Networks, ed. H.G. Schuster, VCH Weinheim, 27-56 (1991).

10. C.E. Shannon, W. Weaver, The Mathematical Theory of Communication, Univ. of Illinois Press (1949).

11. K.Pawelzik, H.-U.Bauer, J.Deppisch and T.Geisel, NIPS92 accepted.

TEMPORAL STRUCTURE OF SPIKE TRAINS FROM MT NEURONS IN THE AWAKE MONKEY

Wyeth Bair[1], Christof Koch[1], William Newsome[2], Kenneth Britten[2]

[1] *Computation and Neural Systems Program, California Institute of Technology, Pasadena, CA 91125*
[2] *Department of Neurobiology, Stanford University School of Medicine, Stanford, CA 94305*

ABSTRACT

We compute the power spectrum associated with single-unit extracellular spike trains recorded in a previous study of 213 neurons in extrastriate area MT of the macaque monkey, a region that plays a major role in processing motion information. The data were recorded while monkeys performed a near-threshold direction discrimination task so that both physiological and psychophysical data could be obtained on the same set of trials [14]. (1) About half of the cells have a flat spectrum with a dip at low temporal frequencies, indicative of a Poisson process with a refractory period. (2) About half of the cells have a peak in the 25–50 Hz frequency band. (3) We show that the peak in the power spectrum is related to the cell's tendency to fire bursts of action potentials and that the shape of the power spectrum can be described without assuming an explicit oscillatory neuronal mechanism. (4) We find no relationship between the shape of the power spectrum, in particular, the amplitude of the peak, and psychophysical measures of the monkeys' performance on the direction discrimination task.

75.1   INTRODUCTION

It is widely held that visual cortical neurons encode information primarily in their mean firing rates. This notion has been challenged recently by reports of stimulus-induced, though not stimulus-locked, semi-synchronous neuronal oscillations in the 30–70 Hz range in the visual cortex of the anesthetized cat, using both multi-unit activity and local field potentials [7][9]. Stimulated by earlier theoretical proposals of Milner [12] and von der Malsburg [20], these experimental findings gave rise to the idea that neurons which signal different features of the same object ("binding problem") oscillate in phase, while neurons which code for properties of different objects (or the "ground") oscillate at a different phase or at random [6][9][18]. However, neuronal oscillations have proven more elusive in the awake macaque monkey, with some authors reporting oscillations in the 25–50 Hz range [8][11][10][13], while others fail to find significant evidence for them [19][21].

The data analyzed in this paper were obtained in a continuing series of experiments attempting to link the responses of neurons in extrastriate area MT, known to be important in processing motion, to the psychophysical performance of trained monkeys [1][14][16]. This experimental paradigm enables us to correlate the temporal structure of the neuronal discharge with the monkey's performance and with the cell's performance as measured in the original study. So motivated, we set out to compute power spectra in search of oscillations. We find that there are peaks in the power spectra near 40 Hz, but that these peaks are indicative of temporal structure related to bursts of action potentials, not to oscillations. In addition, we find that the temporal structure is independent of the stimulus conditions and the performance measures.

75.2 METHODS

Experimental methods for the original data collection are described in detail by Britten *et al.* [1]. Here we only highlight the main features of the experiment. Three adult macaque monkeys were trained to report the direction of coherent motion in a dynamic random dot display in which a fraction, c, of dots moved coherently at a constant speed in one direction while the other dots moved randomly. For $c = 0$ all dots moved randomly, while for $c = 1$ all dots moved coherently. The dot pattern filled the receptive field of a single MT neuron and the direction of coherent motion was either the neuron's preferred direction, or the opposite (null) direction. The monkey held fixation during the 2 sec stimulus and responded afterward. Action potential occurrence times were recorded extracellularly at a 1 msec resolution during this period. For a typical experiment, at least 210 trials were performed as c and motion direction varied.

In the previous analysis [1], a psychophysical threshold, c_{system}, was measured for each experiment, where threshold was taken to be the c value that supported 82% correct performance. Neuronal sensitivity was measured from the responses to preferred and null direction motion obtained over a range of coherence levels; responses were taken to be the total number of spikes that occurred during the two second trial. Using a method based on signal detection theory, a "neurometric function" was computed that expressed the theoretical performance of an ideal observer who judges the direction of motion in the visual stimulus based only on the responses of the MT neuron being analyzed. Computed performance of the ideal observer was plotted as percent correct choices as a function of motion coherency, and "neurometric" thresholds, c_{cell}, were extracted in the same manner described for psychometric thresholds.

For temporal analysis, we performed Fourier transforms on spike trains using a standard Fast Fourier Transform (FFT) algorithm and computed one-sided estimates of the power spectral density using overlapping data segments and Parzen windowing [17]. Each 2 sec long spike train was converted into a sequence of 0's and 1's (1's for action potentials) at 1 kHz resolution and processed as 14 overlapping data segments, 256 msec per segment. This technique yields a one-sided spectrum with 128 entries at equally spaced intervals from 0 to 500 Hz. For short-lived oscillation analysis, power spectra were computed for individual 256 msec windows, and statistics were compiled on the height and location of peaks.

75.3   EXPERIMENTAL RESULTS

Of 213 cells analyzed, we found that about 80 cells had a relatively flat spectrum with a dip at low temporal frequencies and that about 110 cells had a spectrum with a peak in the 25–50 Hz frequency band. In about 10% of all cells, the peak in the power spectrum rose at least twice as high as the baseline. The remaining cells were not considered due to either low spike rates or predominantly transient responses which confounded spectral analysis. The panels on the left in Fig. 1 show data for example cells which have a dip in the spectrum (a) and a peak in the spectrum (c). The power spectra shown are averaged over all trials (all c values) for each cell. Averaging over all trials was performed only after observing that the stimulus conditions had very little effect on the shape of the power spectrum, other than a general scaling effect expected from variations in mean firing rate. In other words, we find that a cell with a peak in its average spectrum also shows a very similar peak if the spectrum is computed by averaging over only trials for (1) a particular c value, (2) preferred (null) direction motion (3) correct (incorrect) response from the monkey, or (4) any combination of these conditions.

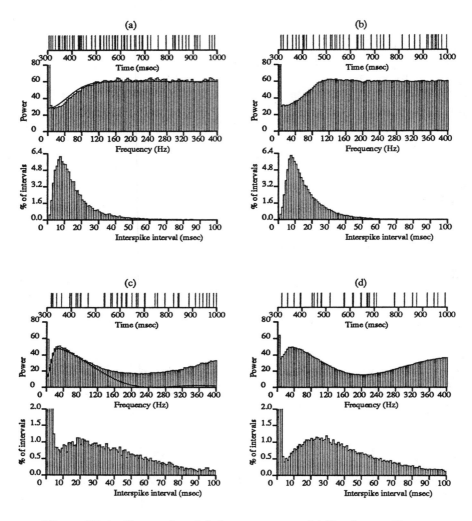

Figure 75.1   Data and models for two neurons. (a) Non-burst cell *e047*. The spike train shows a high firing rate, but spikes rarely cluster. The dip at low frequencies in the power spectrum and the absence of short intervals in the ISI indicate a refractory period. The analytical result for the spectrum (Section 5) is super-imposed (smooth curve). (b) The corresponding computer generated data based on a randomly (Poisson) firing neuron with a refractory period. (c) Burst cell *w190*. The power spectrum has a peak near 30 Hz, and the spike train shows bursting. Bursting is evident in the ISI from the peak (truncated) at short intervals. Again, the analytical curve is super-imposed on the spectrum. (d) The corresponding computer generated data is based on a neuron which fires bursts at Poisson instants but has a burst-related refractory period.

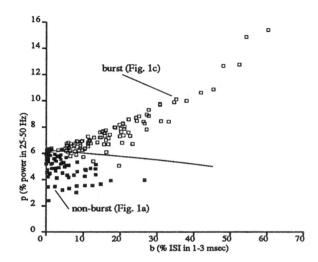

Figure 75.2 The measure p of the peak in the power spectrum (percent power from 25–50 Hz) is plotted against the measure b of burst firing (percent ISI $\leq$ 3 msec). Solid squares indicate cells which have a dip in the power spectrum at low frequencies: *non-burst* cells (Fig. 1a). Open squares indicate cells which have a peak in the 25–50 Hz range: *burst* cells (Fig. 1c). Among burst cells, the correlation between b and p is strong. (Points for cells from Fig. 1 are indicated.)

By comparing observations from the power spectra to observations of the raster spike trains, we discovered a correlation between the peak in the power spectrum and a cell's tendency to fire "bursts" of action potentials, typically 2–4 spikes in 2–6 msec. To quantify this correlation, we define two metrics, one based on the interspike interval (ISI) histogram and the other based on the power spectrum. Our measure of burstiness, b, is the percent of interspike intervals that fall in the first 3 bins of the ISI histogram. Our measure of the peak in the power spectrum is the percent of power in the 25–50 Hz band. These definitions are motivated by the observed interspike intervals within bursts and by the observed locations of peaks in the spectrum for this database. Fig. 2 shows the correlation between the strength of the peak in the spectrum, p, and the measure of burstiness, b. For cells that have peaks in the 25–50 Hz band (open squares in Fig. 2), p increases as b increases. In the modeling section, we show that such a correlation is expected for bursting cells. For cells with a dip at low frequencies (filled squares), there is no strong correlation between p and b. The line in Fig. 2 shows the locations of idealized neurons that fire randomly (Poisson distribution) with mean rates from 1 to 200 spikes/sec. This line separates the continuum of cells into "bursting" cells, which cluster action potentials more often than a randomly firing cell, and "non-bursting" cells, which fire clusters of action potentials less often than a randomly firing cell.

75.4   CORRELATION TO BEHAVIOR

We correlated p_{cell}, the average value of p over all trials for a particular cell, against a number of different variables related to the performance of the cell and the behavior of the monkey during the task: the neuronal threshold c_{cell}, the ratio between neuronal and behavioral threshold c_{cell}/c_{system} and the decision related probability for each individual neuron [15]. The correlation coefficients, r^2, are 0.00025 ($p = 0.82$), 0.002 ($p = 0.54$) and 0.004 ($p = 0.39$), respectively, indicating no correlation between a peak in the power spectrum (i.e. burstiness) and these known behavioral measures. The scatter plots for these correlations show no structure. Thus, the prominence of a peak in the 25–50 Hz region of the power spectrum does **not** correlate with the known behavior of the monkey or performance of the cell.

75.5   STOCHASTIC MODELS

We use two simple analytical models to account for the dips and peaks in the power spectra and to establish the relationship between bursting and a peak in the spectrum. Based on results for the power spectrum of a shot noise process [3], it can be shown that a neuron firing action potentials at random (Poisson distribution), but with a refractory period, has a spectrum with a dip at low frequencies. Fig. 1b shows data from a numerical simulation of this simple two-parameter (mean spike rate λ and a refractory period related σ^2) process with the analytical solution super-imposed on the power spectrum. The second model assumes that a neuron fires bursts, modeled as box-car functions, at random but with a burst-related refractory period. This leads to the results shown in Fig. 1d. Notice that the shape and location of the peak near 40 Hz in the power spectrum is well predicted by both the numerical and analytical models. The analytical result does not match well at high frequencies because it does not account for temporal structure of spikes within bursts. In both models, the parameter values were set based on the neuronal data. Fig. 3 shows the change in the power spectrum when bursts are replaced by single action potentials, giving a convincing demonstration that the burst structure, rather than the timing of bursts, is responsible for the peak near 40 Hz.

75.6   DISCUSSION

We find that about half of the MT cells studied here have peaks in the 25–50 Hz band of their power spectra and that these peaks are related to the cells' tendency to fire bursts of action potentials. We find no correlation between the peaks in the power spectrum and psychophysical or physiological measures of performance on the motion discrimination task. Furthermore, we describe an analytical model which accounts for the peaks in the spectrum without

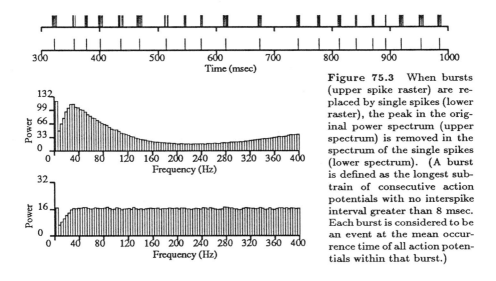

Figure 75.3 When bursts (upper spike raster) are replaced by single spikes (lower raster), the peak in the original power spectrum (upper spectrum) is removed in the spectrum of the single spikes (lower spectrum). (A burst is defined as the longest subtrain of consecutive action potentials with no interspike interval greater than 8 msec. Each burst is considered to be an event at the mean occurrence time of all action potentials within that burst.)

assuming explicit oscillation, and therefore claim that oscillations of the type reported by other labs in visual cortical areas are not present in this database. This conclusion is at odds with a recent study by Kreiter and Singer [10] who report that ca 50% of their multi-unit data in MT of awake monkey show oscillations in response to high-contrast bars. Using their criteria, we find that ca 1% of our cells are oscillatory and that similar multi-unit data (not shown) supports this lack of oscillation. What might cause this discrepancy? Effects of over-training, i.e. learning, may alter the temporal firing patterns in the cortex, or the dynamic dot stimulus may not be as effective in evoking oscillations as high-contrast bars.

Although oscillations are not present here, we find that cells have a temporal signature in that they do not change between bursting and non-bursting firing patterns, regardless of motion coherency c or the absence of stimulation. Whether our bursting cells correspond to the intracellularly identified class of "intrinsically bursting" cells located in layer 5 of rodent slice [4] or whether bursting carries significance to the organism [2][5] is at present not known.

Acknowledgements: We thank Francis Crick, Joel Franklin and Udi Zohary with whom we discussed these findings over the last two years and the Santa Fe Institute for providing a stimulating forum for discussion. This work was supported by the ONR, the James S. McDonnell Foundation, the National Eye Institute, and a NSF Presidential Young Investigator Award to C.K. W.B. received support from a NSF Graduate Fellowship, and K.B. received support from an NIMH training grant.

REFERENCES

[1] Britten KH, Newsome WT, Saunders RC (1992) *Exp. Brain Res.* **88**:292–302.

[2] Cattaneo A, Maffei L, Morrone C (1981) *Exp. Brain Res.* **43**: 115-118.

[3] Champeney DC (1973) Fourier transforms and their physical applications. New York: Academic Press Inc.

[4] Connors BW, Gutnick MJ (1990) *Trends Neur.* **13**:99-104.

[5] Crick F (1984) *Proc. Natl. Acad. Sci. USA* **81**:4586-4590.

[6] Crick F, Koch C (1990) *Seminars Neurosci.* **2**: 263-275.

[7] Eckhorn R, Bauer R, Jordan W, Brosch M, Kruse W, Munk M, Reitboeck HJ (1988) *Biol. Cybern.* **60**: 121-130.

[8] Freeman WJ, van Dijk BW (1987) *Brain Res.* **422**:267-276.

[9] Gray CM, Singer W (1989) *Proc. Natl. Acad. Sci. USA* **86**:1698-1702.

[10] Kreiter, AK, Singer, W (1992) *Eur. J. Neurosci.* **4**: 369-375.

[11] Livingstone MS (1991) *Soc. Neurosci. Abstr.*, **17**:176. (Abstract)

[12] Milner PA (1974) *Psychol. Rev.* **81**:521-535.

[13] Murphy VN, Fetz EE (1992) *Proc. Natl. Acad. Sci. USA* **89**: 5670-5674.

[14] Newsome WT, Britten KH, Movshon JA (1989a) *Nature* **341**:52-54.

[15] Newsome WT, Britten KH, Movshon JA, Shadlen M (1989b) In: *Neural Mechanism of Visual Perception*, ed. D. Lam & C. Gilbert, pp. 171-198.

[16] Newsome WT, Pare EB (1988) *J. Neurosci.* **8**:2201-2211.

[17] Press HP, Flannery BP, Teukolsky SA, Vetterling WT (1988) Numerical recipes in C, the art of scientific computing. Cambridge: Cambridge UP.

[18] Sporns O, Tononi G, Edelman GM (1991) *Proc. Natl. Acad. Sci. USA* **88**: 129-133.

[19] Tovée MJ, Rolls ET (1992) *Neuro. Report* **3**:369-372.

[20] Von der Malsburg C, Schneider W (1986) *Biol. Cybern.* **54**:29-40.

[21] Young MP, Tanaka K, Yamane S (1992) *J. Neurophys.* **67**:1464-1474.

TEMPORAL STRUCTURE CAN SOLVE THE BINDING PROBLEM FOR MULTIPLE FEATURE DOMAINS

Thomas B. Schillen
Peter König

Max-Planck-Institut für Hirnforschung
Deutschordenstraße 46, 6000 Frankfurt 71, Germany

ABSTRACT

We investigate a solution to the binding problem in visual processing using temporal structure in a neuronal network. We, first, demonstrate binding and segregation of assemblies by synchronizing and desynchronizing connections for a single feature representation. Then, we extend this model to multiple feature domains where each member of an distributed assembly is specific for its single particular stimulus feature only. This avoids the combinatorial explosion of multispecific cardinal cells. Besides binding by synchronizing the temporal structure of distributed neuronal responses, our simulations demonstrate that synchronization has to be complemented by a means for stimulus-dependent desynchronization.

76.1 INTRODUCTION

Current theories of the neurobiology of vision assume that local stimulus features like velocity, disparity, colour, etc. are processed by spatially separated neuronal populations in the brain. In the presence of several distinct stimuli, this concept leads to the problem of binding responding cells into assemblies that code uniquely for the different stimuli. A temporal structure of neuronal responses has been proposed as a possible solution to this binding problem [1]. In particular, synchronous oscillatory activity would allow to identify those cells that belong to a particular assembly.

Meanwhile, the formation of stimulus-specific assemblies has been found in oscillatory neuronal activity in the visual cortex of cats and monkeys [2, 3, 4, 5]. In previous work we have demonstrated that networks of simplified excitatory and inhibitory neuronal populations with synchronizing and desynchronizing delay connections can exhibit a stimulus-dependent formation of oscillatory assemblies in close analogy with experimental observations [6, 7, 8, 9]. Based

This work has been supported in part by the *Deutsche Forschungsgemeinschaft*.

on these concepts the current paper investigates the binding problem in an oscillatory network model.

76.2 SINGLE FEATURE MODULE

We use a basic oscillatory element as described before [6, 7, 8]. Within a feature module, each oscillator is considered to respond selectively to stimuli which present the appropriate feature within the oscillator's receptive field. Thus, each feature module can be visualized as a three-dimensional arrangement of oscillators: a two-dimensional topographic map of the visual field versus a one-dimensional representation of the module's feature domain. The module's oscillators are coupled by short-range synchronizing and long-range desynchronizing delay connections along the module's topographic and feature dimensions.

Figure 1 demonstrates the stimulus-dependent assembly formation within a feature module of 16 disparity levels. Stimuli representing an inclined plane and a partially overlapping object in the foreground (Fig. 1A) lead to synchronous activity within the disparity representations of each of the two objects, rsp. (Figs 1B, 1C1, 1C2). The assembly corresponding to the inclined plane comprises oscillators from all disparity levels, which are synchronized by reason of stimulus continuity with respect to feature and topographic dimensions. However, the discontinuity between the two stimuli allows the segregation of the two corresponding assemblies by means of the desynchronizing connections. The activities of the two representations assume a random phase relation with respect to each other (Fig. 1C3). Without active desynchronization the two stimulus representations synchronize completely.

76.3 MULTIPLE FEATURE MODULES

Extending the network to multiple feature modules, synchronizing connections are introduced which couple each oscillator, selective for a particular feature and topographic location, with all the oscillators of any other module's feature domain at the corresponding topographic location. These connections allow the synchronization of assemblies distributed over multiple feature modules where each oscillator is specific for its single particular feature expression only. With this design the required number of connections is approximately proportional to n^2 (n: number of feature modules), thus avoiding the combinatorial explosion of multispecific cardinal cells growing by the n-th power.

In this particular example, we use three feature modules for the representation of colour, orientation, and disparity, rsp. (Fig. 2A). The network's stimulus input consists of two overlapping bars of identical colour, but different orientation and disparity. Figure 2A depicts the realization of this stimulus configuration within the corresponding feature domains. As demonstrated by the cross-correlograms in figure 2C, the network represents each of the two stimulus bars by a distributed assembly of units with synchronized oscillatory activity. In this way the cells representing features of the same stimulus are bound across

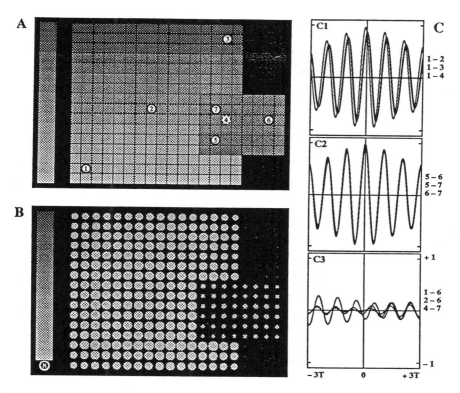

Figure 1: Synchronization and desynchronization in a single feature module.
(A) Stimulus input representing an inclined plane and a partially overlapping
object in the foreground. Disparity level indicated by shading. (B) Activity-
phase map of oscillators stimulated at the respective locations. Oscillation
amplitude is indicated by disk diameter, oscillation phase is coded by shading
$(0 \dots 2\pi)$. The map shows the coherent synchronization within the disparity
representation of each of the two stimuli. With respect to each other, the activities of the two assemblies are out of phase. (C) Mean cross-correlograms
between activities of oscillators at locations numbered in (A). (C1, C2) synchronization within each of the two stimulus representations, (C3) random phase
relation between the two assemblies.

the three modules with near zero phase lag. However, no such synchronization
is observed between the two assemblies.

The presented stimulus configuration illustrates a principle segregation problem which results from spatial overlap: considering the stimulus input to the
colour module only, cells responding to the two different bars would have to
synchronize, resulting in a false binding of the two stimulus assemblies. It is
therefore necessary that this false binding is prevented by the segregation of
the stimulus representations in the other feature domains, which is conveyed

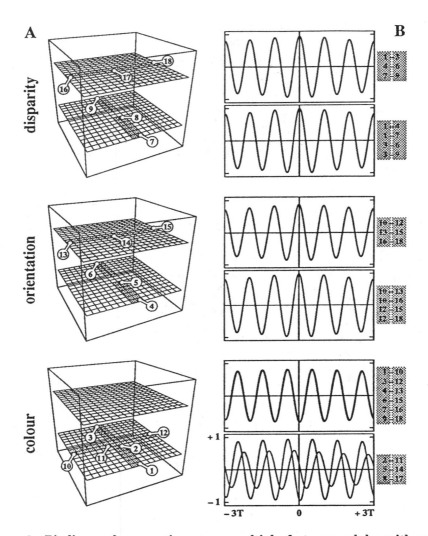

Figure 2: Binding and segregating across multiple feature modules with spatially overlapping stimuli. (A) Stimulus input representing two overlapping bars with identical colour, orthogonal orientation, and different disparities. (B) Mean cross-correlograms between activities of units at locations numbered in (A). The correlograms demonstrate the coherent synchronization within the respective representations of each of the two stimuli. With respect to each other, the activities of the two assemblies are out of phase.

to the colour module via the intermodule coupling. In our model, the necessary desynchronization is suitably provided by the desynchronizing connections already implemented for the function of the single feature module.

Note that the segregation of the two stimulus assemblies in this simulation is not due to asymmetric coupling weights: coupling weights within and between modules are chosen identical. Thus, none of the modules is distinguished. Note also that the demonstrated segregation cannot be accomplished by a suitable noise level in the system: any noise which is able to prevent false binding at the spatial overlap in the colour module will be equally efficient in preventing any synchronization elsewhere.

76.4 DISCUSSION

Within each single module, short range synchronizing connections support coherent activity of oscillators at neighbouring topographic and feature locations. By reason of these connections, any stimulus which is continuous with respect to the module's topographic and feature dimensions will be represented by a population of oscillators responding with synchronized activity. Contrastingly, long range desynchronizing connections along the module's feature dimension establish asynchronous activity of oscillators at neighbouring topographic but different feature locations. This renders responses to several distinct stimuli uncorrelated.

Between modules, synchronizing connections establish coherent activity of oscillators of different feature modules at corresponding topographic locations. This leads to the representation of the different features of a stimulus by a coherently oscillating assembly of cells distributed over the different modules in the network. The desynchronizing connections within each module provide the means for segregating these distributed assemblies across multiple feature domains. Without active desynchronization, stimuli having common features at nearby topographic locations would synchronize completely. This problem cannot be solved in general by increasing the system's noise level.

For related work investigating the synchronization between different feature modules see Sporns et al. [10] and Horn et al. [11].

REFERENCES

[1] C. v.d. Malsburg. *Internal Report 81-2, Max-Planck-Institute for Biophysical Chemistry, Göttingen*, 1981. [2] C.M. Gray, P. König, A.K. Engel, and W. Singer. *Nature* 338:334–337, 1989. [3] A.K. Engel, P. König, and W. Singer. *Proc.Natl.Acad.Sci.* 88:9136–9140, 1991. [4] A.K. Engel, A.K. Kreiter, P. König, and W. Singer. *Proc.Natl.Acad.Sci.* 88:6048–6052, 1991. [5] A.K. Kreiter and W. Singer. *Europ.J.Neurosci.* 4:369–375, 1992. [6] T.B. Schillen and P. König. *Internat.Joint Conf.Neural Networks* II:387–395, 1990. [7] P. König and T.B. Schillen. *Neural Comp.* 3:155–166, 1991. [8] T.B. Schillen and P. König. *Neural Comp.* 3:167–177, 1991. [9] P. König, B. Janosch, and T.B. Schillen. This volume, 1992. [10] O. Sporns, J.A. Gally, G.N. Reeke Jr., and G.M. Edelman. *Proc.Natl.Acad.Sci.* 86:7265–7269, 1989. [11] D. Horn, D. Sagi, and M. Usher. *Neural Comp.* 3:510–525, 1991.

ASSEMBLY FORMATION AND SEGREGATION BY A SELF-ORGANIZING NEURONAL OSCILLATOR MODEL

Peter König, Bernd Janosch and Thomas B. Schillen

Max-Planck-Institut für Hirnforschung
Deutschordenstraße 46, 6000 Frankfurt 71, Germany

ABSTRACT

Experimental evidence demonstrates the stimulus-dependent formation and segregation of neuronal assemblies defined by coherent oscillatory response patterns. In this paper, we investigate whether the self-organization of synchronizing and desynchronizing connections can establish a corresponding temporal response structure using local learning rules. Motivated by recent experimental observations, synchronizing connections are modified according to a two-threshold Hebb-like learning rule, while we generalize this rule to analogous Anti-Hebb-like weight changes for the desynchronizing connections. We show that the resulting network exhibits synchronization and segregation of oscillatory activity in agreement with the experiment.

77.1 INTRODUCTION

Current concepts of visual processing assume that local features of visual stimuli like disparity, orientation, colour, etc. are represented by distributed neuronal populations in the brain. Simultaneous processing of several stimuli in a natural scene leads to superposed responses in each of these feature representations. Several years ago it has been suggested that the temporal structure of neuronal activity could be used to define distinct assemblies, representing different visual stimuli, by the synchronization of their firing patterns [1,2,3]. These hypotheses have found support by physiological experiments in cat visual cortex. It could be shown that coherent stimuli, such as continuous light bars, lead to synchronous activity of oscillatory neuronal responses [4]. In contrast to this, incoherent stimuli, as for example two superposed light bars of different orientations, have been found to lead to synchronous oscillating activity in each of the corresponding assemblies, which were however not synchronized among each other [5].

This work has been supported in part by the *Deutsche Forschungsgemeinschaft*.

77.2 LEARNING THE FORMATION AND SEGREGATION OF ASSEMBLIES

We now investigate the self-organization of a neuronal network, demonstrating assembly formation and segregation in agreement with the experimental evidence. The network's basic oscillatory elements consist of excitatory (E) and inhibitory (I) units, coupled by delayed negative feedback loops [6]. Each such oscillator is interpreted as a population of neurons having a particular orientation specificity. Elementary oscillators are coupled by two types of delay connections. Horizontal connections for coupling different oscillators originate at the excitatory unit of one oscillator and terminate at either the inhibitory (EI) or the excitatory (EE) unit of another. The effect of these connections is to either synchronize (EI) or desynchronize (EE) the activities of the coupled oscillators [6,7].

The experimental situation of Engel et al. [5] is represented by a network of 16 coupled delayed nonlinear oscillators with equidistantly spaced preferred orientations (Fig. 2A). In the network's initial state the weights of synchronizing and desynchronizing connections between the oscillatory units are spatially homogeneous.

During the simulation 16 different input patterns, corresponding to the different preferred orientations of the oscillatory units, are presented in a pseudo-random manner. Based on recent observations in slices from rat visual cortex [8], we use the following two-threshold learning rule (ABS rule) for weight changes of the synchronizing connections in our network:

$$\Delta\omega(t) = \epsilon \, \bar{a}_{pre}(t) \, f_{ABS}(\bar{a}_{post}(t))$$

where t is time, $\Delta\omega$ is the synaptic weight change, ϵ is a rate constant, $\bar{a}_{pre}$ and $\bar{a}_{post}$ are respective mean activities of the pre- and postsynaptic unit, and

$$f_{ABS}(\bar{a}) = \begin{cases} C_0 & : & \bar{a} \leq \theta_1 \\ C_1 & : & \theta_1 < \bar{a} \leq \theta_2 \\ C_2 & : & \theta_2 < \bar{a} \end{cases} \quad \text{with } C_1 < C_0 < 0 < C_2$$

is a two-threshold function of the postsynaptic activity with thresholds θ_1 and θ_2. For the network's desynchronizing connections, we generalize the above modification scheme to a corresponding two-threshold anti-ABS learning rule ($C_2 < C_0 < 0 < C_1$).

Applying the above learning rules to the network while presenting training patterns leads to rapid changes of connection strengths. For synchronizing connections (EI) only the coupling to nearest and next nearest neighbouring oscillators maintains non-zero synaptic weights. All other EI connections degenerate to near-zero weights. The resulting tuning width of synaptic weights (15°) corresponds to only about half the width of the orientation tuning of neuronal activity (30°). Desynchronizing connections (EE) develop mainly between

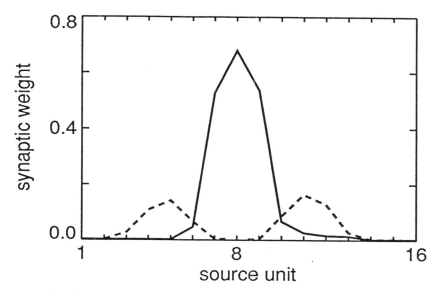

Figure 1: Development of synaptic weights during training. Weight distribution after 4000 stimulus presentations, corresponding to about seven minutes of real time. The figure shows the synaptic weights of synchronizing (EI, solid) and desynchronizing (EE, dashed) connections converging onto the oscillator at position eight.

elements of different preferred, but not orthogonal, orientations. The weight distributions after 4000 stimulus presentations, corresponding to 420 seconds, are depicted in figure 1.

After training was completed we tested the network with single and superposed stimuli in correspondence to the experimental situation. A single stimulus always led to a synchronously oscillating assembly of cells. Two superposed stimuli were represented by two independent assemblies. Within each assembly, all units were oscillating synchronously, but no constant phase relationship was found between the two assemblies (Fig. 2). Thus, the network exhibits a behaviour in agreement with the experimental observations.

77.3 DISCUSSION

The results described above demonstrate that it is possible to develop a network capable of stimulus-dependent assembly formation and segregation by simple local learning rules. Forms of plasticity as found in in vitro preparations of rat visual cortex [8] are well suited to account for the development of synchronizing connections. Furthermore, physiological evidence for the influence of plasticity of intracortical connections on the synchronization of assemblies has recently been found [9].

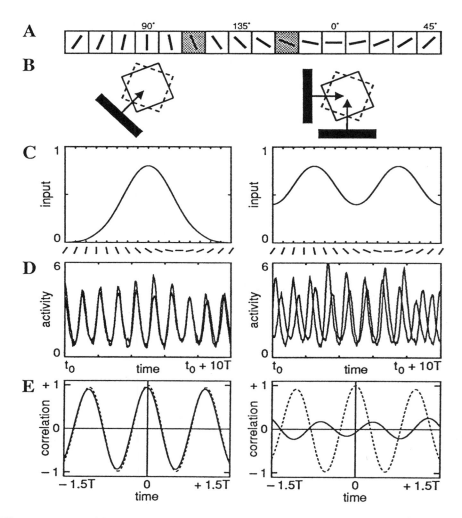

Figure 2: Assembly formation and segregation of temporal structure after learning. (A) 16 oscillators representing orientation-selective cells with identical receptive field locations. (B) Stimulus conditions of a single light bar (left) and two superposed light bars (right). (C) Input profiles corresponding to the stimulus conditions shown in (B). (D) Activity traces from the two units hatched in (A). In the case of a single stimulus (left) both oscillators are well synchronized and thus belong to the same oscillatory assembly. With superposed stimuli (right) each of the monitored oscillators couples to one of the two assemblies representing the two stimuli. Because of the desynchronizing connections the two assemblies are driven out of phase.

The applied modified learning-rule decreases synaptic weight if the postsynaptic neuron fails to exceed a certain threshold, which may be higher than spiking threshold. This leads to synchronizing connections between units of not more than 15° difference in preferred orientation, corresponding to only half the neuron's orientation tuning width. This allows efficient coarse coding while synchronizing only continuous stimuli.

The proposed Anti-Hebb-like rule for the plasticity of desynchronizing connections has not yet been discovered in physiological experiments. However, as most experimental studies concentrate on LTP in pyramidal neurons this may be no surprise. The two-threshold Anti-Hebb-like rule leads to desynchronizing connections (EE) between elements of different preferred, but not orthogonal, orientations. This allows segregation of stimuli which overlap partially in feature space. Furthermore, with respect to binding multiple feature domains, desynchronizing connections allow the segregation of overlapping stimuli which are identical in one feature expression, e.g. colour, by differences in another feature domain, e.g. disparity [10].

Anti-Hebb-like learning rules have been applied in neuronal networks before for the decorrelation of stimulus specificity on long time scales [11,12].

The network described above uses clearly defined objects, e.g. light bars as stimulus inputs. In a natural scene it may not be that clear what constitutes an object. Using an self-organizing approach allows the stimulus environment itself to define the entities of objects. The correlated appearance of their features, e.g. continuity, leads to the development of appropriate synchronizing and desynchronizing connections, which allows later stimulus segregation. This may be the largest benefit of a self-organizing network for segregating different stimuli.

REFERENCES

[1] P.M. Milner. *Psychological Reviews* 81:521–535, 1974. [2] C. v.d. Malsburg. *Internal Report 81-2, Max-Planck-Institute for Biophysical Chemistry, Göttingen*, 1981. [3] H. Shimizu, Y. Yamaguchi, I. Tsuda, and M. Yano. *Complex Systems – Operational Approaches*, Springer Verlag, pp. 225–239, 1985. [4] C.M. Gray, P. König, A.K. Engel, and W. Singer. *Nature* 338:334–337, 1989. [5] A.K. Engel, P. König, and W. Singer. *Proc.Natl.Acad.Sci.* 88:9136–9140, 1991. [6] P. König and T.B. Schillen. *Neural Comp.* 3:155–166, 1991. [7] T.B. Schillen and P. König. *Neural Comp.* 3:167–177, 1991. [8] A. Artola, S. Bröcher, and W. Singer. *Nature* 347:69–72, 1990. [9] P. König, A.K. Engel, S. Löwel, and W. Singer. *Neuroscience Abstracts* 523.2, 1990. [10] T.B. Schillen and P. König. This volume, 1992. [11] H. Barlow. *Neural Comp.* 1:295–311, 1989. [12] J. Rubner and K. Schulten. *Biol.Cybern.* 62:193–199, 1990.

INTER-AREA SYNCHRONIZATION IN MACAQUE NEOCORTEX DURING A VISUAL PATTERN DISCRIMINATION TASK

Steven L. Bressler

Center for Complex Systems, Florida Atlantic University, Boca Raton FL 33431

Richard Nakamura

Laboratory of Neuropsychology, NIMH, Bethesda MD 20892

ABSTRACT

Synchronization among cortical sites was investigated in a macaque monkey performing a visual pattern discrimination task. Single-site power and between-site coherence were computed from transcortical field potentials. At the time of the behavioral response, power and coherence increased for a select set of cortical sites widely distributed in one hemisphere. The increase was not specific to any narrow frequency band, but occurred over the entire observable range between 0 and 100 Hz.

78.1 INTRODUCTION

Although much is known about the properties of single cortical neurons, relatively little is understood about the cooperation of those neurons in large-scale cortical networks. Yet a leading single-cell neurophysiologist has claimed that in order to understand the relation between cortical function and complex behavior, it is necessary to study the dynamic aspects of network activity [1]. In recent years, inter-area synchronization has received attention as a possible mechanism of cortical organization at the network level.

Most of the work in this field has dealt with correlation of oscillatory activity in the γ frequency range (30-80 Hz) [2]. In the rabbit, olfactory perception during a sniff involves an increase in γ correlation between olfactory bulb and cortex that is location-specific in both structures [3]. In the cat visual system, increased γ correlation in response to coherent visual stimuli has been found between sites within striate cortex [4], between homologous striate sites of left and right hemispheres [5], and between striate sites and several extrastriate areas [6].

In monkeys there is still no consensus on the role, or even the existence, of γ oscillations in the cerebral cortex. On the one hand, it has been reported that γ oscillations transiently synchronize between sites in striate cortex of the macaque [7] and squirrel monkeys [8], between sites in the superior temporal sulcus of the macaque [9], and between sites in pre- and post-central sensorimotor cortex of the macaque, particularly during performance of tasks involving fine finger control [10]. On the other hand, one of these reports [7] found that the greater part of the occipital field potential power was concentrated at the low end of the spectrum, and that, although transitory spectral peaks appeared at multiple and variable frequencies in the γ range, the overall appearance of the average spectrum had the form of "1/f noise". Furthermore, the presence of γ oscillations in macaque visual cortex has been questioned by a report [11] which found a broad-band increase in power, and not an increase in γ-band resonant activity, from areas V1 and MT in response to optimally-oriented light bars. In macaque inferotemporal cortex, this report and others [12], also found very little evidence for visual pattern induced γ oscillations. Another group [13] has reported that oscillations are induced in the macaque temporal pole by familiar visual patterns, but in the Δ, not the γ, frequency range.

This report considers the question of inter-area synchronization of local field potentials (LFPs) from widely separate cortical sites in a macaque monkey performing a visual pattern discrimination task. It is important to study a functionally active monkey because synchronization may only occur transiently in conjunction with specific stages of information processing. The issue of frequency is also important. Those reports finding no evidence for γ oscillations in monkey, without considering functional interrelations between cortical sites, do not address the synchronization question. Correlated activity may be an important aspect of cortical function even if it is aperiodic and broad-band.

78.2 METHODS

Experiments were performed at the Laboratory of Neuropsychology at the National Institute of Mental Health in accordance with institutional guidelines. An adult rhesus macaque monkey was trained to perform a visual pattern discrimination task with a GO/NO-GO response paradigm. The monkey was water deprived to provide motivation for task performance. Its water consumption and behavior were carefully monitored to insure adequate hydration. Computer-generated visual stimuli were presented for 100 msec through a silenced piezoelectric shutter. The head was fixed to a headholder located 57 cm from the display screen, the visual angle being 1°/cm. The stimulus set had four diagonal patterns, each formed by four dots (each 9 mm on a side). Each pattern consisted of two dots at opposite corners of an outer square, 6 cm on a side, and two dots at opposite corners of a concentric inner square, 2 cm on a side. Diagonal lines had the outer and inner square dots slanted in the same direction, diagonal diamonds in opposite directions.

The monkey was trained to readily accept chair restraint and to initiate each trial by holding down a lever with the preferred (left) hand. The intertrial interval was randomly varied between 0.5 and 1.25 seconds. The stimulus was presented approximately 115 msec (varied over 25 msec to avoid line frequency locking) following the start of data collection. On GO trials, a water reward was provided 500 msec after stimulus (diagonal lines) onset if the monkey responded correctly by lifting his hand from the lever before 500 msec. On NO-GO trials, the monkey was required to hold down the lever for 500 msec following presentation of diagonal diamonds. GO and NO-GO trials were randomly presented with equal probability in 1000-trial sessions lasting approximately 35 minutes.

All surgical procedures were preceded by ketamine sedation and pentobarbital anesthesia. Surgical preparation was performed in two stages. First, following craniotomy, recording electrode positions were selected based on the sulcal markings found on the inner cranial surface. Stainless steel screws were inserted in the outer cranial surface at these positions and the calvarium was replaced. Additional screws were placed in the same region for grounding. The cranium was allowed to refuse for one month, after which the second surgical stage was performed to implant the recording electrodes. At each recording site, the screw was removed and an opening was made in the exposed dura. Each electrode consisted of a teflon-coated pair of platinum wires with 2.5 mm tip separation and each tip having a 0.5 mm long exposed surface. The electrode was advanced through the dural opening until the less-advanced tip rested against the pial surface. Electrodes were distributed across the right cortical convexity. They were affixed to the cranium with dental acrylic. The monkey was allowed to recover for at least two weeks before proceeding with data collection.

Results are presented from analysis of two sessions' data collected two weeks apart. The data set consisted of 892 correctly performed GO trials and 901 correctly performed NO-GO trials from 11 intracerebral electrode sites in the right hemisphere (Fig. 78.1): one striate, one prestriate, five parietal, two somatosensory, and two motor. Simultaneous transcortical LFPs were recorded by bipolar differential amplification using Grass Model 7P511J amplifiers with passband between 1 and 100 Hz (3 dB points). Digitization was at 200 samples/sec for each channel. Gain settings were individually adjusted for each channel to maximize signal amplitude without clipping. Gain correction was applied to each record to recover the original signal strength in microvolts, and a digital 60 Hz notch filter was applied to remove the effect of line-frequency contamination.

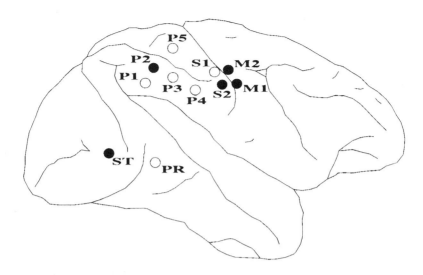

Figure 78.1 Eleven electrode sites were monitored in right cortical convexity of the rhesus macaque monkey. (ST: striate; PR: prestriate; P1 - P5: parietal; S1 - S2: somatosensory; M1 - M2: motor.) Closed circles: 5 sites (group A) in session 1 were synchronized during the response. Open circles: Non-synchronized sites (group B).

78.3 RESULTS

Average Potentials

The peak amplitudes of the post-stimulus average LFPs were in the range of 200-2000 μV. The components generally were of three types: an early negative peak (N1) at or prior to 200 msec, a mid-latency positive peak (P2) prior to 300 msec, and a late broad component after 300 msec. The GO and NO-GO average waveforms were highly similar for at least the first 150 msec post-stimulus at all sites. In session 1, four sites (parietal sites 1,3,4,5) showed only small between-condition differences in waveshape, and no long-lasting between-condition divergence. At the other sites, a distinct divergence began from 165 to 280 msec, a period prior to, and overlapping, the response onset in the GO condition (265 +/- 16 msec). Five of these sites (striate, parietal 2, somatosensory 2, and motor 1 and 2), were distinguished by the GO waveform showing a distinct peak at 345 msec, followed by a slow return toward baseline, that was absent in the NO-GO condition. This group of five sites (group A) thus had important similarities during the response, not shared with the other six (group B). In session 2, the distinction between groups A and B persisted, except that group A expanded to include parietal site 3, which had previously been in group B.

Power

To compare the frequency content of the GO and NO-GO event-related LFPs, the Discrete Fourier Transform (DFT) was computed for the time window consisting of the first 500 msec post-stimulus of each trial. The GO and NO-GO power spectra were averaged for each site and the natural logarithm was applied. These spectra decreased with increasing frequency in a characteristic "1/f" manner. Overall, the power of group A sites was significantly greater than group B by the Student's t-test (session 1: t=2.33, p < 0.05; session 2: t=12.04, p < 0.01; d.f.=9). In addition, power in the GO condition was greater than the NO-GO condition for all group A sites, the difference being significant (session 1: t=4.60, p < 0.01; session 2: t=5.37, p < 0.01; d.f.=9) as compared to group B.

For more precise localization of the time when the spectra of the two conditions differed, average DFTs were again computed, but now for 113 windows, each 80 msec long, centered from 60 msec pre-stimulus to 500 msec post-stimulus. The windows were highly overlapped, each differing from the previous by only one point. The greater temporal resolution resulted in relatively poor frequency resolution, the time window of 80 msec yielding frequency bins at 12.5 Hz intervals. Log power was displayed separately for each frequency bin as a time-series where each point corresponded to the center of one of the 113 windows. (The 0 Hz bin was excluded because of d.c. offset artifact, and the 100 Hz bin was excluded because of poor resolution at the Nyquist frequency.)

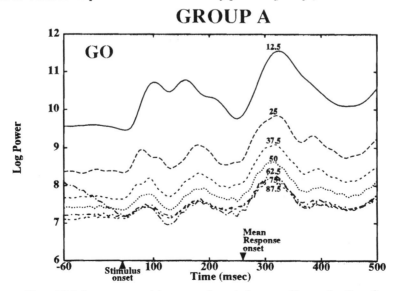

Figure 78.2 Average group A log power for each frequency bin as a function of time. The peak in power (310-325 msec) during the behavioral response was specific to group A sites. Power time-series are labeled in Hz by frequency bin.

The power time-series results were in general agreement with those of the amplitude averages, but showed that the late response-related components seen at group A sites consisted of power across the whole observable spectrum (Fig. 78.2). The major difference between conditions, like the amplitude averages, occurred in group A and consisted of an increase in power with the behavioral response for all frequencies in the GO condition. This change was not observed for group B.

Coherence

The average LFP waveforms indicated that at the time of the response in the GO condition an event took place that was similar at group A sites. Power spectral analysis suggested that this similarity extended across the observable frequency spectrum (0 - 100 Hz). However, single-site amplitude or power measures do not measure similarity directly. For this, the coherence spectrum [14], a measure of waveshape similarity, was employed. Coherence spectra of all 55 pairwise combinations of the 11 sites were computed for each of the same 113 80-msec-long windows that had been used to compute power spectra.

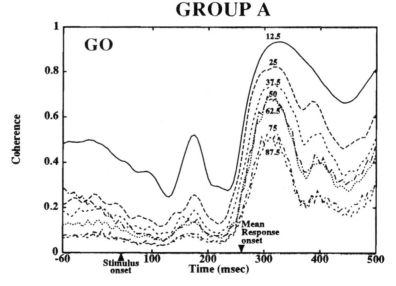

Figure 78.3 The average group A coherence time-series displayed for each frequency bin. The large broad-band increase in coherence with the behavioral response (GO condition) was specific to group A sites. Coherence time-series are labeled in Hz by frequency bin.

From the coherence spectra, coherence time-series were constructed for each available frequency bin. These coherence time-series were averaged for all within-group site pairs, yielding group A and group B average coherence time-series.

Since, by definition, coherence values lie within the range of 0 to 1, the Fisher z-transform, which converts a bounded distribution to an unbounded one, was used before averaging coherence.

The post-stimulus temporal course of coherence conformed to what was expected from the power time-series results. Group B mean coherence was uniformly low (below 0.2) at all frequencies in both conditions. For group A, mean coherence followed the same basic temporal pattern for all frequencies. In both conditions, group A coherence declined following the stimulus. In the GO condition, it then greatly increased beginning shortly before response onset and peaked near 325 msec (Fig. 78.3). NO-GO coherence remained low at this time. For both sessions, the coherence in the GO condition at 325 msec was significantly higher than the pre-stimulus level when compared to the group B pre-/post-stimulus difference (session 1: $t=28.90$, $p < 0.01$; session 2: $t=8.95$, $p < 0.01$; d.f.$=23$). The expansion of group A in session 2 to include parietal site 3 was confirmed by its high coherence with the other group A sites.

78.4 DISCUSSION

The most important finding of this study is that multiple, distributed sites in the monkey neocortex can become functionally linked in large-scale networks during visual task performance. The increase in coherence of group A sites at the time of the response suggests that inter-area synchronization is involved in task performance and that an extended cortical network serves to process task information.

The increase in coherence was spatially selective. Out of the eleven sites available, only five sites (group A) in session 1 became linked at the time of response onset. Spatial specificity was particularly evident in parietal cortex, where only one site out of five in session 1 was linked in the global network. Presumably the same degree of specificity was present in other cortical regions (e.g. striate cortex), although it could not be observed due to the limited number of electrode recording sites. Assuming adequate spatial sampling, the spatial pattern of coherence may be a sensitive indicator, at each successive processing stage, of those sites that are functionally linked in a cortical network.

The coherence peak was spectrally broad rather than being confined to narrow spectral peaks. When group A coherence increased with response onset, it did so over the entire range of observable frequencies, indicating that narrow-band oscillations are not a prerequisite for inter-area synchronization in a cortical network. That synchronization appears even at the low end of the spectrum implies that it is accessible in human scalp-recorded brain potentials, and lends credence to studies which have shown discrete spatial covariance patterns from the human brain related to specific stages of information processing [15].

Acknowledgement

Supported by NIMH Grant 43370 to EEG Systems Laboratory, San Francisco CA. We thank Don Krieger for assistance in data transfer.

REFERENCES

[1] Mountcastle, V.B. In *The Mindful Brain*, G. Edelman (Ed.) MIT Press, Cambridge (1979) 7.

[2] Bressler, S.L. & Freeman, S.L. *Electroencephalogr. Clin. Neurophysiol.* **50** (1980) 19; Bressler, S.L. *Trends Neurosci.* **13** (1990) 161.

[3] Bressler, S.L.. *Electroencephalogr. Clin. Neurophysiol.* **57** (1984) 270; Bressler, S.L. *Brain Res.* **409** (1987) 285; Bressler, S.L. *Brain Res.* **409** (1987) 294; Bressler, S.L. *Behav. Neurosci.* **102** (1988) 740.

[4] Gray, C.M., Konig, P., Engel, A.K. & Singer, W. *Nature* **338** (1989) 334; Gray, C.M., Engel, A.K. Konig, P. & Singer, W. *Visual Neurosci.* **8** (1992) 337; Engel, A.K., Konig, P. & Singer, W. *Proc. Natl. Acad. Sci. USA* **88** (1991) 9136.

[5] Engel, A.K., Konig, P., Kreiter, A.K. & Singer, W. *Science* **252** (1991) 1177.

[6] Eckhorn, R., Bauer, R., Jordan, W., Brosch, M., Kruse, W., Munk, M. & Reitboeck, H.J. *Biol. Cybern.* **60** (1988) 121; Engel, A.K., Kreiter, A.K., Konig, P. & Singer, W. *Proc. Natl. Acad. Sci. USA* **88** (1991) 6048.

[7] Freeman, W.J. & van Dijk, B.W. *Brain Res.* **422** (1987) 267.

[8] Livingstone, M.S. *Soc. Neurosci. Abstr.* **17** (1991) 176.

[9] Kreiter, A.K. & Singer, W. *Europ. J. Neurosci.* **4** (1992) 369.

[10] Murthy, V.N. & Fetz, E.E. *Proc. Natl. Acad. Sci. USA* **89** (1992) 5670.

[11] Young, M.P., Tanaka, K. & Yamane, S. *J. Neurophysiol.* **67** (1992) 1464.

[12] Gawne, T.J., Eskandar, E.N., Richmond, B.J. & Optican, L.M. *Soc. Neurosci. Abstr.* **17** (1991) 443; Tovee, M.J. and Rolls, E.T. *Neuroreport* **3** (1992) 369.

[13] Nakamura, K., Mikami, A. & Kubota, K. *Neurosci. Res.* **12** (1991) 293.

[14] Glaser, E.M. & Ruchkin, D.S. *Principles of Neurobiological Signal Analysis*, Academic Press, New York (1976) 174.

[15] Gevins, A.S. & Bressler, S.L. In *Functional Brain Imaging*, G. Pfurtscheller & F.H. Lopes da Silva (Eds.) Hans Huber Publishers, Bern (1988) 99.

ACETYLCHOLINE AND CORTICAL OSCILLATORY DYNAMICS

Hans Liljenström and Michael E. Hasselmo†

Department of Numerical Analysis and Computing Science
Royal Institute of Technology, S-100 44 Stockholm, Sweden

†Dept. of Psychology, Harvard University
33 Kirkland St., Cambridge, MA 02138, USA

ABSTRACT

Acetylcholine appears to regulate the oscillatory properties of cortical structures. Application of cholinergic agonists induces theta rhythm oscillatory patterns in the piriform (olfactory) cortex [1] and the hippocampus [2]. The effect of cholinergic modulation on cortical oscillatory dynamics was studied in a computational model of the piriform (olfactory) cortex [3]. The model included the cholinergic suppression of neuronal adaptation, the cholinergic suppression of intrinsic fiber synaptic transmission, and the cholinergic suppression of inhibitory synaptic transmission. In particular, we show that the increase in pyramidal cell excitability, due to suppression of neuronal adaptation by acetylcholine can enhance the gamma oscillations in evoked potentials and can induce theta rhythm oscillatory dynamics in the spontaneous potentials.

79.1   INTRODUCTION

Field potential recordings of cortical structures show oscillatory dynamics which appear to depend on intrinsic physiological features such as the time constants of inhibition. Simulations of the piriform cortex, the primary olfactory cortex, show that these oscillatory dynamics can be replicated in computational models based on detailed compartmental simulations of individual neurons [4] or more abstract representations of the transfer function of cortical neurons [3]. In experimental preparations, it has been shown that the cholinergic innervation of cortical regions strongly influences the oscillatory dynamics within those regions. Within the piriform cortex, application of cholinergic agonists increases gamma oscillations in the evoked response to stimulation of the lateral olfactory tract, and causes an eventual increase in theta oscillations in the spontaneous potentials [1]. A wide range of research has shown that the induction of theta rhythms within the hippocampus depends upon cholinergic innervation arising from the medial septum [2]. While some of this modulation may be rhythmic, it has been shown that theta type oscillations can be induced in the hippocampus with tonic, non-rhythmic cholinergic modulation, as shown by microinfusion of cholinergic agonists [5] or by investigation of oscillatory dynamics in hippocampal slice preparations [6].

Computational models of cortical oscillatory dynamics provide an ideal technique for examining how the neuromodulatory effects of acetylcholine influence cortical oscillatory properties. In the research presented here, we have used a previously developed computational model of piriform cortex oscillatory dynamics [3]. This model can closely replicate experimental data from the piriform cortex showing characteristics of the EEG [7] and evoked potentials [8]. The oscillatory dynamics of this model depended solely upon the network circuitry and the characteristic time constants of two types of inhibition. These were based upon experimental data from intracellular recordings in brain slice preparations of the piriform cortex [9]. Here we show how implementation of cholinergic modulation within the model influences the oscillatory dynamics of the network. The cholinergic increase in excitability causes increased oscillatory activity within the model, while the cholinergic suppression of intrinsic synaptic transmission prevents gamma rhythm oscillatory dynamics from predominating.

79.2   METHODS AND RESULTS

We have analyzed the effects of cortical cholinergic neuromodulatory effects in a computational model of piriform cortex oscillatory dynamics [3]. The essential properties of the model have been presented previously, with the basic structure sketched in Fig. 1. In this model, the input-output relationship of populations of neurons is represented as an experimentally determined sigmoid function, with a single parameter Q, which determines threshold, slope and amplitude [10]. This parameter seems to depend on the level of acetylcholine in the system (Freeman, personal communication). The output function of a network unit i, with activity u_i, is given by the function g_i (with C as a normalisation constant),

$$g_i(u_i) = CQ_i\{1 - exp[-(exp(u_i) - 1)/Q_i]\}.$$

The neurons also have a characteristic time constant of decay of activity, different for each cell type. The model includes excitatory afferent input and excitatory intrinsic connections as well as feedback and feedforward inhibitory interneurons with physiologically realistic values for delays due to transmission of activity along interconnecting fibers. These characteristics of the model initially formed the basis for oscillatory properties which mimic the actual oscillatory properties of the piriform cortex, combining the faster "gamma" type oscillations (40-60 Hz) and the slower "theta" type oscillations (5-10 Hz). An additional contribution to the oscillatory properties comes from neuronal adaptation. Here we will focus on the modifications of this model used to represent the neuromodulatory influences of acetylcholine. These modifications mainly involve the input-output function of the network units, and the connection weights of intrinsic excitatory and inhibitory units.

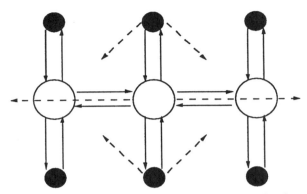

Figure 1 The principal structure of the network model, based on the three layered structure of the olfactory cortex. Open circles are excitatory units, corresponding to populations of pyramidal cells, extensively interconnected. Closed circles correspond to populations of inhibitory interneurons of two different types. (For more details see [3]).

In order to reflect the influences of cholinergic modulation on neuronal excitability, it was necessary to implement neuronal adaptation within the model. This neuronal adaptation arises from voltage-dependent and calcium- dependent potassium currents in cortical pyramidal cells. Depolarization or the influx of calcium due to spiking activity activates these potassium currents, causing a decrease in the frequency of subsequent spiking activity. This phenomenon has been described in piriform cortex [11, 12] and in hippocampus [13]. Neuronal adaptation was modeled by shifting the gain parameter Q of each excitatory unit dependent upon the previous activity of that particular unit. This serves as an approximation to the flattening of the f/I curve observed experimentally at later stages of a sustained depolarization. Here we have

$$Q_i = Q_c exp[- < u_i(t) >^2_{\tau_0} /D],$$

where Q_c is a constant determining the maximum gain, $< u_i(t) >_{\tau_0}$ is the time average of the unit activity over the last τ_0 msecs, and D is an adaptation parameter.

Experimental work in the piriform cortex [11, 12] and hippocampus [13] shows that acetylcholine suppresses neuronal adaptation via activation of muscarinic cholinergic receptors. These receptors shut down the maximal conductance of voltage- and calcium-dependent potassium currents via second messenger systems. This allows the neuron to fire in a more sustained manner in response to input. In the model, this effect of acetylcholine was represented by decreasing the change in Q dependent upon previous activity, i.e. by increasing the adaptation parameter D.

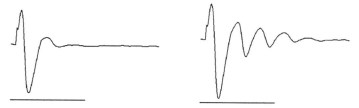

Figure 2   Simulated evoked potential without (left trace) and with (right trace) cholinergic suppression of adaptation. (Time bar is 100 ms).

Acetylcholine also modulates synaptic transmission within cortical structures. It suppresses excitatory synaptic transmission with selectivity for intrinsic but not afferent fiber synaptic transmission within the piriform cortex [14]. In addition, it appears to suppress inhibitory synaptic transmission as well [15]. These effects were represented in the model by decreasing the strength of the different types of synaptic transmission. Thus, cholinergic modulation was represented by selectively decreasing the strength of intrinsic but not afferent excitatory connections, and by decreasing the strength of the two types of inhibitory synaptic transmission.

Introducing neuronal adaptation in the model results in new oscillatory properties, in addition to the network dependent oscillations. It is now possible to get oscillations in the model with only excitatory connections involved, having all inhibitory connections blocked. The frequency of these oscillations depend on the time constant τ_0, and on the adaptation parameter D. Theta rhythm oscillations can be induced by noise with e.g. $\tau_0 = 100ms$ and $D = 10.0$. If D is much smaller, no oscillations will occur.

Implementation of the neuromodulatory effects of acetylcholine in the model caused changes in the oscillatory dynamics analogous to those seen in physiological experiments. In particular, as shown in Fig. 2, the suppression of neuronal adaptation (increasing D) resulted in increased gamma frequency oscillations in the simulated evoked potentials, similar to experimental effects in piriform cortex [1]. Application of the effects of acetylcholine in the model could induce the appearance of theta (and/or gamma) oscillations, even when starting from a state with no clear oscillatory activity. Thus, as shown in Fig. 3, suppression of adaptation and suppression of synaptic transmission could induce theta rhythm oscillatory characteristics in the spontaneous potential, similar to effects seen in experiments on the piriform cortex [1] and the hippocampus [2, 6].

Increasing the adaptation parameter D to represent greater excitability increased the oscillatory properties of the model. Increasing the maximum gain constant Q_c gives a similar result, which acts additively to the increase in D. (It is uncertain whether acetylcholine has this effect *in vivo*). The predominant oscillatory characteristics of the model are also strongly influenced by the selec-

Figure 3 Simulation output showing cholinergic induction of theta rhythm oscillations. At an initially non-oscillatory state the adapatation parameter D is increased and synaptic transmission suppressed, mimicking the neuromodulatory effects of acetylcholine. (Time bar is 100 ms).

tive suppression of excitatory intrinsic fiber synaptic transmission. Suppression of intrinsic fiber synaptic transmission strongly decreases the tendency toward gamma rhythm oscillations, thereby increasing the relative contribution of theta rhythm oscillations. Changing the connection strengths by 10 % can result in quite different dynamic behavior, similar to the change in adaptation or gain parameters discussed above. Thus, while the increase in excitability enhances all oscillatory dynamics, the suppression of intrinsic fiber synaptic transmission appears to push the increase predominantly toward theta rhythm oscillations.

The inhibitory connections also play a role in the oscillatory properties of the system. In addition to the cholinergic effects on intrinsic excitatory connections direct cholinergic modulation of inhibition can further determine the oscillatory state. Thus, if cholinergic modulation also shut down excitatory input from pyramidal cells to feedback inhibitory interneurons, the faster oscillations would be suppressed. There is also some physiological evidence that cholinergic agonists suppress the output of inhibitory interneurons [15].

Thus, the different effects of cholinergic modulation can make the system go from a stationary state to an oscillatory mode, or from one oscillatory mode to another, with either gamma or theta frequency predominating. The final state partly depends on the relative strength of the inhibitory connections. When the connection strengths in the faster, feedback inhibitory loops were strongest, the model would show predominant gamma oscillatory properties. When the connection strengths involved in slower inhibition were strongest (or when suppression of synaptic transmission allowed the time course of neuronal adaptation to predominate), the model would show predominant theta oscillatory properties. When both inhibitory systems were of equal strengths, the interaction of the two oscillatory modes could result in a single oscillatory mode with an intermediate frequency (e.g. 12 Hz instead of 6 and/or 40 Hz). Thus, the final oscillatory characteristics induced by cholinergic changes in excitability depended to some extent on the relative strength of inhibitory processes. This, together with the particular time course of neuronal adaptation, could explain the cholinergic initial increase in gamma oscillations in the evoked potential (Fig. 2) and the subsequent increase in theta oscillations in piriform cortex spontaneous activity (Fig. 3) [1].

79.3 DISCUSSION

The model presented here shows how experimental data on the cholinergic modulation of cortical oscillatory dynamics can be linked to the experimental data on specific neuromodulatory effects of acetylcholine on cortical neurons. Thus, the model accurately reflects how application of the cholinergic agonist carbachol to the piriform cortex alters the evoked potential and spontaneous EEG of this region [1]. These effects on evoked potential and EEG are obtained in the model by implementing experimentally verified effects of acetylcholine, including the suppression of neuronal adaptation, excitatory intrinsic synaptic transmission and inhibitory synaptic transmission. In particular, we have shown that theta rhythm oscillatory behavior can arise from the time course of adaptation, from inhibitory feedback, or from both of these effects combined. The effects are additive and could reflect a general tuning to theta rhythm oscillations in the real systems.

The changes in oscillatory dynamics described here may prove very important for the role of acetylcholine in learning and memory. Acetylcholine has been shown to play a role in memory function in a range of tasks [16], and computational models show that the effects of acetylcholine on cortical structures may enhance their associative memory properties [17]. The effects of acetylcholine on cortical oscillations may also play a role in learning and memory, since it has been shown that theta rhythm oscillatory dynamics appear important in memory behavior [18] and in the induction of long-term potentiation [19]. The model presented here is currently being used to further explore the role of cholinergic modulation of cortical dynamics in memory function. Preliminary results show that the recall time can be greatly reduced and accuracy can be improved in associative memory tasks, when learning occurs with increased gain and reduced synaptic transmission for intrinsic connections [20].

These results may apply to cholinergic modulation of oscillatory dynamics in the hippocampal formation. The piriform cortex and subregions of the hippocampus have a similar three layered structures, and show similar effects of cholinergic modulation. In brain slice preparations of the hippocampus, cholinergic agonists have been shown to induce theta oscillations of extracellular potentials [6, 21]. However, these theta rhythms are not sensitive to blockers of inhibitory synaptic transmission [21]. A computational model of region CA3 suggests that these oscillations are driven by intrinsically bursting neurons [22]. However, we have shown that they could arise from the time course of adaptation. Thus, the mechanisms described here could apply to both in vivo theta rhythms in the piriform cortex and hippocampus, which may be determined by the time constant of inhibition, and to theta rhythms in the hippocampal slice preparations, which may be determined by the time constant of adaptation. The modeling techniques presented here may prove relevant to modeling cholinergic modulation of oscillatory dynamics of neocortical structures as well. Since neuromodulation of cortical structures is influenced by the level of arousal and motivation in an animal, the modeling presented here may allow details of cortical physiology to be understood in the context of specific behavioral states.

79.4 ACKNOWLEDGEMENTS

Financial support from the Swedish Natural Science Research Council (H.L.) and from the French Foundation for Alzheimer Research (M.E.H.) is gratefully acknowledged.

REFERENCES

[1] Biedenbach, M. A., "Effects of anesthetics and cholinergic drugs on prepyriform electrical activity in cats," *Exp. Neurol.*, vol. 16, 1966, pp. 464–479.

[2] Bland, B. H., "The physiology and pharmacology of hippocampal formation theta rhythms," *Prog. Neurobiol.*, vol. 26, 1986, pp. 1–54.

[3] Liljenström, H., "Modeling the dynamics of olfactory cortex using simplified network units and realistic architecture," *Int. J. Neural Systems*, vol. 2, 1991, pp. 1–15.

[4] Wilson, M. and Bower, J., "Cortical oscillations and temporal interactions in a computer simulation of piriform cortex," *J. Neurophysiol.*, vol. 67, 1992, pp. 981–995.

[5] Rowntree, C. and Bland, B., "An analysis of cholinoceptive neurons in the hippocampal formation by direct microinfusion," *Brain Research*, vol. 362, 1986, pp. 98–113.

[6] Konopacki, J., MacIver, M., Roth, S.H., and Bland, B., "Carbachol induced EEG 'theta' activity in hippocampal brain slices," *Brain Research*, vol. 405, 1987, pp. 196–198.

[7] Bressler, S., "Spatial organization of EEGs from olfactory bulb and cortex," *Electroencephalogr. Clin. Neurophysiol.*, vol. 57, 1980, pp. 270–276.

[8] Bressler, S. and Freeman, W., "Frequency analysis of olfactory system EEG in cat, rabbit and rat," *Electroencephalogr. Clin. Neurophysiol.*, vol. 50, 1980, pp. 19–24.

[9] Tseng, G.-F. and Haberly, L., "Characterization of synaptically mediated fast and slow inhibitory processes in piriform cortex and in vitro slice preparation," *J. Neurophysiol.*, vol. 59, 1988, p. 1352.

[10] Freeman, W., "Nonlinear gain mediating cortical stimulus-response relations," *Biol. Cybern.*, vol. 33, 1979, pp. 237–247.

530

[11] Constanti, A. and Sim, J., "Calcium-dependent potassium conductance in guinea-pig olfactory cortex neurones in vitro," *J. Physiol.*, vol. 387, 1987, pp. 173–194.

[12] Hasselmo, M. and Barkai, E., "Cholinergic modulation of the input/output function of rat piriform cortex pyramidal cells," *Soc. Neurosci. Abstr.*, vol. 18:220.9, 1992, p. 521.

[13] Nicoll, R., "The coupling of neurotransmitter receptors to ion channnels in the brain," *Science*, vol. 241, 1988, pp. 545–551.

[14] Hasselmo, M. and Bower, J., "Cholinergic suppression specific to intrinsic not afferent fiber synapses in rat piriform (olfactory) cortex," *J. Neurophysiol.*, vol. 67, 1991, pp. 1222–1229.

[15] Pitler, T. and Alger, B., "Cholinergic excitation of gabaergic interneurons in the rat hippocampal slice," *J. Physiol.*, vol. 450, 1992, pp. 127–142.

[16] Hagan, J. and Morris, R., "The cholinergic hypothesis of memory: A review of the animal experiments," in *Psychopharmacology of the Aging Nervous System* (L.L. Iversen, S. I. and Snyder, S., eds.), New York: Plenum Press, 1989, pp. 237–324.

[17] Hasselmo, M., Anderson, B., and Bower, J., "Cholinergic modulation of cortical associative memory function," *J. Neurophysiol.*, vol. 67, 1991, pp. 1230–1246.

[18] Winson, J., "Loss of hippocampal theta rhythm results in spatial memory deficit in the rat," *Science*, vol. 201, 1978, pp. 160–163.

[19] Larson, J. and Lynch, G., "Induction of synaptic potentiation in hippocampus by pattern stimulation involves two events," *Science*, vol. 232, 1986, pp. 985–988.

[20] Liljenström, H., "Oscillations and associative memory: brain and model," in *Proc. Symp. Brain and Mind* (Cotterill, R., ed.), Royal Danish Academy of Sciences and Letters, August 16-21 1992 (manuscript).

[21] MacVicar, B. and Tse, F., "Local neuronal circuitry underlying cholinergic rhythmic slow activity in CA3 area of rat hippocampal slices," *J. Physiol.*, vol. 417, 1989, pp. 197–212.

[22] Traub, R., Miles, R., and Buzsaki, G., "Computer simulation of carbachol-driven rhythmic population oscillations in the CA3 region of the in vitro rat hippocampus," *J. Physiol.*, vol. 451, 1992, pp. 653–672.

DISSIPATIVE STRUCTURES AND SELF-ORGANIZING CRITICALITY IN NEURAL NETWORKS WITH SPATIALLY LOCALIZED CONNECTIVITY

Robert W. Kentridge

Department of Psychology, University of Durham, Durham DH1 3LE, U.K.

1 LOCALIZED CONNECTIVITY MODELS

Artificial neural network (ANN) models suggest principles which may underly cortical information processing. Nevertheless, differences between the cortex and the architectures and unit characteristics of ANNs imply that quite different collective behaviors may be used for information processing in the brain. The cortex neither contains symmetric interconnections between neurons nor conforms to the constraints of ANN feedforward architectures. Although cortex could be treated as a 'dilution' of a symmetrically connected network this ignores its spatial structure, treating it as a randomly connected network. The introduction through structure of a meaningful length scale to cyclically connected systems allows a class of behavior to arise between simple attractor dynamics and chaos—self-organized dissipative structure formation. Dissipative structures are spatiotemporally organized patterns of energy dissipation in systems far from equilibrium. Their formation depends on a balance between interactions of spatial and temporal instabilities in a system. Coupling between diffusive spatial processes and localized state-determining processes allow particular states to become dominant over a wide ranges. In neural networks with spatially localized connectivity activity in one group of neurons will tend to be transmitted to neighboring cells diffusively. Competition between excitatory and inhibitory processes within a localized group of cells and intra-neural processes of membrane decay and refractoriness can produce localized selection of sustained activity, sustained quiescence or even more complex behaviors. This paper investigates the likelihood that dissipative structure formation in networks containing these features could provide a basis for information processing in the brain.

In the networks described here the probability of neural interconnection was proportional to a gaussian with standard deviation sd of the distance between neurons. The strength of connections between neurons were randomly set between 0 and 1. The neurons were spiking leaky integrators with an absolute refractory period r.

$$\text{if} \quad v_j(t) > th_j \qquad x_j(t) = ap_j \quad \text{and} \quad v_j(t+1+r) = \Sigma w_{ij} x_i(t+r)$$

$$\text{if} \quad v_j(t) \le th_j \qquad x_j(t) = 0 \quad \text{and} \quad v_j(t+1) = \Sigma w_{ij} x_i(t) + d_j v_j(t)$$

t denotes time, $v_j(t)$ the 'potential' of neuron j, th_j its firing threshold, $x_j(t)$ its activity, ap_j the strength of an action potential, d_j the time decay constant of potential and w_{ij} the synaptic strength between neurons i and j. The network parameters were average threshold th, average membrane decay-constant d, standard deviation of neurons' connectivity gaussian sd and average connectivity of each neuron n. These variables were selected from gaussian distributions. The refractory period and the strength and sign of the action potential ap were fixed for each type of neuron.

2 STRUCTURE AND DYNAMICS

The conditions under which one-dimensional networks could support the simplest dissipative structures, propagating pulses of activity, were investigated. The networks were initially quiescent, with all membrane potentials set to zero. A group of adjacent neurons at the bottom of the one-dimensional networks were then fired and the subsequent behavior was studied. The networks consisted of 1005 neurons of which 75% were excitatory and 25% inhibitory. Parameters were $n=75$, $sd=75$, $ap=-10$, $d=0.95$, $th=50$, $r=1$ for the inhibitory neurons and $n=75$, $sd=75$, $ap=1$, $d=0.95$, $r=1$ for the excitatory ones. The average threshold th of neurons varied between 10.4 and 10.8. 100 additional excitatory neurons fired together at the bottom of the net provided the initial stimulus.

Three results are possible, either a pulse travels across the entire network without losing coherence, or the pulse may decay, or it may lose coherence and spread through the net chaotically. These three different behaviors were found in networks differing by less than 5% in the average value of their average neural membrane time-constants, moreover, the boundary between those behaviors were not abrupt, but rather 'islands' of each behavior appear and disappear as parameter values vary. Such sensitivity appears to make a functional role for dissipative structures unlikely in biological systems. This parameter sensitivity problem has been well well known since the classic work of Beurle [1] on nets of this type, we will now consider its solution.

In the previous example the stimulus has two separate roles to play, providing the coherent seed for a larger scale structure, and putting the system into a state near a phase-transition. The choice of parameter values may be less critical if the network is maintained near a phase-transition by diffuse background activity. Studies of two-dimensional networks in which each neuron in the network had a small amount of potential added to it each time-step were undertaken. A variety of behaviors in which coherent non-linear wave fronts formed were found. These wave fronts formed target waves typical of dissipative structures in chemical systems being able to annihilate one-another in nets in which multiple centers of activity formed (see [2] for further details).

The assumption that diffuse low-intensity stimulation maintained networks in a critical regime and therefore facilitated dissipative structure formation is now examined in more detail. Diffusely driven neural networks with spatially localized connectivity and forest fire models used to study critical phenomena in percolation theory share many features. In some forest fire models a critical state can be maintained over a relatively wide range of conditions though an active relaxation process of self-organized criticality [3]. Production of criticality over a similar wide parameter range would increase the biological plausibility of a role for dissipative structures in cortical information processing. The procedures used in [3] were applied to examine networks for evidence of self-organizing criticality. The spatial hallmark of criticality is the wide range of scales over which components in a system interact. This is reflected in the power-law distribution of the sizes of clusters of components sharing similar energies (for neurons, membrane potentials). The relationship between cluster size and their probability of occurrence can be measured objectively using the structure function:

$$S_q(r) = < [|v_i - v_{i+r}|^q]$$

v_i is the membrane potential of neuron i, v_{i+r} is the potential of another neuron at a distance r from i, square brackets indicate averaging over a shell of all neurons at the distance r, angle brackets indicate time-averaging. The structure function of a network of 2304 units was measured over 50,000 time steps. Network parameters were $n=15$, $sd=3$, $ap=1$, $th=7.5$, $d=0.999$, $r=1$. 0.05 units of potential were injected into each unit each time step. Figure 1 (left) shows membrane potentials (proportionate to the diameter of the circles, firing neurons are shown as asterisks) in the network at one time. The range of cluster sizes is similar to that expected at criticality (cf. figure 1 in [3]). The method used to calculate the membrane potential structure function (figure 2 right) was very computationally intensive which limited the length of the run. More conclusive results may have been obtained had a longer run been possible. The structure function obtained does appear to show a power law relationship, how-

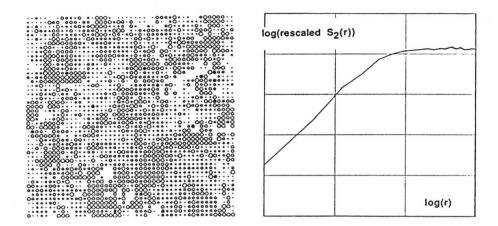

Figure 1 Membrane potentials (left) and structure function (right) in a 2 dimensional network.

ever, the scaling range was well below system size. Nevertheless, the average range of connections in the network was approximately 2 units and 90% of all connections were over less than 5 units so the scaling relationship cannot be directly attributed to local connectivity.

If one of a pair of otherwise identical systems is subject to a very small perturbation then the growth of the difference between the states of the two systems characterizes its dynamics. In chaotic systems the perturbation grows exponentially in time, in non-chaotic ones it eventually stops growing, in the critical regime growth is as a power of time rather than exponential. Perturbation growth was measured in neural networks by making small changes to the membrane potentials of neurons in one of a pair of networks and then measuring the growth of this difference between corresponding units. Both membrane and action potential variable difference were measured in a series of simulations. In order to differentiate between noisy exponential and power-law growth a control condition in which a network whose range of connectivity was so large as to be essentially space independent, but which was identical to the experimental network in all other parameters was used. Figure 2 shows results produced by toroidal networks of 8,100 neurons with the following parameters: $n=15$, $ap=1$, $th=15$, $d=0.999$, $r=1$. The connectivity range was $sd=3$ in the experimental condition and $sd=900$ for the control. On each step each unit was injected with 0.15 units of potential in one series of experiments and 0.075 units in a second series. After 500 steps the potential in each unit was perturbed in one net of each pair by an average of 0.0015 units. The results from one experiment (0.075 unit driving rate, potential difference measure) are shown in figure 2.

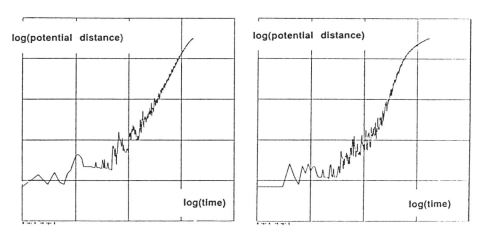

Figure 2 Perturbation growth in experimental (left) and control (right) networks.

The experimental case shows clear power law growth while the control data show chaotic exponential growth. Similar results were obtained with the action potential difference measure and at the higher 0.15 unit driving rate. The fact that critical behavior was not disrupted by a doubling of the driving rate is indicative of self-organizing criticality. These studies show that it is quite possible that dissipative structures of coherent activity may form in cortex-like networks. Self-organizing criticality may play a role in increasing the likelihood that such structures can form. Evidence for $1/f$ power spectra in the cortex [4] which are only readily explained in terms of self-organizing criticality, support the notion that the processes investigated in this paper may indeed be taking place in the cortex.

REFERENCES

[1] Beurle, R.L. (1956) Phil. Trans. R. Soc. B **240** 55-94.

[2] Kentridge, R.W. (1993) *Critical dynamics of networks with spatially localised connections.* In M. Oaksford & G. Brown (Eds.) *Psychology & Neurodynamics.* London: Academic Press.

[3] Chen, K.; Bak, P. & Jensen, M.H. (1990) Phys. Lett. A **149** 207-10.

[4] Young, M.P.; Tanaka, K. & Yamane, S. (1992) J. Neurophysiol. **67** 1464-74.

INDEX

A

acetylcholine 268, 273, 377, 459, 523
adaptation 363, 464
alpha function 63, 133
Alzheimer's disease 461, 467
amacrine 159
ambiguus, nucleus 90
apparent motion 159
ART classifier 224
associative memory 261, 273, 431, 443
attention   215, 221
auditory system
  development 438
  peripheral 229, 241
auditory cortex   see cortical areas,
auditory

B

backpropagation 252, 297, 337
baroreceptor 82, 89
Bayesian statistics 55, 432
bifurcation, Hopf 263
binding problem 487, 496, 503
boundary element method 283
brainstem 241, 307
burst 261, 495

C

CABLE 98, 115
CAJAL 331, 482
cell assembly 431, 449, 503, 509
central pattern generator 295
cerebellum 67, 331, 343, 349, 355
channel, in membrane see ionic current
chaotic system 43, 264, 369, 481, 531
chemoreceptor 249, 255
coding hypothesis 43
color blobs 171, 173
compartmental model   see model,
compartmental
complex cells 171, 173
conditioning 173, 331
correlation function 207, 488
correlations 27, 68, 207, 210, 390, 437

cortex
  areas of see cortical areas
  oscillation in 475, 481, 487, 495,
    503, 509, 515, 523
cortical areas,
  extrastriate 495, 515

maps 395, 403, 409, 415
MT 455, 495
olfactory 267, 273, 481, 515, 523
sensorimotor 475, 516
V1 171, 215, 221, 389, 395
visual 171, 201, 207, 383, 395, 403,
    409, 481, 487, 495, 503, 509, 515
current clamp 81, 229, 291
current,
  calcium 5, 38, 116, 133, 308, 343,
    356, 432
  potassium 5, 10, 22, 32, 38, 91, 110,
    116, 153, 308, 345, 356, 432, 476
  sodium 10, 22, 32, 91, 110, 116,
    153, 308, 356, 476

D

delay line 73, 450
dendritic spines 345, 356, 425
development, in visual system
    see visual system, development
directional selectivity 159, 201
distributed processing 337, 340
dopamine 369
dynamic organization 532

E

EEG 487, 515, 523, 524
electric organ 281
electroreceptor 281
enteric nervous system 313
entropy 43
EPSP 97, 358, 475
evoked potentials 524
eye blink 331

F

fast-most model 165
feature detection 189, 438, 503
feedback 215, 256, 383, 389, 437,
    456, 510
field potential   see EEG
finite-element method 283
firing patterns 49, 265
Fourier series 367, 497

G

GABA 221
GABA-A receptor 159, 202, 274, 357
GABA-B receptor 202, 274, 458
Gabor filter 171
ganglion cells, in retina 159, 165

Genesis 38, 110, 115, 134, 273, 324, 356, 426, 476
genetic algorithm 295
grandmother cell 25

H
habituation 103
hair cell 229
Hebb rule 3, 207, 262, 416, 431,
  see also learning, Hebbian
hippocampus 103, 445, 523
homunculus 55

I
ideal observer 55
information, in neural signal 43, 165, 487
intersegmental coordination 301
ionic current see current
ISIH 499
isocortex 455

K
Kohonen, feature maps of 404

L
lamprey 301
lateral geniculate nucleus 38, 98, 171, 202, 215, 389
leaky-integrator neuron 61, 532
learning
  backpropagation see backpropagation
  covariance 377, 443
  Hebbian 3, 25, 172, 207, 273, 366, 377, 383, 397, 415, 439, 509
  in conditioning 332
  reinforcement 364
  unsupervised 25, 363, 437
LGN see lateral geniculate nucleus
limulus 151
LTP 268, 377, 416, 513

M
MacNeuron 116
matched filter 27
model
  compartmental 4, 6, 11, 19, 38, 97, 115, 159, 251, 274, 331, 343, 356,
426, 476, 481,523
connectionist see neural networks
Hodgkin-Huxley 9, 19, 32, 81, 89, 109, 116, 237, 289, 307, 432, 476, 482
integrate and fire 61, 255, 296
Jeffress 235, 241
network 89, 115, 151, 215, 274, 307, 531
of development 383, 389, 395, 403, 415
resonance 229
single compartment 19, 81
motor unit 325
motorneurons 10, 291, 314, 325, 337, 433
movement 369
MSO (medial superior olive) 241
multicomputer 121

N
NeMoSys 11
network, recurrent 337, 431, 467
neural
  assembly see cell assembly
  code 9, 55
  networks 20, 49, 67, 115, 121, 224, 295, 307, 323, 363, 431, 437, 455, 461, 467, 531
neuromodulation 273, 433, 524
NEURON 38, 134
nictitating membrane 332
nitric oxide 377
NMDA 202, 378, 425, 432
noise process, Poisson 476, 495

O
ocular dominance columns 171, 208, 377, 389, 395, 409, 415
olfactory system 255, 261, 267, 523
orientation columns 395, 409
oscillation 251, 289, 301, 308, 481, 488, 495, 503, 509, 523

P
parallel fibers 331, 349, 357

peristaltic reflex 313
photoreceptor, invertebrate 152
Purkinje cell 67, 331, 343, 349, 355
pyramidal cell 97, 201, 221, 267, 431

R
Rallpack 133
receptive field 165, 171, 189, 395
Renshaw cell 324
resonance 289
respiration 307
retina
  lateral interactions 152
  processing 106, 143, 157, 159, 165,
    171
rhythm 295, 307

S
SABER 115, 426
self-organization 3, 403, 509, 531
serotonin 291, 433
SFINX 146, 192
sigmoidal function 230, 369, 452, 482
silicon neuron 127
simulation, Monte-Carlo 222
simulator see model
solitarius, nucleus 89
speech production 363
spike-initiating zone 11
stretch reflex 323
SWIM 432
synaptic
  compensation 467
  integration 97, 355
synchrony 475, 481, 487, 495, 503,
  509, 515

T
telencephalon 104
texture 189
tuning characteristic 216, 229

U
undulatory locomotion 301
user interface 118

V
velocity storage 337
vestibulo-ocular reflex (VOR) 337
vision, cortex see cortical areas, visual
visual system, development 171, 207,
  383, 389, 395, 403, 415
VLSI 127
voltage clamp 32, 37, 111

X
X-cell, in retina 165

Y
Y-cell, in retina 165